New Tools for Environmental Protection

Education, Information, and Voluntary Measures

Committee on the Human Dimensions of Global Change

Thomas Dietz and Paul C. Stern, *Editors*

Division of Behavioral and Social Sciences and Education
National Research Council

NATIONAL ACADEMY PRESS
Washington, DC

NATIONAL ACADEMY PRESS • 2101 Constitution Avenue, N.W. • Washington, DC 20418

NOTICE: The project that is the subject of this report was approved by the Governing Board of the National Research Council, whose members are drawn from the councils of the National Academy of Sciences, the National Academy of Engineering, and the Institute of Medicine. The members of the committee responsible for the report were chosen for their special competences and with regard for appropriate balance.

The study was supported by Purchase Orders 9W-3489-NANX and 1W-2501-NANX from the Environmental Protection Agency to the National Academy of Sciences. Any opinions, findings, conclusions, or recommendations expressed in this publication are those of the author(s) and do not necessarily reflect the view of the organizations or agencies that provided support for this project.

Suggested citation: National Research Council (2002) *New Tools for Environmental Protection: Education, Information, and Voluntary Measures.* Committee on the Human Dimensions of Global Change. T. Dietz and P.C. Stern, eds. Division of Behavioral and Social Sciences and Education. Washington, DC: National Academy Press.

Additional copies of this report are available from National Academy Press, 2101 Constitution Avenue, N.W., Washington, DC 20418

Call (800) 624-6242 or (202) 334-3313 (in the Washington metropolitan area)

This report is also available online at **http://www.nap.edu**

Printed in the United States of America

THE NATIONAL ACADEMIES

National Academy of Sciences
National Academy of Engineering
Institute of Medicine
National Research Council

The **National Academy of Sciences** is a private, nonprofit, self-perpetuating society of distinguished scholars engaged in scientific and engineering research, dedicated to the furtherance of science and technology and to their use for the general welfare. Upon the authority of the charter granted to it by the Congress in 1863, the Academy has a mandate that requires it to advise the federal government on scientific and technical matters. Dr. Bruce M. Alberts is president of the National Academy of Sciences.

The **National Academy of Engineering** was established in 1964, under the charter of the National Academy of Sciences, as a parallel organization of outstanding engineers. It is autonomous in its administration and in the selection of its members, sharing with the National Academy of Sciences the responsibility for advising the federal government. The National Academy of Engineering also sponsors engineering programs aimed at meeting national needs, encourages education and research, and recognizes the superior achievements of engineers. Dr. Wm. A. Wulf is president of the National Academy of Engineering.

The **Institute of Medicine** was established in 1970 by the National Academy of Sciences to secure the services of eminent members of appropriate professions in the examination of policy matters pertaining to the health of the public. The Institute acts under the responsibility given to the National Academy of Sciences by its congressional charter to be an adviser to the federal government and, upon its own initiative, to identify issues of medical care, research, and education. Dr. Kenneth I. Shine is president of the Institute of Medicine.

The **National Research Council** was organized by the National Academy of Sciences in 1916 to associate the broad community of science and technology with the Academy's purposes of furthering knowledge and advising the federal government. Functioning in accordance with general policies determined by the Academy, the Council has become the principal operating agency of both the National Academy of Sciences and the National Academy of Engineering in providing services to the government, the public, and the scientific and engineering communities. The Council is administered jointly by both Academies and the Institute of Medicine. Dr. Bruce M. Alberts and Dr. Wm. A. Wulf are chairman and vice chairman, respectively, of the National Research Council.

COMMITTEE ON THE HUMAN DIMENSIONS
OF GLOBAL CHANGE

THOMAS DIETZ *(Chair)*, Department of Environmental Science and Policy, and Department of Sociology and Anthropology, George Mason University

BARBARA ENTWISLE, Department of Sociology, University of North Carolina

MYRON GUTMANN, Inter-university Consortium for Political and Social Research, University of Michigan

RONALD MITCHELL, Department of Political Sciences, University of Oregon

EMILIO MORAN, Department of Anthropology, Indiana University

M. GRANGER MORGAN, Department of Engineering and Public Policy, Carnegie Mellon University

EDWARD PARSON, John F. Kennedy School of Government, Environment and Natural Resources Program, Harvard University

ALAN RANDALL, Department of Agricultural, Environmental, and Development Economics, Ohio State University

PETER J. RICHERSON, Division of Environmental Studies, University of California, Davis

MARK R. ROSENZWEIG, Department of Economics, University of Pennsylvania

STEPHEN H. SCHNEIDER, Department of Biological Sciences, Stanford University

SUSAN STONICH, Department of Anthropology and Environmental Studies Program, University of California, Santa Barbara

ELKE U. WEBER, Department of Psychology, Columbia University

THOMAS J. WILBANKS, Oak Ridge National Laboratory, Oak Ridge, TN

CHARLES KENNEL *(Ex Officio, Chair-Committee on Global Change Research)*, Scripps Institution of Oceanography, University of California, San Diego

ORAN R. YOUNG *(Ex Officio, Liasion to International Human Dimensions Program)*, Institute of Arctic Studies, Dartmouth College

PAUL C. STERN, *Study Director*
DEBORAH M. JOHNSON, *Senior Project Assistant*

Preface

There has been increasing concern among environmental protection officials in the federal government about the problem of diminishing returns from regulation. Many believe that the quick environmental fixes from command-and-control regulation mainly have been achieved and that the balance of pollution sources is shifting from large "point sources" to more diffuse sources that are more difficult and expensive to regulate. In addition, changes in the political climate have made it increasingly difficult to use command-and-control regulations. Consequently, there has been a search for alternatives to regulation, including a shift to market-based approaches such as tradable emissions permits, to informational approaches, and to voluntary measures.

The Office of Environmental Education of the U.S. Environmental Protection Agency (EPA), responding to these concerns, asked the National Research Council (NRC) to organize a workshop to examine these issues. At the workshop, held November 29-30, 2000, participants examined the belief that changed conditions call for increased use of alternatives to regulation and economic measures. They also presented and discussed scientific evidence on the efficacy of education, information, and voluntary measures for achieving environmental protection objectives. The chapters of this volume include revised versions of presentations at the workshop, comments from discussants, and overviews of the issues by the editors and other workshop participants.

Since its birth in 1989, the Committee on the Human Dimensions of Global Change of the NRC has recognized the importance of understanding the individual and organizational behaviors responsible for environmental degradation both in order to anticipate environmental outcomes and to inform policy decisions intended to improve those outcomes. A previous committee effort, *Environmentally Significant Consumption: Research Directions* (NRC, 1997), examined the determinants of some of those behaviors. This volume examines some of the policy tools that are being used to change them.

We believe the result of our project is a rich series of contributions that review what we know about the potential importance and effectiveness of education, information, and voluntary measures in environmental protection; brings together what have been somewhat disparate literatures; and points directions for the future. We hope this volume achieves several goals. First, we hope it provides a sound grounding in what we have learned about the effectiveness of the "new" tools, both individually and in combination with other policy instruments. Second, we hope it provides a broad state-of-the-art review and shows connections and gaps in knowledge that may not have been obvious in the past. Third, for researchers and those funding research, we believe it conveys a sense of what has been learned and indicates priorities for future work. Finally, although not a management handbook, we hope it provides some guidance to those who design and manage policies and programs that employ education, information, and voluntary approaches.

On behalf of the committee, I wish to thank the EPA for its support of this project and Ginger Keho of the EPA's Office of Environmental Education for having the foresight to envision this project. The committee's gratitude goes to Brian Tobachnick, who managed the logistics of the project during its early stages; to Cecilia Rossiter, who provided additional organizational help at early stages; and to Deborah M. Johnson, who carried it the rest of the way. We also owe a debt to Laura Penny, who did the copy editing, and to Kirsten Sampson Snyder and Yvonne Wise, who managed the review and editorial processes.

I wish to thank the following individuals for their participation in the review of the papers in this volume: Clint J. Andrews, Rutgers University; Richard N.L. Andrews, The University of North Carolina at Chapel Hill; Lynton K. Caldwell, Indiana University, Bloomington; Doug McKenzie-Mohr, McKenzie-Mohr Associates, Ontario, Canada; James Meadowcroft, The University of Sheffield, United Kingdom; Joanne Nigg, University of Delaware; Stuart Oskamp, Claremont Graduate University, Claremont, California; Paul Sabatier, University of California, Davis; Lynnette Zelezny, California State University, Fresno.

Although the individuals listed provided many constructive comments and suggestions, they were not asked to endorse the final draft of the report before its release. The review of this report was overseen by Barbara Entwisle, University of North Carolina at Chapel Hill to whom we are most grateful. Appointed by the NRC, she was responsible for making certain that an independent examination of this report was carried out in accordance with institutional procedures and that all review comments were carefully considered. Responsibility for the final content of this report rests entirely with the authoring committee and the institution.

Thomas Dietz, *Chair*
Committee on the Human Dimensions
of Global Change

Contents

PART I

INTRODUCTION

1

Exploring New Tools for Environmental Protection

Thomas Dietz and Paul C. Stern

Many believe that the nature of environmental policy is changing. In much of the world during the last third of the 20th century, environmental policy was dominated by "command-and-control" approaches.[1] Under command and control, government agencies develop a set of rules or standards. These determine technologies to be used or avoided; amounts of pollutant that can be emitted from a particular waste pipe, smokestack, or factory; and/or the amounts or kinds of resources that may be extracted from a common pool such as a fishery or forest. The agencies then issue commands in the form of regulations and permits to control the behavior of private firms, other government agencies, and/or individuals.[2] This approach is venerable. It can be found in the 4,000-year-old Code of Hammurabi, which prescribes penalties for faulty construction.[3]

In the past fifteen years, experiments with market-based environmental policies have proliferated. This change came in response both to theoretical developments in economics and to the continued resistance to command-and-control policies by those regulated. In market-based approaches, instead of detailing what can and cannot be done, government places a constraint or tax on pollution or resource extraction and lets those targeted by the policy decide how best to economize on those activities. One of the best known market-based strategies is the tradable environmental allowance (Tietenberg, 1985, 2002; Rose, 2002). Government determines an appropriate level of emissions and issues permits to emit that are limited to that level.[4] Permits can be bought and sold in the market. The theory is that individual firms and plant managers are in a better position than government regulators to determine how to meet the targets most economically. Because the permits can be sold, firms that are especially efficient at reducing emissions can actually profit from their efforts at preventing pollution.

Another market-based approach flows from the insight that markets do not normally include environmental impacts in the costs of production or the prices of goods and services. When this happens, society, which shares the costs of the environmental impacts, is providing a hidden subsidy for these products. The subsidy can be countered with an environmental impact or pollution tax that would compensate society as a whole and provide an incentive for producers to reduce environmental impacts.[5] Current discussions of a carbon tax to reduce emissions of carbon dioxide follow this logic.

TOWARD NEW TOOLS

Although command-and-control and market-based approaches have dominated U.S. environmental policy in recent decades, other approaches have also been employed. Environmental education efforts aimed at both the public and at students have been used since the 1960s. Information-based efforts for energy conservation, such as home energy audits and appliance labeling programs, began in the aftermath of the energy crises of the 1970s. The environmental impact assessment provisions of the U.S. National Environmental Policy Act of 1969 provided a wealth of new information on proposed policies and projects for stakeholders to evaluate. In much the same spirit, the Emergency Planning and Community Right-to-Know Act of 1986 required private firms to provide the federal government with information on releases of toxic substances. A major goal of the effort was to inform the public about toxics. Starting in the 1990s, the U.S. Environmental Protection Agency (EPA) and the U.S. Department of Energy initiated several plans for voluntary action by industry, while as early as 1989 the Chemical Manufacturers Association (now the American Chemistry Council) began the Responsible Care program—a voluntary effort conducted by the chemical manufacturing industry without direct government involvement.

In this volume we refer to approaches that are neither command and control nor market based as "new tools." What do we mean by "new tools?" As will be clear in this and subsequent chapters, a strict taxonomy of environmental policy tools is not possible and perhaps not necessary.[6] Our analysis has been based on a fivefold classification of policy types:

- Command and control,
- Market-based policies,
- Education,
- Provision of information, and
- Voluntary measures.

The first two approaches constitute the "old tools" that have been most prominent over the past quarter century; the last three are the "new tools" that are the subject of current experiments.[7] With the old tools, explicit external controls are placed on behavior: Those who do not do as prescribed face specific tangible sanctions. The new tools rely more on implicit sources of behavioral control, so that the resulting behavior is likely to be perceived as voluntary. Education includes the provision of information in a systematic and structured way, but usually goes further, encouraging deeper understanding and, perhaps, values and norms regarding behaviors. Simple provision of information offers "just the facts." The line dividing these two categories is not always distinct, but the contrast is useful for comparing, for example, a school- or community-based program of education on toxic substances in the environment with the maintenance of the Toxics Release Inventory on the World Wide Web by the EPA. Voluntary measures include agreements between regulatory agencies and private firms, agreements among firms in an industry, and voluntary actions across industries, such as when firms set environmental requirements for their suppliers.

Some other taxonomies also include plans (e.g., Andrews, 1994). Clearly plans can play an important role in defining general expectations and setting goals. But we have not included them in our taxonomy because plans require implementation methods that will usually involve the five policy types in our taxonomy. We also do not include the development of new technologies that lead to reduced environmental impact. Policy to encourage technological development is as complex as policy intended to change the environmental behavior of individuals, communities, or organizations, and thus deserves separate treatment. Encouraging technological development may be one of the most effective ways of reducing environmental impact, and that technological innovation may be driven by a range of policy approaches: new tools, direct regulation, and market forces, as well as technological policy per se. Finally, we include codes and norms of "best professional practices" established by professional or industry groups within the broad category of voluntary measures. In engineering and management, such practices can do a great deal to reduce environmental impact. They deserve more attention than they have been given to date as a means of reducing environmental impact.

The new tools are an evolving set of supplements to command-and-control and market-based methods. They take many forms, as is obvious from the diverse policies considered in subsequent chapters. But they all have one or both of two features. They use education and the provision of information to try to change behavior, and the changes in behavior are voluntary in the sense that they are not driven by specific regulatory directives, externality taxes, or permit markets. Of course, concerns about market threats, opportunities, and risks, such as consumer boycotts, may provide indirect financial incentives.

Categorizing an approach to environmental protection as "command and control," "market based," or based on "education, information, and voluntary measures," although useful analytically, overlooks the fact that every tool is actually a hybrid of all these forms. Individual and organizational response is normally a function of prices, the lure of economic opportunities, the threat of sanctions, the availability of useful information, concern with reputation, and various intrinsic motivations. For example, market considerations often influence compliance with regulations. Few major command-and-control regulations in the United States are implemented without lawsuits from both environmental groups and industry. The targets of regulation certainly weigh the economic costs of a lawsuit and its chances of success against the costs of complying with a new regulation. Noninstrumental goals also may drive responses to command-and-control regulations. Some corporate managers operate on a strict profit maximization principle, but many others have serious environmental concerns and want to do as good a job as possible at minimizing their environmental impact within the fiscal limits they face. Given the apparent increase in green consumerism, such motivations are not unrelated to concerns about market share and profitability.

Just as command-and-control approaches engage market incentives, so too do market incentives involve some of the characteristics of command and control. Government, often in cooperation with stakeholders, must design the institutions that will implement tradable permits or pollution taxes. They must set the level of pollution allowed or the tax rate, as well as penalties for breaking the rules. They may also require market participants to provide accurate information on their resource use or pollutant emissions. All these activities involve command and control.

New tools based on education, information, and voluntary measures are present in every command-and-control and market-based policy as well. New measures, whether command and control or market based, always involve a learning curve in which those affected must learn how to operate efficiently in the face of the changed environment. The cost of information needed to comply with a new regulation or to strategize effectively in the face of a market-based policy may be high. Those affected educate themselves, sometimes by trial and error, sometimes by imitation of others, sometimes by discussion with those implementing the new regulatory regime. Governments provide information as a part of every command-and-control strategy. One of the major arguments in favor of market-based schemes is the view that markets are fast, accurate, and efficient transmitters of information. So although the distinction among "command-and-control," "market-based," and "new tools" approaches to environmental protection is useful analytically, analysis also will benefit from attention to the degree to which each approach is embedded in, and embeds within itself, the others.

Calling education, information, and voluntary measures *new* tools is something of a misnomer. Certainly, command-and-control and economic instruments are very old, dating to the earliest states. We alluded earlier to a safety "regulation" in the Code of Hammurabi; taxes to provide public goods or discourage undesirable behavior are probably about as old. But the "new tools" based on education, the provision of incentives, reputation, and peer pressure are even older. Before the state emerged, humans lived in groups with relatively little hierarchy, and the market was not a feature of daily social life. Societies of food foragers and early horticulturalists usually had no permanent political leadership and traded mostly for things not produced locally. Governance involved discussion, ritual, tradition, and peer pressure. Although debate continues about how well preagricultural societies managed the parts of the environment that supported their lives, the management tools they used were surely closer to what we are calling "new tools" than to the "old tools" of command-and-control and market incentives.[8] Thus, although these approaches may be considered innovations in the contemporary policy system, they have an ancient lineage.

WHY NEW TOOLS?

What changes in the past decade or so have led to the interest in new tools? This is a question that has not attracted as much careful scholarship as it has speculation, so we don't have a clear answer. But a number of hypotheses are available.

New Targets

One hypothesis is that the rise in interest in new approaches is a result of a shift in the sources of pollution that need attention. Proponents of this view (reviewed in Rejeski and Salzman, this volume, Chapter 2) argue that command-and-control regulations were effective with the major manufacturing and resource-extractive corporations to which they were applied from the 1960s on.[9] Once their legal resistance to a regulation ceased, these corporations could afford the capital investments required to comply with the regulations and could hire technically competent staff from within or from consulting firms to interpret and implement them.[10] According to this argument, complying with command-and-control regulations became a normal part of doing business for these firms, and they responded with major decreases in pollution. This arrangement could work because any regulatory agency had a relatively modest number of firms with which to deal, making the tasks of contacting, negotiating, and monitoring manageable.

As large firms were regulated, attention turned to other forms of pollution (or in a few cases, the same pollutants) that were emitted by thousands or even millions of sources. These small and "nonpoint" sources were hard to identify and difficult to monitor. Moreover, those responsible for their emissions often lacked the ability to understand or comply with regulations because of cost and technical capacity. Although applying command-and-control approaches to such sources is not possible, it makes sense to try alternative approaches.

Proponents of the new-target view also note a growing frustration with the fragmented character of U.S. environmental policy, which involved dozens of separate legislative mandates that were not well coordinated (Andrews, 1999:358-362; National Academy of Public Administration, 1995; Esty and Chertow, 1997). A single manufacturing plant might be regulated under a dozen different statutes and have to deal with that many or more offices at the EPA and other federal agencies. They sometimes note that other industrial nations are developing innovative, integrative, and seemingly effective approaches to comprehensive environmental policy that rely on additional tools (Janicke and Weidner, 1996). There is hope the new tools will provide for a more integrated and coordinated approach to environmental policy by encouraging responses that go beyond compliance with assorted regulations to address underlying problems.

In parallel with an interest in more integrated approaches to regulation has come exploration of new technologies that allow for multipollutant, multimedia emissions control. Work under the general heading of "industrial ecology" attempts to find technologies that reduce overall environmental impacts of production, consumption, and waste disposal, often by reengineering entire product life cycles rather than focusing on "end of pipe" control technologies. Because, as noted, most environmental regulations are designed to regulate specific kinds of pollutant emissions to specific media, they can sometimes create obstacles to this more holistic approach.[11]

New Attitudes

A complementary explanation for the increased interest in new tools is that attitudes about the environment have changed substantially since around the first Earth Day in 1970. Although survey research on national samples does not provide unambiguous evidence of striking attitude change among individuals, there is evidence of aggregate change, perhaps because older birth cohorts have been replaced by newer ones whose members hold more proenvironmental views (Jones and Dunlap, 1992; Kanagy et al., 1994). So it is reasonable to propose that cohorts who have received their education since about 1970 are far more aware of and concerned with environmental issues than prior cohorts.

The increased environmental consciousness of the public corresponds with the rise of green consumerism, which is actively promoted by many environmental organizations. In response, some firms may be seeking a niche market defined in terms of minimal environmental impact from their products. Even firms that do not see environmentalism as part of their marketing strategies acknowledge that environmental impacts play some part in the purchasing decisions of many consumers. In addition, many firms are sensitive to the possibility of boycotts orchestrated by environmental groups.

Not all of the shift in corporate concern should be attributed to external pressures. Some firms have accepted the idea from the emerging field of industrial ecology that pollutant emissions are a result of inefficient processes. The precursors to many pollutants are expensive, and the release of the pollution into the environment is thus a waste of money.[12] Furthermore, the current generation of corporate environmental managers and a growing number of senior executives are members of environmentally aware cohorts.[13] Although managers always must be attentive to the bottom line, many are also anxious to protect the environment because they believe it is important and ethical to do so.

Such a shift in attitudes can provide new opportunities for environmental protection. It may be possible to see many firms not as recalcitrant actors who must be dragged to better environmental practice, but as ready partners with regulatory agencies, environmental groups, and local stakeholders in designing approaches that go beyond what is minimally required by command-and-control regulations. This idea has led to a call for more emphasis on some of the new tools and less on regulation.

New Distribution of Power

Michel's notion of agency capture offers another hypothesis about the move toward new tools. This view emphasizes the strong resistance from the private sector to the wave of environmental regulation in the late 1960s and early 1970s. The rise of conservatism to power in U.S. politics during the 1980s and 1990s made the highest levels of federal agencies less antagonistic and more sympathetic to industry concerns. Evidence includes the observation that many of the political appointees in environmental agencies under the Reagan and two Bush administrations were drawn directly from industry or from think tanks, research institutes, or lobbying firms highly sympathetic to industry. In the view of those who emphasize the shift to conservatism, complete deregulation was not politically viable because of the strong opposition from the public and environmental movement organizations to efforts to eliminate or weaken key federal command-and-control regulations. However, a shift from command-and-control strategies to market-based approaches and to the new tools was a viable way for the new regulatory authorities to make environmental policy less antagonistic to industry concerns.[14]

Nothing New

Yet another possible explanation for the increased interest in new tools is that it is not really a recent phenomenon. Since the start of the new wave of environmental regulation in the late 1960s, considerable emphasis has been placed on education, information, and voluntary measures. Environmental education directed at individuals has been an important element of both government and environmental organization strategy for decades. Though the relationship between the regulated and the regulators often has been stormy, it also has always involved elements of the education, information flow, and voluntary cooperation that characterize the new tools. What is new is simply the evolution of understanding. More experience has led to more sophisticated approaches. In addition, the past decade has seen a substantial move toward using education, information, and voluntary strategies not just with individuals, but also with communities and especially with private firms.

Toward a Synthesis

Each of the hypotheses described has some validity. The new tools are of increased interest now as the limits of command-and-control and market-based controls become better understood (see Rejeski and Salzman, this volume, Chapter 2). Partly because of new attitudes, some of the new tools may be more applicable than before to traditional regulatory targets. The new distribution of power has induced regulators to seek alternatives to command and control, and the new tools are attractive options. Whether they are effective for achieving traditional environmental policy goals is still a matter of controversy, however. Reasonable arguments also suggest that the new tools are not a better way to protect the environment or a good way to handle new priorities, but rather the result of increasing resistance to traditional approaches. In this view, they are a weak compromise at best.

All these developments in environmental policy offer good reasons for looking closely at the potential of the new tools at the turn of the millennium. Command-and-control regulations may be reaching a point of diminishing returns, and their tendency to be monolithic and to slow innovation limits their value. If there is increasing heterogeneity in the kinds of sources generating environmental impact, approaches that allow flexibility of response are required. Market-based mechanisms often are put forward as the way to meet the needs for flexibility and for accommodating heterogeneity, but such mechanisms are not always practicable—and when they are, experience with their actual operation shows they do not work well for all environmental problems (Tietenberg, 2002; Rose, 2002). The new tools offer hope that strategies might be tailored to specific contexts.

The new tools need to be considered not only on their own, but also in combination with each other and with traditional policy tools. Few propose eliminating regulation entirely—in fact, it is a central element of all policy tools. What changes when new tools are adopted is what is regulated and the balance of emphasis across policy tools. For example, tradable allowance policies still require regulations of the maximum allowable extraction or pollution, though not of the way the limit is met. Information-based policies like the Toxics Release Inventory include a regulatory requirement—not to limit releases, but to provide information to the public on those releases. Policies based on voluntary agreements normally are presented as a way to decrease pollution faster or further than regulations require. The backdrop of regulation is what motivates participation in voluntary agreements.

With these considerations in mind, examining the new tools provides a new way to think about environmental policy in general. Instead of debating which policy strategy is best or most cost-effective, policy analysts can consider the best policy package for a particular purpose, activity to be controlled, and actor. They can consider a combination of tools and evaluate outcomes from various standpoints, including environmental quality, economic efficiency, and equity. To do this, it is necessary to build an understanding of the behaviors and the individuals or organizations whose behavior is to be changed (see National Research Council, 1997). It is also necessary to understand how each available tool works, when it works more or less effectively, and the conditions (including target behaviors, actors, and the use of complementary tools) under which it can produce the best results. This volume begins to undertake this task for education, information, and voluntary measures.

THE GOALS OF THIS VOLUME

We do not focus here on controversies about the best policy strategy. Whether one views education, information, and voluntary measures as an important step forward, evidence of a retreat in commitment to environmental protection, or a set of complementary approaches to the old tools of command and control and market mechanisms, the research agenda remains the same. Careful taxonomies of the new tools are needed, including consideration of whose behaviors they are intended to affect, which behaviors, and how the change is brought about. It is important to examine the theories of individual and organizational behavior that underpin the tools, to view each use of a new tool as an experiment in policy design, and to analyze these experiments to provide for as much social learning as possible.

Our goal in this volume is to bring together state-of-the-art research on the new tools for environmental protection. The chapters examine empirical research on new tools, extract lessons from adjacent policy and research traditions that can inform work on new tools, and consider the conceptual and theoretical issues that must be addressed. A great deal of excellent research already has been conducted, and much can be learned from it. We hope this volume will provide practical guidance to those who are working to make the new tools more effective. We also hope it will foster the theoretical and empirical research needed to improve new approaches to environmental policy.

Many key research questions remain open. There is a real need for conceptualization, taxonomies, and theorizing, as well as analysis of historical and contemporary innovations. As is evident from the chapters that follow, research on new tools also addresses a number of important basic social and behavioral science questions about information processing, learning, behavioral and organizational change, and other topics. This research also requires attention to some of the central theoretical and methodological questions of the social sciences.

The effort to better understand the tools for environmental protection, new and old, will benefit not only efforts to protect the environment, but also the science of human-environment interactions.

The book is divided into four sections. Part One, which includes this chapter, provides a context for the emergence of interest in new tools. Chapter 2 by Rejeski and Salzman, explores some of the shifts that are (and are not) occurring in pollution sources and outlines some emerging technological developments that are transforming manufacturing and transportation processes and that have profound implications for environmental policy. Chapter 2 suggests that we will need even newer tools to meet the challenges and opportunities that will accompany technological and organizational changes.

Parts Two and Three are organized around targets and strategies. Part Two examines the potential of information and education campaigns to change the behavior of individuals, households, and communities. Part Three focuses on the private sector and the potential of voluntary agreements between government and industry and among firms. Part Four consists of a concluding chapter of reflections on what we know and need to know about the new tools.

NOTES

1 Andrews (1999) provides an excellent history of American environmental policy. Hays (1987) covers most of the period of command-and-control regulation in the United States. Many important experiments in environmental policy have been and are being conducted outside the United States; some chapters in this volume draw on these experiences. Ultimately, understanding of environmental policy must be comparative. This book, however, focuses on the U.S. case as a necessary prelude to a more robust analysis that draws on global experience. International comparisons are beginning to appear (tenBrink, 2002).

2 In the United States, most discussion of command-and-control and market-based approaches to environmental policy assumes the federal government develops regulations or sets up market mechanisms to control the behaviors of private firms. In fact, most U.S. federal command and control regulations apply to federal agencies and to state and local governments as well. In some cases, the state or local government has the responsibility for implementing regulations developed by federal agencies. These complexities are consequential in developing practical strategies and theoretical understanding. For simplicity of language, in this chapter we refer to federal regulation of private firms, but the other regulators and the others being regulated should also be kept in mind.

3 For example, Articles 229-233 read:

If a builder build a house for some one, and does not construct it properly, and the house which he built fall in and kill its owner, then that builder shall be put to death.

If it kill the son of the owner the son of that builder shall be put to death.

If it kill a slave of the owner, then he shall pay slave for slave to the owner of the house.

If it ruin goods, he shall make compensation for all that has been ruined, and inasmuch as he did not construct properly this house which he built and it fell, he shall re-erect the house from his own means.

> If a builder build a house for some one, even though he has not yet completed it; if then the walls seem toppling, the builder must make the walls solid from his own means."

There are comparable penalties for flooding a neighbor's field, grazing sheep in a neighbor's pasture, and other offenses against property. This text is available online at http://www.yale.edu/lawweb/avalon/hammenu.htm.

4 As Tietenberg (2002) and Rose (2002) note, determining how the permits will be allocated initially is a political decision, and one that is often critical to the acceptability of tradeable environmental allowance policies.

5 Freeman (1993:93-140) provides an overview of neoclassical welfare economics as applied to environmental problems. Kneese and Bower (1968) were early advocates of the impact tax approach. Of course, hidden subsidies benefit those who profit from the factory's output. If one factory is receiving such social subsidies by being allowed to generate adverse environmental impacts, it has a competitive advantage over a factory that does not receive such subsidies. It is interesting to note that although neoclassical economists and Marxists use different language to describe this situation, their analysis follows the same logic. For a Marxist analysis of environmental problems see Foster (1999, 2000) or Anderson (1976).

6 Kaufmann-Hayoz et al. (2001) offer a thoughtful discussion on the logic of taxonomies of environmental policies.

7 Kaufmann-Hayoz et al. (2001) suggest a parallel fivefold classification of environmental policy instruments: command-and-control, economic, service and infrastructure, collaborative agreements, and communication and diffusion policies. Their first two categories exactly match our first two. Their category of service and infrastructure provision includes everything from mass transit infrastructure to recycling centers to databases on pollutant emissions. Although this is an important class of policies, we do not emphasize it here except when the service or infrastructure in question involves the provision of information. Their category of collaborative agreements matches our voluntary measures. Their concept of communication and diffusion instruments overlaps closely with our concepts of education and information provision. Neither our typology nor that of Kauffmann-Hayoz et al. (2001) has a separate place for policies of institutional design, such as those that are central to a large literature on institutions for managing common-pool resources (e.g., Ostrom, 1990; National Research Council, 2002). Although changes in institutions do not fit within any of the categories just listed, they often embody several categories at once. For example, tradeable environmental allowances are a market-based institutional innovation that involves much more than economic incentives: government control and information diffusion are both critical to their operation, and infrastructure and voluntary agreements also may be necessary for measuring and monitoring emissions.

8 Krech (1999) provides an overview of the controversy and an analysis of the sustainability of Native American ecological practices.

9 See Commoner (1992) for a dissenting view regarding how effective such efforts were.

10 Dietz and Rycroft (1987) show there has been a high degree of job mobility among federal regulatory agencies, law and consulting firms, and corporations, so the corporate employees often have experience in regulatory agencies and vice versa.

11 Andrews (1994) provides a seven-nation comparison of policies to encourage clean technologies. See also Allenby (1999).

12 Socolow et al. (1994) introduce the key ideas of industrial ecology. Esty and Chertow (1997) suggest that ideas from industrial ecology can provide a basis for the next generation of environmental regulation. Some theorists have suggested that the insights of industrial ecology along with increased societal demand for environmental quality, are leading to a new form of development termed "ecological modernization" (Mol and Sonnenfeld, 2000). Rosa et al. (2001) offer a somewhat skeptical view of ecological modernization.

13   The extent to which corporate managers can act with some autonomy from profit maximization has been a subject of debate at least since Galbraith. It is an important area for research if sound assumptions are to underpin new policy tools.

14   Several of the key innovations in EPA policy that are considered part of the new tools and discussed in subsequent chapters were initiated under the first Bush administration and continued under the Clinton administration (Andrews, 1999).

REFERENCES

Allenby, B.R.
    1999   *Industrial Ecology: Policy Framework and Implementation.* Upper Saddle River, NJ: Prentice-Hall.
Anderson, C.H.
    1976   *The Sociology of Survival: Social Problems of Growth.* Homewood, IL: Dorsey Press.
Andrews, C.J.
    1994   Policies to encourage clean technologies. Pp. 405-422 in *Industrial Ecology and Global Change,* R. Socolow, C.J. Andrews, F. Berkhout, and V. Thomas, eds. Cambridge, Eng.: Cambridge University Press.
Andrews, R.N.L.
    1999   *Managing the Environment, Managing Ourselves: A History of American Environmental Policy.* New Haven, CT: Yale University Press.
Commoner, B.
    1992   *Making Peace with the Planet.* New York: New Press.
Dietz, T., and R.W. Rycroft
    1987   *The Risk Professionals.* New York: Russell Sage Foundation.
Esty, D., and M. Chertow
    1997   *Thinking Ecologically: The Next Generation of Environmental Policy.* New Haven, CT: Yale University Press.
Foster, J.B.
    1999   Marx's theory of metabolic rift: Classical foundations for environmental sociology. *American Journal of Sociology* 105:366-405.
    2000   The ecological tyranny of the bottom line: The environmental and social consequences of economic reductionism. Pp. 135-153 in *Reclaiming the Environmental Debate: The Politics of Health in a Toxic Culture,* R. Hofrichter, ed. Cambridge, MA: MIT Press.
Freeman, A.M., III
    1993   *The Measurement of Environmental and Resource Values.* Washington, DC: Resources for the Future.
Hays, S.P.
    1987   *Beauty, Health and Permanence: Environmental Politics in the United States, 1955-1985.* Cambridge, Eng.: Cambridge University Press.
Janicke, M., and H. Weidner
    1996   *National Environmental Policies: A Comparative Study of Capacity Building.* New York: Springer Verlag.
Jones, R.E., and R.E. Dunlap
    1992   The social bases of environmental concern: Have they changed over time? *Rural Sociology* 57:28-47.
Kanagy, C.L., C.R. Humphrey, and G. Firebaugh
    1994   Surging environmentalism: Changing public opinion or changing publics? *Social Science Quarterly* 75:804-819.

Kaufmann-Hayoz, R., C. Battig, S. Bruppacher, R. Difila, A. DiGiulio, U. Friederich, M. Garberly, H. Gutscher, C. Jaggi, M. Jegen, A. Muller, and N. North
 2001   A typology of tools for building sustainable strategies. Pp. 33-107 in *Changing Things— Moving People: Strategies for Promoting Sustainable Development at the Local Level*, R. Kaufman-Hayoz and H. Gutscher, eds. Basel, Switz.: Birkhäuser.

Kneese, A.V., and B.T. Bower
 1968   *Managing Water Quality: Economics, Technology and Institutions.* Baltimore, MD: Johns Hopkins University Press.

Krech, S., III
 1999   *The Ecological Indian: Myth and History.* New York: W.W. Norton.

Mol, A.P., and D.A. Sonnenfeld
 2000   *Ecological Modernisation Around the World.* London: Frank Cass Publishers.

National Academy of Public Administration
 1995   *Setting Priorities, Getting Results: A New Direction for EPA.* Washington, DC: National Academy of Public Administration.

National Research Council
 1997   *Environmentally Significant Consumption: Research Directions.* Committee on the Human Dimensions of Global Change. P.C. Stern, T. Dietz, V.W. Ruttan, R. Socolow, and J. Sweeney, eds., Commission on Behavioral and Social Sciences and Education, National Research Council. Washington, DC: National Academy Press.
 2002   *The Drama of the Commons.* Committee on the Human Dimensions of Global Change. E. Ostrom, T. Dietz, N. Dolšak, P.C. Stern, S. Stonich, and E.U. Weber, eds., Division of Behavioral and Social Sciences and Education, National Research Council. Washington, DC: National Academy Press.

Ostrom, E.
 1990   *Governing The Commons.* Cambridge, Eng.: Cambridge University Press.

Rosa, E.A., R. York, and T. Dietz
 2001   *Modernization and the Environment: Modeling the Impacts of Economic Development.* Pullman, WA: Department of Sociology, University of Washington.

Rose, C.
 2002   Common property, regulatory property and environmental protection: Comparing common pool resources to tradable environmental allowances. Pp. 233-257 in National Research Council, *The Drama of the Commons,* Committee on the Human Dimensions of Global Change, E. Ostrom, T. Dietz, N. Dolšak, P.C. Stern, S. Stonich, and E.U. Weber, eds., Division of Behavioral and Social Sciences and Education, National Research Council. Washington, DC: National Academy Press.

Socolow, R., C. Andrews, F. Berkhout, and V. Thomas
 1994   *Industrial Ecology and Global Change.* New York: Cambridge University Press.

tenBrink, P.
 2002   *Voluntary Environmental Agreements: Process, Practice, and Future Use.* Sheffield, Eng.: Greenleaf Publishing.

Tietenberg, T.
 1985   *Emissions Trading: An Exercise in Reforming Pollution Policy.* Washington, DC: Resources for the Future.
 2002   The tradable permits approach to protecting the commons: What have we learned? Pp. 197-232 in National Research Council, *The Drama of the Commons,* Committee on the Human Dimensions of Global Change, E. Ostrom, T. Dietz, N. Dolšak, P.C. Stern, S. Stonich, and E.U. Weber, eds., Division of Behavioral and Social Sciences and Education, National Research Council. Washington, DC: National Academy Press.

2

Changes in Pollution and the Implications for Policy

David W. Rejeski and James Salzman

A sk people to describe the archetypal pollution problems we face today and they may well recount a Dickensian vision—a dirt-streaked factory shrouded in smoke, leaking effluent, churning out drums of waste. And for good reason. When the drafters of our pollution control statutes surveyed the landscape in the 1970s, their regulatory landscape was filled with smokestack industries. But what if this vision of environmental threats, still resonant today, has become largely irrelevant? What if we have transformed from a manufacturing-based to a service-based economy? What if manufacturing itself is being transformed radically, if we are entering a new industrial revolution?

This is no idle speculation, for big changes are afoot in both the service and manufacturing sectors. In this chapter, we will begin to explore these changes and transformations and try to tease out their implications for environmental protection and policy. We begin with the transformation in services.

The service sector now dominates America's economy, supplying more than three-quarters of our Gross Domestic Product (GDP) and four-fifths of our employment (*The Economist,* 1994). Over the past few decades, manufacturing's relative economic importance has dramatically declined (a phenomenon known as "deindustrialization"). In 1970, roughly one in four workers was employed in manufacturing. By 2005, that number will drop to less than one in eight (Bureau of the Census, 1997; Rowthorn and Ramaswamy, 1997). Over the same period, employment in services has increased correspondingly, and most often the new service jobs have been knowledge based, marking a shift from material-processing to information-processing activities (Stewart, 1993). Just think of the transformation of Pittsburgh from dirty center of steel production to hub of clean high-tech services. As *The Economist* has asserted bluntly (*The Economist,*

1994:91): "It is still common to refer to [Organization for Economic Co-operation and Development (OECD)] members as the 'industrialized economies.' Common, yet quite wrong."[1]

It has become commonplace for commentators to speak of a fundamental transformation now shaping our economy. The labels vying to capture this era include the "service economy" and the "postindustrial society," but the most commonly used name is the "information revolution"—hailed as the third great economic revolution of human history.[2] The agricultural revolution generated wealth from plowed fields, the industrial revolution from the mechanized production of material goods. In the information revolution, its observers claim, wealth derives from the management, creation, and ownership of knowledge (Carnoy et al., 1993). Famed management guru Drucker has succinctly described such an economy as one where "the basic economic source . . . is no longer capital, nor natural resources nor land. It is and will be knowledge (Drucker, 1994:8).[3] To be sure, the term "information revolution" is a trendy label, suggesting the increasingly central role of information in how we think of ourselves and our society, but it also describes very real transformations. For our purposes, regardless of the label, if the rise of services signals a fundamental change in means of production and patterns of consumption, then the law must adapt accordingly. Otherwise environmental law's focus on smokestack sources risks becoming a Maginot Line: "strong, powerful, bristling with legalistic weaponry, providing comfortable but illusory control and dominance—and increasingly irrelevant" (Allenby, 1997:36).

What are the environmental implications of this transition? Does the rise of services pose important new challenges, or perhaps powerful opportunities, for environmental protection? Surprisingly, no one seems to know. More surprisingly, almost no consideration has been given to these questions. Although literally thousands of books and articles have explored the implications of smokestack industries for environmental law and policy, a mere handful have considered the service sector. To begin to provide answers, we need to rethink our basic assumptions of pollution sources and, as a consequence, environmental protection strategies. This requires understanding better the current economic and environmental trends and their underlying causes.

DEINDUSTRIALIZATION

What is the evidence for this new economy? Most suggestive is the process of deindustrialization—the dramatic decline of manufacturing's relative economic importance. The unrelenting growth of the service sector and the apparently corresponding decline of the manufacturing sector has been taking place for decades in the United States, Europe, and Japan, engendering heated debate over the consequences. The service sector has expanded in all but one quarter over the past 50 years (Rejeski, 1997). Between 1955 and 1980, the U.S. economy added

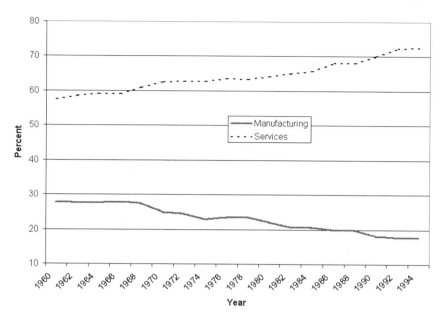

FIGURE 2-1 Value added by sector as a share of U.S. Gross Domestic Product (GDP) at current prices.
Source: Rowthorn and Ramaswamy (1997).

40 million jobs, yet only 1 in 10 of these was in manufacturing (see Figures 2-1 and 2-2). Over the same period, the health sector added more jobs than did all of manufacturing combined (Cohen and Zysman, 1987). Most services, such as communications, wholesale trade, finance, insurance, and real estate, have grown steadily. In recent years, the health care and computer systems fields have been among the fastest growing sectors in the entire economy for both employment and revenue.[4]

In considering these impressive figures, one must keep some points in mind. First, note that Figure 2-2 shows employment data. Because of the way the Bureau of the Census defines manufacturing and service employment, the statistics tend to overstate service employment (Salzman, 1999:429). Second, because labor productivity has risen faster in the manufacturing sector than in services, employment has fallen while production has increased. Much of this increased productivity has been due to greater reliance on the services sector, which has not increased productivity at the same rate (Salzman, 1999:434). Finally, Figure 2-1 shows sectoral contribution to GDP *as a percentage*. During this period, though, GDP has been growing as well.

As a result, a close analysis of economic indicators reveals two broad trends at work in the past few decades. First, there *has* been sustained growth in the service sector such that in relative terms it now dominates our nation's economic

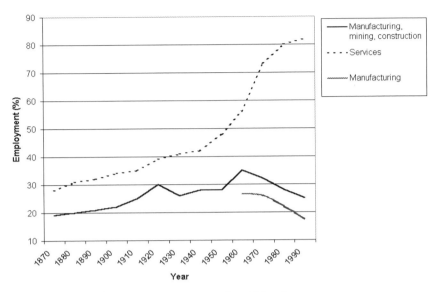

FIGURE 2-2 Employment as a percentage of total labor force.
Source: Rowthorn and Ramaswamy (1997).

activity. Despite overestimates of its growth, the service economy is for real. Second, this rise of services has masked significant productivity gains and an absolute increase in manufacturing activity. It is *not* the case that services have grown while manufacturing has disappeared. Rather, the growth of services has outpaced manufacturing's growth, despite the fact that we are producing more than ever (Bureau of the Census, 1995:748, 759). These results should not, on reflection, be surprising. The need for food did not go away at the end of the agricultural revolution nor has industrial activity dimmed in the brilliance of the information revolution's dawn.

Even if the smokestack economy is still alive and well, albeit diminished in stature compared to services, one might still expect environmental benefits. The core thesis of Drucker's (1994) and others' writings on the information revolution has claimed that knowledge is supplementing natural and human made capital as factors of production. Intuitively, this makes sense. One would expect, for example, that increasing use of e-mail would reduce the environmental impact from overnight express and postal mail, that telecommuting and videoconferencing would reduce the transport impacts from traveling to work, and that bioengineered crops would reduce the need for pesticides and fertilizer. (von Weizsacker et al., 1997). Such examples surely suggest that as the information revolution advances, there will be an "environmental bonus." But is this happening? The best data to assess this question comes from a series of studies conducted by the World Resources Institute (WRI) that examined material flow through the United States economy (Adriaanse et al., 1998). WRI sought to quantify all the natural resources directly and indirectly consumed by economic activity in four major industrialized nations (the United States, the Netherlands, Germany, and Japan). Based on the industrial ecology principle of material flow accounting, the study tracked the consumption of natural resources in the economy, from the extraction of raw materials through to their ultimate disposal. Importantly, the study sought to track the entire lifecycle, capturing material flows overseas as well as domestically.

Figure 2-3 shows the results for U.S. material intensity, measuring Total Material Requirement (TMR) per unit Gross National Product (GNP).[5] If the economic infrastructure is changing, moving toward more information processing than material processing, then this should be reflected in less material consumption *per unit* of economic activity. In mathematical terms, the measures of material intensity should show decreasing slopes. The study found that TMR material intensity has, in fact, decreased, as has the measure of direct material intensity (which included traditional material inputs such as oil, copper, or water, but not the hidden material flows captured in TMR). Less comprehensive studies have reached similar conclusions. These data therefore are consistent with the thesis that knowledge is replacing physical inputs as factors of production and, that services are replacing resource-intensive activities.

These results can be explained by a number of other factors as well. The first of these is *input substitution,* the use of new materials as efficient replace-

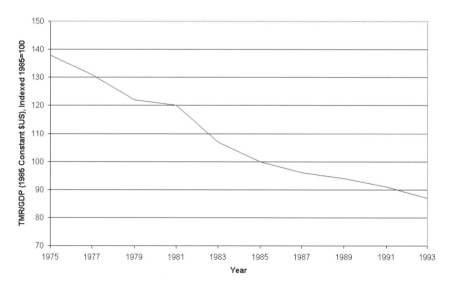

FIGURE 2-3 Overall material intensity: Total Material Requirement/Gross Domestic Product (TMR/GDP) Index.
Source: Adriaanse et al. (1998). Reprinted with permission of World Resources Institute.

ments for current materials. Fiber optics, for example, are replacing old copper wire communication lines, using less material and increasing the carrying capacity by 30 to 50 times (Cleveland, 1985). Similarly, the amount of steel in a car has decreased by more than a third since the early 1970s, while plastics and composites have increased (Wernick et al., 1996). A second factor is *increased production efficiency* that conserves materials. This can occur through redesign of the process, closed-loop recycling, and other pollution prevention techniques that contribute to improved manufacturing efficiency. Finally, *product design* has helped to reduce material consumption. Changes as simple as "light-weighting," or reducing the weight of a product, have led to dramatic differences in material consumption. Beverage cans, for example, have become much smaller and lighter, first moving from glass to steel to aluminum, and then reduced in weight an additional 25 percent. As in the case of pollution reductions, these types of changes also may be driven by command-and-control regulations, by market prices reflecting scarce resources, or by environmental regulations that implicitly or explicitly change relative prices.

Focusing only on the material intensity slope misses the central point, however, for there has been little improvement in the measure of material consumption per capita. In fact, in Japan, Germany, and the Netherlands, material consumption per capita has increased. Put another way, GNP has grown faster than population. Thus measures of material intensity will be more impressive than measures of per capita consumption. What matters for the environment, of course, is total consumption of physical units (Stern, 1997). The important corollary is that because of population growth and increasing economic activity, absolute resource consumption has actually *increased,* despite reductions in material intensity. As the WRI study concluded, "meaningful dematerialization, in the sense of an absolute reduction in natural resource use, is not yet taking place" (Adriaanse et al., 1998:2). These findings have been confirmed by other research in the field.[6]

If services are substituting for manufacturing, if knowledge is in certain instances replacing inputs of natural capital, we would expect to see improvements in material intensity, and we do. The observed improvements in material intensity, though, largely may be due to other factors such as increased production efficiencies and input substitution. The data also suggest that rising absolute consumption is offsetting improvements in dematerialization and efficiencies. In fact, the data raise the possibility of a counterthesis—that the information revolution and rise of services have a net *negative* environmental impact because they increase overall economic activity and thus overall resource consumption (Ehrlich et al., 1999). This may occur in two related ways. First, as knowledge becomes a more important factor of production in some sectors, reductions in the cost of obtaining that knowledge stimulate economic growth, leading to *increased* environmental impacts through increased resource flow and conversion. Second, services may serve as complements to, rather than substitutes for, traditional production factors such as labor and resources, simply increasing their efficiency, rather than replacing them. In both cases, technical advances decrease the cost of an activity and, as a result, increase the overall level of activity. Thus advances in telecommunications and data processing technologies, by making relevant information cheaper and transactions easier, have increased the total number of transactions.

ENVIRONMENTAL PROTECTION AND THE SERVICE SECTOR

Despite the relative growth of the service sector and decline of manufacturing, the data clearly show that these factors have not led to a decrease in resource consumption. Hence, although the service economy may not mark a clear pathway toward sustainable development, it surely merits explicit consideration in environmental policy both because services are important sources of pollution and because they pose different challenges than traditional smokestack sources. Overlooking the role of the service sector in environmental protection is myopic, for it produces environmental impacts in its own right. But we know remarkably little about either the environmental impacts of services or the appropriate policy tools. The few writings seriously examining the environmental impacts of services have identified important themes using anecdotes, but they have not set out a coherent framework for thinking about services' impacts and, depending on their severity, the appropriate governmental response.

This is no easy task, for the service sector comprises a remarkably heterogeneous grouping of economic activities as varied in their function as in their environmental impact. They include transportation and public utilities, wholesale and retail trade, finance, insurance, real estate, business services, health services, legal services, and government services. To develop effective policy recommendations, we must first delineate services into categories meaningful for environmental protection. To do so requires distinguishing between services that cause high direct impact per facility and low (smokestack services), those that do not cause significant environmental harm at the level of individual operation but collectively have large impact (cumulative services), and those that act as leverage points, influencing behavior both upstream and downstream (leverage services). It is important to note that these categories of services are not mutually exclusive. A sector such as the electric utilities, for example, is both a strong smokestack service and a strong leverage service. The following sections briefly explain these categories and their policy implications.

Smokestack Services

As set out in Figure 2-4, smokestack services have high direct environmental impact per facility. For environmental policy analysts, smokestack services are the most obvious of the three categories because their activities *already* are regulated. Sulfur dioxide emissions from power plants are heavily regulated, the

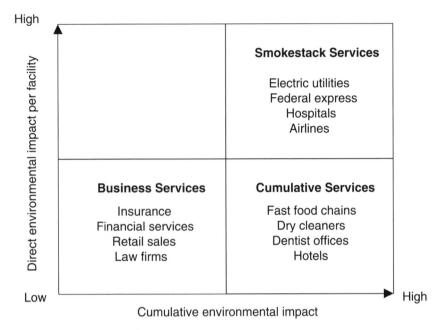

FIGURE 2-4 Categories of service.

subject of the entire trading program of the 1990 Clean Air Act amendments (U.S. Code, 1998a). Air pollution from the Federal Express fleet of delivery vans is subject to requirements under the mobile sources provision of the Clean Air Act (U.S. Code, 1998b). Biomedical waste from hospitals is regulated by the federal Resource Conservation and Recovery Act (RCRA) (U.S. Code, 1998c).

If smokestack services do warrant further attention from environmental lawmakers, it stems from the historical fact that many of the applicable laws were *not* drafted with service industries in mind. The net result can be inefficient governance, requiring the regulated entity to devote quite significant resources to compliance. Although this is, of course, a general problem of regulatory design, inefficient regulation of smokestack services can significantly impede innovative environmental protection measures. A recent study in the *Harvard Business Review* of productivity in the service sector made a similar point, concluding that regulation of services is very inefficient. One of the most important ways "government can help the service sector is not to overregulate it . . . The point is that regulation should be carried out in both spirit and practice to minimize the demands made on [service] businesses' attention and resources" (van Biema and Greenwald, 1997:87-88). As an example, consider the situation of the telecommunications provider in the Northeast, BellAtlantic (now known as Verizon Corporation).

Although BellAtlantic does not produce large amounts of hazardous waste, its diffuse operations constitute innumerable small sources that must be individually regulated. This includes wastes from maintaining a fleet of more than 18,000 vehicles, treating sediment from 113,000 manholes, and managing the use and disposal of more than 2.5 million utility poles treated with wood preservatives (of the 170 million poles in the country). The manhole sediment is typical of the mismatched regulatory burdens facing BellAtlantic. When repairing cables, BellAtlantic employees often work in manholes that contain water and sediment from the street. To get at the cables, it may be necessary to remove some of the water and sediment from the manhole. If they contain more than 5 parts per million (ppm) of lead, however, the water and sediment must be treated as RCRA hazardous waste. BellAtlantic tests have shown that the sediment is below 5 ppm about 55 percent of the time. Yet, in practice, BellAtlantic routinely treats the sediment as hazardous waste (complying with all the attendant RCRA Subtitle C requirements) in order to save time. This means the company must obtain a separate U.S. Environmental Protection Agency (EPA) hazardous waste identification number for every manhole treated. The ID system, required for waste manifests, was designed with smokestack sites in mind because it was assumed there would be one site, and therefore one source of hazardous waste generation. Perhaps not surprisingly, BellAtlantic has the largest number of waste ID numbers in the country.

Similarly, when BellAtlantic designed a mobile treatment unit that would eliminate the toxicity characteristics of the sediment, it found itself prevented from improving environmental performance by a regulatory system that had not anticipated the application of regulation to this service industry. New Hampshire refused to permit the process, stating that mobile on-site treatment only was allowed for manufacturing companies. Because BellAtlantic's Standard Industrial Classification (SIC) code identified it as a service company, it could not apply for the permit. Another example is BellAtlantic's use of emergency standby generators. BellAtlantic has more than 1,800 emergency diesel generators to provide power for the phone system in the event of a power failure. The generators run an average of 29 hours per year. The 1990 Clean Air Act amendment's "potential to emit" clause requires hundreds of permits or exceptions annually because it is assumed the generator runs constantly in a factory setting. In addition to the permits, there is considerable paperwork required for the company to report the presence of and risk management plans for the lead acid batteries in every BellAtlantic building.[7]

The point in raising these brief examples is not to argue that the regulation of smokestack services is unnecessary but, rather, that such regulation warrants special attention because of the potential for poor fit. RCRA, for example, was not written with services in mind. BellAtlantic's operations simply do not fit the model situation the law was drafted to address. Indeed, smokestack services provide an excellent opportunity for innovative regulatory strategies. Large transport services such as Federal Express, Hertz, and Allied Van Lines, for example, might be willing to reduce their overall emissions if they could "bubble" their vehicle fleet, treating it as one larger source of pollution, or obtain other forms of regulatory relief. One would think such possibilities should be attractive to the Common Sense Initiative and Project XL, the EPA's flagship reinvention initiatives to develop smarter, more effective, and cheaper alternatives to traditional regulation. The Department of Energy's well-funded Industries of the Future initiative also would seem appropriate. These initiatives receive more than $100 million in support, but have ignored services. None of the implemented Project XL initiatives have focused on services, none of the Industries of the Future include a service, and only one of the six Common Sense Initiative sectors is considered a service industry.

Cumulative Services

This category contains the largest number of services and is in many respects the most difficult to address because it brings into play the problem of cumulative impacts. In describing the history of environmental protection efforts, Caldwell (1990) described two generations of environmental problems. The first generation consisted of traditional point source emissions of local or, at worst, regional pollutants. These were classic smokestack industry problems of air, water, and soil pollution. Their impacts were reduced by a series of 1970s statutes and what has become known as command-and-control regulation. The second generation introduced transboundary and global threats such as ozone depletion, trade in hazardous wastes and climate change, problems requiring coordination among nations and therefore problems that are poorly suited for first generation command-and-control policies and institutions focused on domestic concerns.

The rise of the service sector may well coincide with the advent of a third generation of environmental problems, the challenge of atomized sources. These sources create, from a policy perspective, a "nonpoint" world where the cumulative impacts of small diffuse sources become significant and begin to resemble unmanageable runoff, potentially overwhelming traditional regulatory approaches.

Many cumulative services may be viewed as simply concentrating everyday activities, such as those at a hotel or restaurant. The environmental impacts do not differ in kind from those of a household; they are simply magnified. Consider the little placards discreetly placed in hotel rooms asking whether you want your towel washed daily. The energy and wastewater impacts of washing the towel at a hotel are little different than if you did so at home. The impact from washing a thousand rooms' towels, however, differs greatly. Although the environmental impacts from a single McDonalds drivethrough are minor, the cumulative impacts of 22 million meals served daily are significant.

A similar concern arises from cumulative services with more direct causal links to specific environmental harms. The services' pollutant emissions individually are negligible, but cumulatively significant and identifiable. The contribution of dry cleaners' volatile organic compounds to smog formation provides one example. Perhaps surprisingly, dentist offices provide another.

In the early 1990s, the San Francisco Bay Regional Water Quality Control Board started detecting significant levels of the heavy metal silver in the water, in sediment, and in tissues of fish and marine mammals in the Bay (Rejeski, 1998). But where was the silver coming from? No silver mines were anywhere near the Bay's watershed. A material flow analysis provided a surprising result, pointing a finger directly at dentist offices. Indeed, the 90,000 dentist offices in the United States account for roughly half of the more than 3,800 metric tons of silver consumed annually. The silver dissolves in fixer solutions used to develop x-rays and goes down tens of thousands of drains and eventually into bays and other watersheds. The small amount of fixer used at each office (less than 5 gallons per month at 80 percent of the sites) provides too little silver to offset the costs of recovery equipment, and RCRA presents serious regulatory burdens to on-site and off-site recovery efforts.

Thus cumulative services pose significant administrative challenges to regulation. This plays out first as an informational challenge. Using the silver discharges by dentists as an example, it is no simple task to link such diffuse emissions with an identifiable harm. Assuming the link has been established, however, how much silver should each office be allowed to discharge? There is a significant difference between regulators allocating SO_2 emissions among 3 smokestacks in an airshed and 1,200 dentist offices in a watershed. Determining equitable and efficient levels can be done, but at a high cost.

Compliance and monitoring expenses may be even higher. For pollutants with clearly identifiable impacts of concern, such as dental offices, auto repair shops, and dry cleaners, the traditional regulatory response has been local command-and-control regulation. Although the idea of a meaningful point source permit for every dentist office seems horribly resource intensive, it can work. Palo Alto's water district, one of the best funded and most sophisticated in the country, routinely regulates small services and inspects their premises. Its ordinance on photoprocessors and medical offices reduced silver levels in the Bay by more than 90 percent in 5 years (Palo Alto Regional Water Quality Control Plant, 1998). In the face of such informational, compliance, and enforcement costs, however, a more common response to cumulative services has been *no* response at all. As the environmental manager for Palo Alto's treatment plant observes, "People [in wastewater treatment] look to industrial sources and aren't used to thinking about services or residential activities as the source of the problem" (K. Moran, Manager, Palo Alto Pollution Prevention Program, personal communication, April 14, 1998). Although cumulative services' concentration of activities provides a more accessible target for permit-based regulations, the sheer number of sources overwhelms enforcement and compliance monitoring capacity.

Beyond a command-and-control approach, two complementary policies therefore deserve close consideration:

- Economic instruments such as taxes are particularly well suited to the harms posed by cumulative services because they direct diffuse behavior throughout a complex system with little need for permitting oversight. This capacity for self-regulation in a decentralized setting greatly reduces administrative oversight costs. Put simply, if you get the prices right, services will look after themselves in an environmentally responsible manner. Because most cumulative services concentrate individuals' environmental impacts, there is no need for "service-specific" economic instruments.
- A second approach relies on information dissemination. Unlike smokestack sources, cumulative services by definition do not face significant environmental regulation and therefore expend few management resources on the subject. That is not to say, however, that services have no interest in improving their environmental performance. In fact, there are a large number of voluntary initiatives in the service sector (Rejeski, 1997:31). Government can play a key role in supplementing and fostering these information exchanges. This can take the form of informing cumulative services that their activities cause environmental impact and providing guidance on proven practices to reduce these impacts.

Leverage Services

To a large extent, services are interstices of the economy, acting as market mediators to provide the commercial link between primary production (mining, agriculture, fishing), manufacturing, and end-users. A class of these services, known as *leverage services,* acts as a funnel through which products must flow. These include retailers, utilities, financial services, and fast-food chains.

In recent years a number of sectors have witnessed a shifting concentration of market share to a small number of companies. In the retail sector, this has resulted in a shift of commercial influence from producer to retailer. A small number of retailers such as Walmart, K-Mart, and Sears now account for roughly 10 percent of retail sales throughout the country (Guile and Cohon, 1997).[8] Toys-R-Us is twice the size of its two largest suppliers put together (Rejeski, 1997). Other leverage services have long dominated their product chain. In the restaurant sector, for instance, serving 22 million meals daily, McDonalds funnels enormous amounts of products to the consumer, influencing agriculture, ranching, and pulp and paper manufacture. Similarly, between the many coal mining and oil companies and the millions of electricity customers stand a small number of utilities.

Before considering the merits of extending laws to govern the leverage services for environmental protection, it should be recognized that this extension of environmental stewardship already occurs daily in the marketplace with no governmental intervention at all. As an alternative to vertical integration strategies, a growing number of major corporations (both services and manufacturers) have taken control over their supply chain, exercising leverage upstream in their product or service lifecycle. This is occurring voluntarily in response to three market forces: anticipation of consumer demands, direct consumer pressure, and secondary boycotts.

Seeking competitive advantage by anticipating consumer demands, major corporations have established environmental purchasing requirements. Most of these are based on product content, with companies requiring suppliers to provide recycled products such as paper or rerefined oil. The U.S. government, the biggest service provider of all and the world's largest consumer, has followed suit in its "green" procurement standards. In 1993, for example, President Clinton issued an executive order requiring every executive agency to practice waste prevention and recycling as well as to promote the market for recovered materials through its procurement process (Executive Order No. 12,873, 1993). Likewise, a number of companies set requirements for suppliers' process and production methods. The major home improvement retailers in the United States and Britain, for example, have committed to sell only tropical timber products certified from sustainably managed forests (The ENDS Report, 1998). Kinkos purchases 36,000 tons of white paper annually. It has considerable influence over its supplier paper mills and requires that they exercise sustainable natural resource management policies, including a commitment not to purchase wood or paper from old-growth forests (Kinkos, 2001). Major manufacturers have launched similar initiatives, because their large raw material and component requirements give them considerable influence upstream over their suppliers.

In response to direct consumer pressure, major corporations have interceded directly upstream to minimize adverse publicity. The most publicized examples have concerned labor practices. For example, charges of unsafe working conditions, child labor, and pitifully low wages have energized sneaker companies such as Reebok and Nike to improve the labor practices of their suppliers in Asia. Often these campaigns are directed at leverage services. The Campaign for Labor Rights' most recent campaign against Nike, for example, called for a boycott of the sport retailer Footlocker because "(1) Footlocker is Nike's largest retail outlet; (2) Nike is Footlocker's largest supplier; and (3) Footlocker is owned by Woolworth Corporation, which is concurrently embroiled in another sweatshop scandal involving the manufacturing facilities of its wholly-owned clothing subsidiaries in Canada."

The action proposed by the Campaign for Labor Rights against Footlocker, known as a secondary boycott, presents another common means to exert pressure indirectly. The Rainforest Action Network has followed this type of strategy in opposing Mitsubishi Corporation's tropical logging practices (Mitsubishi Motor Sales of America, 1997). This nongovernmental organization has organized a boycott of Unionbancal, formerly the Bank of California and Union Bank but now 80 percent owned by Mitsubishi, hoping to pressure the multinational to change the practices of its logging subsidiary (Crockett, 1996). Socially responsible investing and e-mail campaigns also are being used to influence the company.

What are the triggers for change in this group of services? Though the government does not have a large role to play, it can foster these voluntary developments. In particular, the EPA's voluntary programs for smokestack industries, such as the 33/50 initiative and Design for the Environment, may be worth adapting to the service sector. Under the 33/50 initiative, in 1990 EPA Administrator Bill Reilly sent letters to 1,300 companies operating 6,000 facilities in the United States. Reilly listed 17 priority chemicals and challenged the companies to reduce their emissions of these chemicals 33 percent by 1993 and by 50 percent by 1995. Participation was high, and the EPA's 50-percent goal was achieved a year early. By the end of 1995, the EPA reported emission reductions of more than 750 million pounds.[9]

The Design for the Environment program provides funding to promote companies' integration of environmental considerations into the design and redesign of products, processes, and technical and management systems (U.S. EPA, 2000). The EPA could direct a similar initiative to integrate environmental considerations into the practices of leverage services in reducing lifecycle impacts, whether they involve retailers in their selection of goods, fast-food chains in their purchasing practices, or banks in their lending practices.

The fusion of extended producer responsibility with reflexive law (i.e., laws that require generation, disclosure, and, hopefully, consideration of information) suggests a further intriguing potential development. Interviews with leverage service providers reveal equal parts interest and frustration. They understand their pivotal role in the lifecycle, but bemoan their lack of information and expertise to make decisions. Home Depot, for example, spent a number of years trying to establish a corporate policy regarding the arsenate in pressure-treated wood and ultimately found it too difficult a problem. As the discussion on cumulative services explained, because most services do not confront significant environmental issues in their daily operations, their institutional competency is weak, often with neither the in-house capacity to make decisions on sourcing nor access to much of the information they would need.

Reflexive law provides a means to overcome this barrier, relying on a disclosure rather than sanctioning approach. Laws such as the National Environmental Policy Act, the European Union's Eco-Management and Audit Scheme, and the Toxics Release Inventory are intended to enhance the information content of decisions. The goal of these statutes is not to constrain or dictate behavior, but rather to generate information and ensure its meaningful consideration.

Building off the categories already described, Box 2-1 sets out the threshold ques-

---

BOX 2-1
Regulating Services

For each service activity, determine:

Categories → *What is the environmental problem?*
- The direct impacts of the service itself? (smokestack services)
- The cumulative impact of small actions? (cumulative services)
- Significant impacts upstream or downstream? (leverage services)

Regulatory Action → *Do the environmental impacts warrant governmental intervention?*
- If so, do current regulations address the problem?
- If so, do they address it inefficiently, hindering additional improvements?

Regulatory Targets → *Whom should the intervention target?*
- Within the product's life cycle, which market actor is positioned to reduce the greatest environmental impact at least social cost?
- What are the equitable and legal constraints to placing this responsibility on the least-cost provider?

Policy Instruments → *Which form of intervention is most appropriate?*
- Should the instrument be voluntary or regulatory?
- If the service is administered inefficiently, should the action be deregulatory?
- Which instrument or combination of instruments is most efficient—e.g., command-and-control standards, subsidies, education, liability, information collection and dissemination, and so on?

---

tions to determine whether services should be regulated and, if so, in what manner. We have reached a point today where our existing approaches to environmental protection may not fit the dominant and growing service sector. But what if these same tools are becoming less applicable to the industrial sector they were originally designed to address? What if manufacturing and services are becoming more similar, not less?

MANUFACTURING AS A SERVICE

Most people have never heard of Selectron, Celestica, Flextronics, or SCI Systems but these companies make a majority of personal computers and PC peripherals today. In fact, so-called *contract manufacturers* produce nearly 100

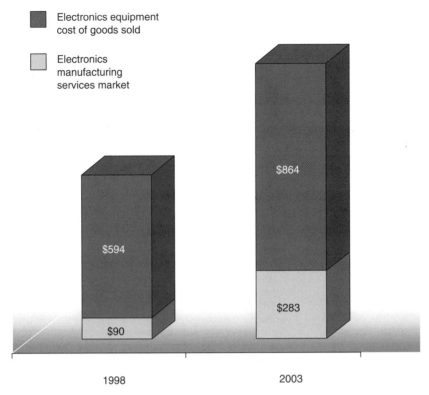

FIGURE 2-5 The growth of outsourcing.
Source: Clancy and Rejeski (2000). Reprinted with permission of RAND.

percent of all Hewlett Packard's personal computers and about 75 percent of their ink-jet printers. These companies represent the emergence of manufacturing as a service, a service that is becoming increasingly globalized. Revenue growth in the contract electronics-manufacturing sector has been exceeding 30 percent per year consistently since 1992 (see Figure 2-5). In the pharmaceutical industry, contract manufacturing of key chemical inputs accounted for 50 to 60 percent of production in 1998 and is projected to reach 60 to 70 percent by 2005 (Van Arnum, 2000).

If environmental policymakers are looking for emerging industrial sectors, contract manufacturing is one that will have increasing importance. It also serves as an indicator of larger changes in the manufacturing landscape.

Some people have viewed this trend as the emergence of a new model of industry organization, one reliant on the development of *turnkey production networks* (Sturgeon, 1997). This is a large departure from early organizational models where companies were concentrated in one geographical area, focused on one piece of the value chain, and were vertically integrated (Cohen, 2000). By using networked models, companies can now decouple production from innovation, thereby reducing manufacturing overhead and inventory/logistics costs, and focus on core values around product design and marketing. What began with IBM's decision to outsource its microprocessors and operating system has changed our industrial landscape.

Flexible, networked manufacturing will allow companies to effectively "deconstruct" their value chains and reassemble them close to cheap labor, large markets, and key customers (Evans and Wurster, 2000). Firms can shift to open-source models for manufacturing and postpone various aspects of the production process to the point of final assembly or use. This actually may transform the geography of production and shift new production away from traditional industrial corridors. For example, in 1980, 50 percent of auto production employment in the United States was concentrated in 16 counties. By 1996, only a third of manufacturing was concentrated in these counties (Helper et al., 1997). Much of this new manufacturing activity moved into new areas in the Southeast United States (see Figures 2-6a, 2-6b).

In a highly networked and deconstructed world, manufacturing does not look like manufacturing anymore; it begins to take on the characteristics of both mobile and nonpoint sources.   Right now, it is possible to purchase or lease turnkey production miniplants that will fit into 20- or 40-foot containers, transport these plants to nearly anywhere in the world, and make everything from baked goods to roofing materials, medical equipment, or mufflers. Imagine this scenario. Two German-made robotic-manufacturing modules are air lifted to Mexico and produce cell phones, one for an American firm and one for a Japanese firm. After 6 months, they are moved to Ireland and reprogrammed to produce parts for personal digital assistants for two firms, one in England and one in Thailand. Who is responsible for the environmental performance and compliance of these systems and their products?

The other possibility that has emerged is to completely decouple production codes from production. Design verification software now allows a three-person firm in California to design logic chips and ship the production code anywhere, such as to a silicon wafer fabrication plant in a jungle in Borneo (see Doler, 2000). This scenario is likely to become more and more common, especially for low-weight/high-value items that can be moved rapidly from far-flung production facilities to markets via airfreight.

Maybe the ultimate service will be the ability to manufacture at a personal level. Neil Gershenfeld at the Massachusetts Institute of Technology (MIT) Media Lab makes the point that fabrication today is where computation was 20 years ago (Clancy and Rejeski, 2000). It tends to occur in large, centralized facilities and it is only now finding its way out into the wider world (as the personal computer did) at smaller scales that allow customized production of short runs (lot-size-of-one). Take a look at what has happened to that workhorse

(a) Counties representing 50 percent of allied automobile employment in 1980

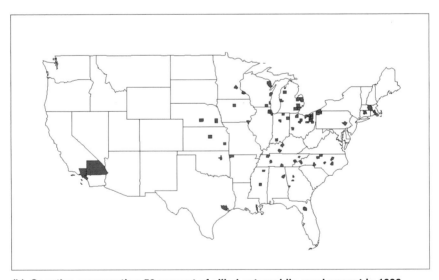

(b) Counties representing 50 percent of allied automobile employment in 1996

FIGURE 2-6 Allied automobile employment. Reprinted with permission of RAND.
Source: Clancy and Rejeski (2000).

FIGURE 2-7 Computer-driven, powder metallurgy press (foreground) with traditional press behind. Reprinted with permission of Mii Technologies.

of the first industrial revolution, the press. New powder metallurgy presses can generate twice the pressure in a fraction of the space and can produce parts 50 percent faster than traditional presses (Kluger, 2000). We now have a high-volume, computer-controlled production system that can almost fit on a desktop (see Figure 2-7). But change often moves in two directions. Take the workhorse of the information revolution, the printer, and turn it into a production machine. There are a wide range of desktop systems that allow very complex objects to be printed using polymer-based powders (see Figure 2-8).

We can begin to see the outlines of a world where production can take place nearly anywhere (see "Manufacturing Anywhere," in Clancy and Rejeski, 2000). In a recent book that explores manufacturing in the year 2020, the authors suggest that "steel manufacturing that could only be performed in Cleveland will be everywhere. Autos produced only in Detroit's mile-long factories will emerge from knockdown garage assembly shops in the Amazon and East Eighty-sixth Street in New York" (Moody and Morley, 1999).

It is not far fetched to imagine 10 to 20 years in the future, systems that store production codes on servers and allow the code to be downloaded to small-scale and personal fabrication devices, much in the way we download music today (we might describe this as an MP 3-D system). Another possibility is to upload production code directly from desktop computer-aided design and verification systems (see Figure 2-9).

FIGURE 2-8 Object with seven articulating joints from a 3-D printer. Courtesy of the MIT Media Lab.

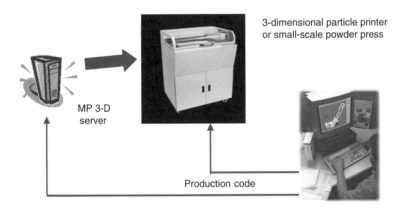

FIGURE 2-9 Schematic diagram of a system for the deconsruction and personalization of production.

From an environmental standpoint, the positive aspect is that we could produce where needed, moving bits (production code) to atoms (production process), and avoiding a significant transportation and logistics penalty. On the other hand, production then could take place anywhere, in thousands of unregulated, and largely nonregulatable, environments. That means that environmental considerations would have to be integrated into the production codes and operations of the fabricators (we would also need a corresponding capacity to defabricate).

In the more distant future, it may be possible to combine the capacity to self-fabricate with autonomous design based on evolutionary computation. Such production could be set in motion with the specification of a set of outcomes or characteristics we would desire from a yet-to-be designed or produced device. This process would result in one-of-a-kind products that have evolved to meet our specifications in Darwinian-like process (Lipson and Pollack, 2000). Production truly begins to replicate nature.

Environmental policy was not set up to handle highly dynamic and mobile production systems, systems that may become increasingly autonomous. The EPA has struggled for years to develop facility identification codes based on the premise that production stays put, or at least does not change faster than the phone book. The rules of the environmental protection game change if production begins to operate more like a service; if it can be moved, reprogrammed, and reconfigured; if it is organized using networks, not hierarchies. From a policy standpoint, it is important to understand that these emerging networks may require very different strategies than those applied to the hierarchies or markets where most environmental policy traditionally has focused (Powell, 1990).

More than 30 years ago, the modern environmental movement began by focusing on the byproducts of production. More recently, policies have advanced to consider the products of production (for instance, European take-back laws or the E.U. Integrated Product Policy). The challenge we now face is to focus on production itself, including the intimate relationship between production and services.

An incredible opportunity is appearing on the horizon. Let us assume that the management gurus and industrial researchers are right in their assessments of a rapidly globalizing service sector, a second industrial revolution, the deconstruction of value chains, an explosion in contract manufacturing, the *personalization* of fabrication, and the emergence of a digital economy. It would be like creating an environmental protection agency in the late 1800s when the first industrial revolution was occurring, when we had an opportunity to proactively shape the system rather than simply react to its adverse impacts for the next 100 years. Maybe our existing system of regulations will work in this rapidly changing world, but maybe not.

Management guru Drucker has made the comment that the theory of business is no more than a hypothesis that must be examined and tested continually (Drucker, 1999). The same is true of environmental policy. Built into our regulatory system and environmental policy institutions must be ways to continually test our system of regulation and the models of business on which our regulations are based. In today's rapidly changing world, the greatest danger posed to effective environmental policy will be unchallenged assumptions about the nature and dynamics of business. We need to put every regulation, every policy, and every assumption about the causes and effects of environmental damage on trial for life. Otherwise, we will face a radically transformed future both ill informed and underequipped.

NOTES

1   OECD is an international governmental organization dedicated to the promotion of policies that expand growth in market-based economies. Its 29 members include all of the major industrialized modern economies.

2   A December 1997 LEXIS/NEXIS database search found 125 separate newspaper and magazine stories contrasting the information and industrial revolutions. The following passage from *Foreign Affairs* is embellished, but typical of these references (Wriston, 1997:172):

> We are now living in the midst of the third great revolution in history. When the principle of the lever was applied to make a plow, the agricultural revolution was born, and the power of nomadic tribal chiefs declined. When centuries later, men substituted the power of water, steam, and electricity for animal muscle, the Industrial Revolution was born. Both of these massive changes took centuries to unfold. Each caused a shift in the power structure. Today, the marriage of computers and telecommunications has ushered in the Information Age, which is as different from the Industrial Age as that period was from the Agricultural Age. Information technology has demolished time and distance.

3   Consider the central role of information in the following descriptions:

> [I]n the changed world economy, the sources of higher productivity are increasingly dependent on knowledge and information applied to production, and this knowledge and information is increasingly science-based. Production in the advanced capitalist societies shifts from material goods to information processing activities that focus on symbol manipulation in the organization of production and in the enhancement of productivity (Carnoy et al., 1993:5).

> With rare exceptions, the economic and producing power of a modern corporation or nation lies more in its intellectual and systems capabilities than in its hard assets of raw materials, land, plant, and equipment (Quinn et al., 1997:20).

> A pre-industrial society is primarily extractive, its economy based on agriculture, mining, fishing, timber and other resources such as natural gas or oil. An industrial sector is primarily fabricating, using energy and machine technology, for the manufacture of goods. A post-industrial sector is one of processing in which telecommunications and computers are strategic for the exchange of information and knowledge (Bell, 1976:xi-xiii).

4   An additional 44 million jobs were added between 1980 and 1999 (Bureau of the Census, 2000). The *Statistical Abstract* lists the fastest growing occupations as home health aides, computer engineers and analysts, physical therapists, systems analysts, and correction officers (Bureau of the Census, 1995).

5   WRI researchers developed a new measurement unit of Total Material Requirements. This quantifies both the direct and indirect use of natural resources flowing through an economy. Direct material requirements include feedstock resources in the production process such as grain, copper, coal, and gas. Indirect material requirements include "hidden flows." These are natural resources that are not sold as commodities and never enter the economy, such as overburden and waste from

extractive activities, biomass from crop harvesting and logging, soil erosion from agriculture, and earth moved during construction.

6 A study at Rockefeller University concluded (Wernick, 1996:5):

[A]n assessment of consumption per unit of economic activity shows a dematerialization in physical materials of about one-third since 1970 . . . [I]ndividual items in the American economy may be getting lighter but the economy as a whole is physically expanding . . . We see no significant signs of net dematerialization at the level of the consumer or saturation of individual material wants.

Since 1950, per capita consumption of copper, steel, energy, timber, and meat has doubled; consumption of plastic has increased fivefold and aluminum sevenfold. Although America has the highest per capita consumption levels in the world, the resource consumption in Western Europe and Japan is only slightly less (Durning 1992:29, 38).

A 1997 study examined the consumption of a range of metals, minerals, agricultural chemicals, and pertroleum products in 32 countries over 21 years. They concluded that a general reduction in resource consumption was not evident in the most developed countries (Jänicke et al., 1997:467). The most exhaustive literature survey (Cleveland and Ruth, 1998:45) on the subject similarly concluded that "[d]espite claims to the contrary, there is no compelling macroeconomic evidence that the U.S. economy is decoupled from material inputs." OECD studied the *global* material intensity for steel and wood from 1970 through 1992. Throughout this period, although the material intensities of wood and steel showed a negative slope, the "total world materials consumption rose by 38%" (Organization for Economic Co-operation and Development, 1998:64-65). The linkage of economic growth and resource consumption also was confirmed by a recent government study (Interagency Working Group on Industrial Ecology, 1998).

Analyses of energy consumption and waste generation show similar results. Ausubel (1996:4) notes, "Although the soaring number of products and objects, accelerated by economic growth, raised municipal waste in the United States annually by about 1.6% per person in the last couple of decades, trash per unit of GDP dematerialized slightly."

7 The information about BellAtlantic is drawn from a consultant's report written for NYNEX (a corporate predecessor to BellAtlantic) (MacDonald, 1999), and personal communication with Roy Deitchmann (former environmental manager at NYNEX) on March 5, 1999.

8 Prior to becoming a law professor, the author served as the environmental manager for a major consumer products company. The company had a special sales office in a small town in Arkansas for the simple reason that Walmart's purchasing office was located there. If, for whatever reason, Walmart requested a change in product formulation or packaging, there was an immediate and compliant response. This represents a sharp break from the earlier balance of power in the consumer goods market. The retail trade traditionally has been highly fragmented. As a result, companies such as Procter & Gamble or Coca Cola, because of the importance of their products to consumers, generally hold the upper hand in negotiating with retailers.

9 Reviews of the 33/50 program have been mixed. Contrast Mazurek (this volume, Chapter 13) and Harrison (this volume, Chapter 16) with Karkkainen (2001).

REFERENCES

Adriaanse, A., S. Bringezu, A. Hammond, V. Moriguchi, E. Rodenburg, D. Rogich, and H. Schultz
  1998   *Resource Flows: The Material Basis of Industrial Economies.* Washington, DC: World Resources Institute.
Allenby, B.
  1997   Clueless. *Environmental Forum* (September):36.

Ausubel, J.
    1996   The liberation of the environment. *Daedelus* 125:1.
Bell, D.
    1976   *The Coming of Post-Industrial Society: A Venture in Social Forecasting.* New York: Basic Books.
Bureau of the Census
    1995   *Statistical Abstract of the United States.* Available from the Superintendent of Documents, U.S. Government Printing Office. Washington, DC: U.S. Department of Commerce.
    1997   *Statistical Abstract of the United States, 2000.* Available from the Superintendent of Documents, U.S. Government Printing Office. Washington, DC: U.S. Department of Commerce.
    2000   *Statistical Abstract of the United States.* Available from the Superintendent of Documents, U.S. Government Printing Office. Washington, DC: U.S. Department of Commerce.
Caldwell, L.K.
    1990   *International Environmental Policy: Emergence and Dimensions.* 2nd ed. Durham, NC: Duke University Press.
Carnoy, M., M. Castels, S.S. Cohen, and F.Cardoso, eds.
    1993   *The New Global Economy in the Information Age: Reflections on Our Changing World.* University Park: Pennsylvania State University Press.
Clancy, N., and D. Rejeski, eds.
    2000   *Our Future - Our Environment.* Santa Monica, CA: RAND. [Online]. Available: http://www.rand.org/scitech/stpi/ourfuture/ [Accessed: March 12, 2002].
Cleveland, H.
    1985   The twilight of hierarchy. In *Information Technologies and Social Transformation,* National Academy of Engineering, B. Guile, ed. Washington, DC: National Academy Press.
Cohen, M.
    2000   Survey—mastering management: All change in the second supply chain revolution. *Financial Times,* October 2.
Cohen, S.S., and J. Zysman
    1987   *Manufacturing Matters: The Myth of the Post-Industrial Economy.* New York: Basic Books.
Crockett, B.
    1996   Unionbancal faces boycott over Mitsubishi ties. *The American Banker,* May 10, p. 4.
Doler, K.
    2000   Jungle Fab. *Electronic Business,* November 1.
Drucker, P.
    1994   The age of social transformation. *The Atlantic Monthly* (November):53-56.
    1999   *Management Challenges for the 21st Century.* New York: Harper Business.
Durning, A.
    1992   *How Much Is Enough?* Washington, DC: Worldwatch.
The Economist
    1994   The manufacturing myth. *The Economist.* March 19, 91.
Ehrlich, P., G. Wolff, G.C. Daily, J.B. Hughes, S. Daily, M. Dalton, and L. Goulder
    1999   Knowledge and the environment. *Ecological Economics* 30(2):267-284.
The ENDS Report
    1998   Interest grows in label for products from sustainable forests. *The ENDS Report,* January, 276, 27.

Evans, P., and T. Wurster
2000 *Blown to Bits: How the New Economics of Information Transforms Strategy.* Boston: Harvard Business School Press.
Executive Order No. 12,873
1993 58 Fed. Reg. 54,911.
Guile, B., and J. Cohon
1997 Sorting out a service-based economy. In *Thinking Ecologically,* M. Chertow and D. Esty, eds. New Haven, CT: Yale University Press.
Helper, S., et al.
1997 Pollution Prevention Assistance in the Automotive Supply Chain. Unpublished report of the Weatherhead School of Management, Case Western Reserve University.
Interagency Working Group on Industrial Ecology
1998 *A Report of the Interagency Workshop on Industrial Ecology, Material and Energy Flows.* Washington, DC: U.S. Department of Energy. Office of Industrial Technologies.
Jänicke, M., M. Binder, and H. Mönch
1997 Dirty industries: Patterns of change in industrial countries. *Environmental and Resource Economics* 9.
Karkkainen, B.
2001 Information as environmental regulation: TRI and performance benchmarking, precursor to a new paradigm? *Georgetown Law Review* 89:257.
Kinkos
2001 Commitment to Action. [Online]. Available: http://www.kinkos.com/website/aboutus_module.jsp?suite= kinkenvironment&screen=6 [Accessed March 21, 2001].
Kluger, J.
2000 A New Factory for a New Age. *Time* 156(23). Available: http://www.time.com/time/magazine/article/0,9171,89548,00.html.
Lipson, H., and J.B. Pollack
2000 Automatic design and manufacture of robotic lifeforms. *Nature* 406:974-978.
McDonald, K.
1999 Rethinking environmental regulations impacting a telecommunications company. Unpublished consultant report written for NYNEX.
Mitsubishi Motor Sales of America
1997 The Agreement between Rainforest Action Network, Mitsubishi Motor Sales of America, and Mitsubishi Electric America, November 12.
Moody, P.E., and R.E. Morley
1999 *The Technology Machine: How Manufacturing Will Work in the Year 2020.* New York: Free Press.
Office of Industrial Technologies
no date Industries of the Future. [Online]. Available: http://www.oit.doe.gov/industries.shtml [Accessed March 21, 2001].
Organization for Economic Co-operation and Development
1998 *Economic Globalization and the Environment.* Paris: Organization for Economic Co-operation and Development.
Palo Alto Regional Water Quality Control Plant
1998 *Clean Bay Plan No. 56.* Palo Alto, CA: Regional Water Quality Control Plant.
Powell, W.W.
1990 Neither market nor hierarchy: Network forms of organization. *Research in Organizational Behavior* 12:295-336.
Quinn, J.B., J.J. Baruch, and K.A. Zien
1997 *Innovation Explosion.* New York: Free Press.

Rejeski, D.
    1997   An incomplete picture. *Environmental Forum* 14(5):26-34.
    1998   Mars, materials and morality lays. *Journal of Industrial Ecology* 1(4):13-18.
Riddle, D.
    1987   The role of the service sector in economic development: Similarities and differences by development category. In *The Emerging Service Economy,* O. Giarini, ed. New York: Free Press.
Rowthorn, R., and R.O. Ramaswamy
    1997   *Deindustrialization—Its Causes and Implications.* International Monetary Fund Working Paper WP/97/42. Washington, DC: International Monetary Fund.
Salzman, J.
    1999   Beyond the smokestack: Environmental protection in the service economy. *UCLA Law Review* 47:411-489.
Stern, P.C.
    1997   Toward a working definition of environmentally significant consumption. In *Environmentally Significant Consumption: Research Directions.* Committee on the Human Dimensions of Global Change, P.C. Stern, T. Dietz, V.W. Ruttan, R. Socolow, and J. Sweeney, eds. Washington, DC: National Academy Press.
Stewart, T.
    1993   The new era: Welcome to the revolution. *Fortune,* December 13, p. 76.
Sturgeon, T.
    1997   *Turnkey Production Networks: A New American Model of Industrial Organization?* Working Paper 92A. Cambridge, MA: MIT Center for Technology, Policy, and Industrial Development.
Thurow, L.
    1997   New rules. *Harvard International Review* 20:54.
U.S. Code
    1998a   42 U.S.C. § 401-416.
    1998b   42 U.S.C. § 202-216.
    1998c   42 U.S.C. § 6922 et seq.
U.S. Environmental Protection Agency
    2000   Design for the Environment. [Online]. Available: http://www.epa.gov/dfe [Accessed March 21, 2001].
Van Arnum, P.
    2000   Bulls or bears? Outlook in contract manufacturing. *Chemical Market Reporter,* February 14.
van Biema, M., B. Greenwald
    1997   Managing our way to higher service-sector productivity. *Harvard Business Review* July/August:87-88.
von Weizsacker, E., A.B. Lovins, and L.H. Lovins
    1997   *Factor Four: Doubling Wealth, Halving Resource Use.* London, Eng.: Earthscan Publications.
Wernick, I.K., R. Herman, S. Govind, and J.H. Ausubel
    1996   Materialization and dematerialization: measures and trends. *Daedalus* 125(3):171-198.
Wriston, W.B.
    1997   Bits, bytes, and diplomacy. *Foreign Affairs* (September/October):172.

PART II

INFORMATION AND EDUCATION FOR INDIVIDUALS, HOUSEHOLDS, AND COMMUNITIES

Introduction

In this part of the volume, the contributors examine the use of "new tools" to influence the behavior of individuals, households, and communities. We find it useful to distinguish between two general strategies for employing the new tools of communication and diffusion discussed in this part: social marketing and public education.

Chapters 3-8 examine influence attempts that follow a logic of social marketing (McKenzie-Mohr and Smith, 1999). A target behavior is identified on the basis of its presumed environmental benefits, and communication and diffusion instruments are mobilized to increase the prevalence of the target behavior in a target population. Social marketing interventions may use the full range of communication and diffusion instruments. They may appeal to the target group's values and beliefs, try to shape those values and beliefs, provide information or skills, elicit commitments, promote social norms and expectations, create partnerships with organizations that might be influential with the target population, and so forth. Like other kinds of marketing, social marketing works within and does not attempt to change the context set by social institutions, financial incentives, and existing infrastructure. It normally focuses on behaviors that have fairly direct impacts on environmental quality—behaviors such as recycling of household wastes, use of private or public transport, and household appliance purchases and maintenance, rather than on behaviors that may affect the environment indirectly by influencing public policy.

Proenvironmental social marketing often has been controversial in the United States. This is because people sometimes disagree sharply about whether it is proper for government agencies to use communication and diffusion instruments stronger than mere information provision for environmental policy purposes. In

Chapter 3, Lutzenhiser discusses some of the political debates since the 1970s over the social marketing of energy conservation. The extent to which governments are willing to use the more intrusive communication instruments—those involving persuasion, appeals to values, or efforts to change social norms—probably depends on the urgency of the behavioral objective and the strength of public support for it. These factors probably account for the long history of vigorous social marketing to promote disaster preparedness and public health measures such as vaccination and "safe sex" behaviors (see Chapters 6 and 7). That history may hold lessons for environmental social marketing, which has a shorter history and a sparser record of evaluation research.

Chapters 3, 4, and 5 review knowledge about the most extensively studied types of environmental social marketing—efforts aimed at decreasing household energy use, increasing participation in recycling programs, and increasing the market share of "green" household commodities. Some of these programs have been government sponsored, while others have relied partly or exclusively on nongovernmental organizations. It is worth noting that the target behaviors of these programs are not the most important ones in terms of direct environmental impact. Decisions about the size and location of one's dwelling unit, the purchase of motor vehicles, and the frequency and method of travel are more significant in environmental terms than most of the behaviors targeted by the programs reviewed here. We report on the well-studied cases in the hope that they can illuminate more general issues as well.

Chapters 6 and 7 complement the environmental chapters with summaries of lessons learned from social marketing in the areas of public health and disaster preparedness. These chapters are included not because the target behaviors are believed to have significant environmental impacts, but because the programs share some common elements with environmental social marketing. The extent to which these lessons may transfer to the environmental context is discussed in Chapter 8. It is worth noting that social marketing in the areas of public health and disaster preparedness has sometimes used communication and diffusion instruments in more aggressive ways than they have been used in environmental social marketing. The lessons of these efforts may be useful for governments or communities that attach sufficient urgency and importance to changing environmentally relevant behaviors to warrant adopting strong measures of communication and diffusion.

Public environmental education is a very different strategy conceptually from social marketing. As Ramsey and Hungerford define environmental education in Chapter 9, its main goal is to promote responsible citizenship behavior. The presumption is that if people develop solid knowledge about environmental processes and conditions and the skills necessary for effective citizenship, they will move the society in ways that will tend to provide the environmental protection that people want. Public education, defined in this way, does not try to change specific behaviors that have direct environmental impact. Rather, its aim is to increase the prevalence of effective citizenship behaviors that affect the environment only indirectly. The particular citizenship behaviors cannot be defined in advance because well-educated citizens will differ in how they participate, and even in the environmental goals they favor. Thus, the best test of environmental education as defined here is the level and sophistication of public involvement in environmental decision making at all levels of government and outside government. Environmental impact is only an indirect effect.

Public environmental education, like social marketing, is sometimes controversial in the United States. Some of this controversy can be attributed to the perception, correct or incorrect, that environmental education programs as actually implemented are disguised social marketing. This potential for confusion makes it useful to maintain a sharp conceptual distinction between the different logics of environmental education and social marketing, even if the distinction is sometimes blurred in practice. For example, educational organizations sometimes engage in aggressive social marketing with broad public support, as they do when they advocate against the use of illegal drugs. The conditions under which educational organizations are used for social marketing are probably similar to those under which other public organizations are used for this purpose: perceived urgency of the behavioral objective and strongly supportive social norms.

Chapters 9 and 10 discuss interventions that involve environmental education. Ramsey and Hungerford (Chapter 9) examine research on school-based environmental education programs, with a major focus on citizenship behavior as an outcome variable. Andrews, Stevens, and Wise (Chapter 10) develop a concept of "community-based environmental education" that is actually a hybrid of the educational and social marketing strategies. The ethical issues sometimes raised by combining education and marketing presumably are addressed because the interventions are aimed at adult members of the communities that create the programs. Thus, the targets of social marketing have had the opportunity to participate in its design.

Andrews and colleagues' community-based environmental education model uses many of the influence techniques common in integrated community-based environmental programs that do not describe themselves as educational. Community recycling programs (see Chapter 4) are a frequently studied example. Community-based programs also have been organized to clean up polluted rivers, decrease greenhouse gas emissions, and achieve other environmental objectives. Community-based environmental programs, whether or not described as educational, have not yet received systematic research attention. Nevertheless, some researchers and practitioners have examined available knowledge to identify program characteristics that seem to promote success in these programs (e.g., McKenzie-Mohr and Smith, 1999; Gardner and Stern, 1996: Chapter 7). These characteristics are discussed further in Chapter 12.

Chapter 11 examines community-based environmental programs through a

wider lens, focusing on their social and political contexts. It is commonly observed that certain communities are environmental and civic innovators across
many different areas. Chapter 11 provides some empirical grounding and a
theoretical framework to go with these observations. It presents a policy capacity framework for thinking about characteristics of communities and their contexts that enable them to take effective environmental action. It also considers
what governments at higher levels might do to provide favorable conditions for
local initiatives. Chapter 12 offers a conceptual framework and some tentative
conclusions regarding the usefulness of communication and diffusion instruments for changing behavior in individuals, households, and communities.

REFERENCES

Gardner, G.T., and P.C. Stern
    1996   *Environmental Problems and Human Behavior.*   Needham Heights, MA: Allyn and
           Bacon.
McKenzie-Mohr, D., and W. Smith
    1999   *Fostering Sustainable Behavior: An Introduction to Community-Based Social Market-
           ing.*   Gabriola Island, British Columbia, Can.: New Society Publishers.

3

Marketing Household Energy Conservation: The Message and the Reality

Loren Lutzenhiser

Τhis chapter explores what we know about the social marketing of energy conservation, a well-studied proenvironmental behavior that once again has appeared as a policy focus in energy-constrained parts of North America. Energy conservation (EC) requires changes in behavior that are often difficult to accomplish. Underlying patterns of energy use are hard to recognize and their impacts (on budgets, prices, ecosystems, and energy systems) are distant in time and space. Support for EC has been sporadic and intervention outcomes have been mixed. As a result, EC is hardly a poster child for social marketing success in the environmental arena.[1]

I first consider why the conservation of energy periodically has been identified as a public good and policy goal. Understanding shifting motivations helps us understand why the serious social marketing of EC actually has been quite rare in the United States and why its effects have been highly variable. Drawing on research and policy related to U.S. household nontransportation energy use—the area where most of the literature is centered—I then offer an overview of what we know about successes and failures of EC marketing. I conclude by considering what else we need to know to make conservation interventions more effective in the future.[2]

APPROACHES TO ENERGY CONSERVATION

In different historical periods, different views of energy have led to quite different reasons for pursuing EC (National Research Council, 1984). Prior to the 1970s, utility marketing was focused on promoting the expanded use of energy for personal benefit (comfort, convenience, luxury) and social good

49

(progress, prosperity, modernity). However, beginning in 1973, energy shortages and price spikes led to concerns about energy dependence, national security, and the economy. The social marketing of EC by the federal government and civic organizations began in this period, using television and print media advertising to promote simple measures to save energy, such as turning down the heat, using less hot water and lights, and installing weatherstripping and storm windows. The rationale was national interest—the situation being, in President Carter's words, "the moral equivalent of war."

With the passing of the crises, however, the marketing of EC almost entirely disappeared. The exceptions were a national appliance efficiency labeling program that continues today and localized efforts based on concern about the costs and impacts of energy supply system growth. In the 1980s, "demand side management" (DSM) programs were initiated in New England, the West Coast, and parts of the Midwest to secure EC reframed as "energy efficiency." Persons were asked to install insulation in housing, to purchase more efficient heating and cooling equipment, and to install upgraded appliances and lighting. Marketing contact with consumers shifted to the use of "bill stuffers" and other direct appeals, including the provision of free "weatherization" services for low-income households by community agencies. Under DSM, the policy logic saw EC as the "least cost" source of energy supply—with efficiency investments essentially allowing repurchase of energy from existing users at costs lower than new sources of supply. Social marketing under DSM focused on self-interest and bill reduction, sometimes with financial incentives included. Consumer appeals contained little mention of system-level goods or environmental benefits.

During the 1990s, as the prices of conventional fuels fell, investment in DSM declined. The little remaining social marketing of EC consisted of scattered environmental movement and government appeals to reduce greenhouse gas emissions (primarily on late-night public service announcements and Web sites). However, the appliance labeling requirement continued in place, and the U.S. Environmental Protection Agency (EPA) expanded its designation and labeling of a wide variety of products under the EnergyStar™ brand. The EPA's motivation was primarily to reduce the impacts of greenhouse gas emissions, and the agency worked quite effectively (and essentially on its own) with business groups to promote and bring to market more efficient technologies. In the early 1990s, similar programs were launched by some utilities, along with state governments and environmental groups (usually in regional consortia), to pursue this strategy—now termed "market transformation" (MT)—in an effort to get the greatest impact from shrinking DSM funding (Geller and Nadel, 1994; York, 1999). Consumer marketing strategies by MT initiatives have included television and print advertising, along with subsidies for lighting and appliance replacement. The bases of MT appeals have been environmental as well as economic. With emergent energy "crises" in California, the Northwest, and New

York in 2001, serious attention once again has been given to communicating the need for conservation directly to consumers on a mass basis.

In sum, during the past 25 years, the social marketing of EC has taken place only for brief periods. Comprehensive social marketing of the sort used for AIDS prevention or antismoking campaigns never has been tried in the case of energy in the United States, with the range of policy instruments used (efficiency appeals, advertisements, incentive payments, labeling) generally being quite restricted and applied only in selected settings.[3] Yet, as both a near-term imperative in some parts of the United States and a long-term environmental good on a global scale, the greater conservation of energy is clearly necessary, desirable, and quite feasible.[4]

MARKETING CONSERVATION AND CHANGING BEHAVIOR

Given its history, it is probably not surprising that the social marketing of energy conservation—even accompanied by incentives and subsidies—has been highly variable in its effects. Stern (2000) and others have argued for some time that various conservation program approaches have tended to overlook a variety of barriers and limiting factors, leading to spotty success (e.g., Gardner and Stern, 1996; Schultz, this volume, Chapter 4; Thøgersen, this volume, Chapter 5). Stern (this volume, Chapter 12) offers a good overview of our current knowledge of how social marketing and policy approaches can be better combined to confront barriers of different sorts.

As noted, over the past two decades, social marketing of EC has received little serious, well-funded, carefully targeted, or persistent attention in the United States. But with that said, and despite continuing general pessimism among energy policy analysts about the potential for household conservation response, we have observed recent significant reductions in energy use in California (as much as 10 to 12 percent) in response to a combination of conservation appeals, critical events, and policy interventions (e.g., California Energy Commission, 2002). A few earlier interventions, such as the Hood River Project (Hirst, 1987), that aggressively targeted whole communities also led to impressive responses, with nearly universal public adoption of energy efficiency innovation. But considerably more than simple mass advertising was involved in those efforts.

In Hood River, Oregon, in 1983, marketing messages were accompanied by Bonneville Power Administration financial incentives, a pervasive rationale (civic participation plus energy savings), and free technical assistance to households to help them evaluate EC alternatives and potential benefits (Keating and Flynn, 1984; also see Gardner and Stern, 1995:153-156 for an overview). Sociological research in the community prior to the project helped to develop a picture of community culture and social structure to use in the design of program offerings and marketing messages. A variety of regional and local groups provided support that ranged from dissemination of official program information and word-of-mouth accounts of project success to ongoing feedback about possible program improvements. Various mass media outlets and advertising strategies were used, and a menu of services was offered to the community's 15,000 residents. The result was a 15-percent decrease in electricity consumption, achieved by what we might call a comprehensive, multipronged social marketing strategy (see Andreasen, 1995, for a discussion of optimal social marketing design).

Unfortunately, the Hood River model was not reproduced elsewhere during the ensuing two decades. Furthermore, despite energy efficiency improvements in housing and appliances resulting from tighter regulations—and even with the support of a variety of DSM and MT initiatives—energy use in the United States continued to rise through the 1980s and 1990s (Energy Information Administration, 2002).

Beliefs and Background Knowledge

Why might this be? Public opinion polls in the 1970s and 1980s—and again in 2001—found that most persons believe energy problems are real, but generally attribute them to utility and government failures, rather than consumer behavior. At the same time, the public has consistently shown support for renewable energy and conservation, and has viewed energy use as something that could be better managed by consumers (e.g., Farhar et al., 1980; Farhar, 1993). During the 1970s and early 1980s, more than 80 percent of households reported that they had "cut back" on heating or air conditioning "to conserve energy" (Farhar, 1993:xviii). Recent polling shows that EC involving "real changes" is seen as a social necessity, even in parts of the United States with no prospects of energy shortages (Gallup Poll News Service, 2001).

But in-depth interviews have shown that the public's understandings of the connections among environment, energy, behavior, and conservation are quite limited (Kempton et al., 1995). Although appeals for energy conservation were widespread during the two 1970s energy crises, not all consumers were willing to conserve, and many likely conserved very little—a situation that recently has been observed in California as well (Lutzenhiser, 2001b).

Attitudes and Behavior

In part because of the uncertainties in predicting consumer behavior, household EC became an object of interest of psychologists in the late 1970s and early 1980s (for reviews, see Olsen, 1981; Costanzo et al., 1986). At the center of this work was an effort to link persons' *attitudes* to their likelihood to pursue conservation behaviors—that is, testing whether one's concerns about energy shortages, feelings of obligation to the community and society, or trust in scientific opinion would lead to changes in thermostat settings or fewer trips to the store (e.g., Heberlein and Warriner, 1982; Seligman, et al., 1979).

Because attitudes can be elicited, measured, and shaped by targeted appeals, EC attitude change has been the object of a good deal of investigation. Many researchers grounded their work in the Fishbein and Ajzen (1975) model—an approach that treats intentions to act, as well as the actions that follow from them, as outcomes of a cognitive balancing of the actor's attitudes with the influences of his or her social environment. The conservation message is inserted into this system in order to elicit a combination of attitude, intention, and behavior changes.   Some studies in this tradition reported significant positive relationships between attitudes and subsequent action (Seligman et al., 1979; Becker et al., 1981).   But others found situational factors (e.g., price, energy supply, weather, knowledge, income) to be more important (Stutzman and Green, 1982; Wilhelm and Iams, 1984).   Ester's (1985) rigorous test of the Fishbein-Ajzen model found it to be an overall weak predictor of energy conservation behavior.   Strong, favorable attitudes about conservation—whether the person already has these or whether they are developed as a result of social marketing—may have little to do with what he or she subsequently does.   It turns out that the circumstances of consumer choice are often more important than attitudes in predicting outcomes (Black et al., 1985; Stern, 2000).

Other psychologists, the behavior analysts, have focused their attention on how to change actual behaviors through selective use of stimuli and reinforcement. This view of social marketing cares little about attitude-altering messages, but is concerned with crafting appeals, information, and motivators that encourage and reward specific conserving actions (turning off the lights, limiting the use of hot water, closing unoccupied rooms), and discourage nonconserving actions.   Work in this vein focused on EC behavior change has been reviewed by Katzev and Johnson (1987).   Geller (2001) offers a current assessment of the state of behavior analysis related to the broader topic of environmentally significant behaviors in general.   He argues that quite different kinds of approaches and appeals are necessary to induce proenvironmental behaviors among different groups of consumers (who vary both in terms of their *consciousness* of behavioral impacts and in the *competence* of their actions).

Incentives and Prices

Stern (1992) points out that both attitudes and behaviors interact with a large number of other factors in EC (see also Stern, this volume, Chapter 12). Several of these—particularly incentives and knowledge—have been extensively studied.   During the 1980s, a number of DSM initiatives subsidized the purchase of energy-saving technologies (often through low-interest loans and rebates) and they marketed their programs with well-funded campaigns stressing expected monetary savings.   But customers often didn't buy into these DSM deals.   During this same period, related efforts to improve consumer background knowledge through appliance labeling programs, utility public service ads, and governmental appeals often had little or no effect (Dyer and Maronick, 1988). Research does suggest that "prompts" (i.e., "do this" or "don't do that" messages) had less effect than information accompanied by "feedback" about a person's own recent consumption.[5] More "humanized" information provided by video images seemed to be more effective, as were community role models and interactive, rather than "one-way," communications (e.g., via personal contacts with consumers and information delivered through local networks [Ester and Winett, 1982]).

Large financial incentives seem to have a marked effect, but even these were remarkably slow to penetrate the residential market. The frequent failure of incentives alone to induce conservation behavior has been traced to poor economic/energy information and the fact that different subgroups of consumers are differentially attracted by various inducements, leading to the ironic conclusion that "the stronger the financial incentives are, the more important the nonfinancial factors—especially marketing—become to a program's success" (Stern et al., 1986:162). But virtually none of these EC interventions were undertaken for long enough or with enough impetus to reveal the upper bound of possible effect.

If incentives present a mixed picture, then shouldn't high and rising energy *costs* have a demonstrable effect on conservation behavior—even granting limited energy knowledge (Kempton and Montgomery, 1982)? This is a bedrock assumption of the rational/economic model of action assumed in energy policy since the 1980s (Stern, 1986; Lutzenhiser, 1993). Once a month the utility presents a "price signal" to the consumer in the form of an energy bill from which he or she is required to gauge the effects of behavior and consider alternatives. In support of this "signal," utilities also have tried to fill in knowledge gaps, helping customers calculate the savings potentials of different efficiency investments.

However, different consumers react in quite different ways to price levels and price changes (Dillman et al., 1983; Schwartz and True, 1990). Research has suggested "price consciousness" itself to be a variable condition—and only one of a number of ways of relating to energy use (Heslop et al., 1981). Even if significant opportunities to save energy (and money) are present, only those with certain rationalistic styles may be able to appreciate that fact (Kempton, 1984). Also, consumers claiming to be well informed about energy and believing themselves to be acting in an economically rational fashion may often be mistaken. Several studies have found lack of consumer energy knowledge, including "information indispensable to even gross cost calculation" (Archer et al., 1986:78; see also Kempton and Layne, 1988).

In the end, the two most widely used theories of energy and behavior ("attitude change" and "economic rationality") generally have been unable to predict EC behavior (Archer et al., 1984). A few researchers have proposed alternative models focused on social networks and technology diffusion processes (Costanzo et al., 1986) or the interaction of economic, psychological, and social variables (Stern, 1986). Stern and Oskamp (1987), for example, offer a model that treats attitude-behavior processes as part of larger, nested systems of beliefs, events, institutions, and influential "background factors" (e.g., income, education, family size, and ecological variables) that shape and constrain action. However, apart from Stern's continuing interest in this area (e.g., Stern, 2000), there has been a dramatic decline in social psychological EC studies—despite the fact that these are obviously useful in developing policy (Coltrane et al., 1986).[6]

Some have suggested that, even in the absence of a well-developed model of behavior, it still might be possible to at least consider the relative EC effects of various program, incentive, and information combinations. Although the necessary data are not readily available, it is fair to say that EC social marketing efforts in the 1970s and more recently in 2001, when coupled with other policy initiatives and motivating events, might be expected to produce energy savings in the neighborhood of 10 percent. However, the persistence of these savings is widely doubted by energy analysts, and it is reasonable to suppose (again, see Stern, this volume, Chapter 12) that weakening of the combined effects of social marketing, incentives, and context factors ought to probably result in a reduction of EC efforts.

WHY PERSISTENT NONRESPONSE?

So, behavior is resilient, consumption is expansive, and standard models are of limited value. Is the problem intractable? Perhaps not. There is another way to look at it. From a sociological point of view, the important "contextual factors" (and even the "personal capabilities") that Stern (2000) and others argue should be used to improve psychological models also can be seen as elements of highly organized social systems—systems that actively shape and constrain individual behavior. From this point of view, the question "Why haven't we had greater success with our efforts to promote energy conservation?" is best addressed by considering a set of system characteristics, including the social *embeddedness* of energy use, the *constrained* nature of household choice, the *countermarketing* of consumptive lifestyles and behaviors, and the lack of *impetus* for change.[7]

Embeddedness

Social life is inherently *energetic* (Cottrell, 1955; Rosa et al., 1988), which means that the energy flows we might want to alter are both ubiquitous and invisible—hidden in walls and machines, habits and routines. What's more, both energy and EC behaviors (such as the investment in more efficient energy-using equipment) are embedded in cultural systems with logics that have little to do with either energy or investment. Here we're talking about the logics and behaviors involved in homemaking, childrearing, cooking and cleaning, getting by, and doing well.[8]   As a result, consumption is deeply embedded in social routines and personal habits (Lutzenhiser, 1988; Hackett and Lutzenhiser, 1985).

Social actors are competent masters of their routines and habits. But as such, there is no reason to expect them to have much energy "literacy," to know how their houses are built or equipped (Singh et al., 1989), or to know how to think and act responsibly about efficiency choices (Kempton and Montgomery, 1982). The energy flows that sustain them are also usually a collective accomplishment of households and follow a social division of labor (Kempton and Krabacher, 1984), with divided responsibilities for matters involving appliances, energy bills, and conservation (Wilhite and Wilk, 1987; Wilhelm and Iams, 1984; Klausner, 1979; Ritchie and McDougall, 1985).

Often the cultural categories useful for understanding social worlds not only conceal energy flows, but demand actions that are nonconserving. Social identities are at stake in those worlds, and serious issues of respectability, authority, and stigma are implicit in even the most routine energy-using behaviors such as cleaning, bathing, or entertaining. To "conserve" may be an action associated with poverty, and loss of comfort, convenience, cleanliness, or pleasure, and with the threat of social costs incurred if these losses are noticed by others.

Finally, because social status often is achieved and displayed through possession of the largest houses, cars, appliances, or bathtubs, high rates of consumption in the United States are actually a *constitutive feature* of the social matrix within which the actor and his or her social appliances are enmeshed (Lutzenhiser and Gossard, 2000). Energy use tends to be hierarchical and embedded in lifestyle requirements that are socially imposed, rather than a matter of individual choice or preference (Electric Power Research Institute, 1990; Lutzenhiser, 1993; Schipper et al., 1989; Wilhite et al., 2001; Lutzenhiser and Gossard, 2000).

Constraints on Choice

Energy conservation also is rendered difficult by the physical characteristics of buildings, systems, and even infrastructures. There is often a good deal of uncertainty about just how much EC might be possible because of difficulties associated with building orientation, layout, and major systems. Even when persons know what can and should be done (e.g., in terms of insulation, windows, furnaces, landscapes), they are routinely constrained by lack of financial and professional/commercial resources. Only innovations that are allowable under code, and that the utility agrees to connect to its system, can be adopted. One can only choose what's available in the marketplace and what can be installed and maintained. These choices are often severely constrained by lack of enthusiasm for more efficient products by the trades and businesses in the technology supply chain who are risk averse and resistant to innovation.

Countermarketing

At the same time, would-be social marketers of EC find themselves in the midst of many other social marketing efforts, including the American cult of newness and continuous bombardment by "bigger is better" and "shop 'til you drop" messages. These cultural values are not universally held, of course, and a variety of lifestyles are not oriented to excess (some are even oriented to frugality, simplicity, and sustainability). Certainly there is no one-to-one correlation between the amounts that people spend and their associated environmental impacts.[9] However, the core beliefs of consumerism are widely shared, and they provide fertile ground for consumption messages.

The marketing of expanded consumption is incessant in media representations of ways of life that also offer advertiser opportunities for "product placement" on television programs and in movies. Furthermore, the merchandising of goods in retail settings, coordinated with marketing, media, and consumer culture, not only shapes choices and encourages buying, but creates new "needs" on the spot (Ewen, 1979; Stein, 1979; Underhill, 1999). All of these processes, in the hands of skilled agents of change, work together to effectively counteract social marketing messages about conservation.[10]

Impetus

Finally, for social marketing in support of EC to be successful, there has to be some impetus for change—some sponsorship for conservation. With ample energy supplies, low prices, little interest in the global climate consequences of expanded energy use, lack of a "green" political party, and stalemate on environmental policies between the executive and legislative branches, there has been a lack of federal sponsorship for EC (Lutzenhiser, 2001a). EPA efforts to build the EnergyStar™ program and brand (recently with the assistance of the U.S. Department of Energy) are clear exceptions. But in the context of overall federal inactivity, those efforts must be seen as quite modest investments.

In some states, environmental interests and utility regulators have had the political will and leverage to mount EC programs under DSM and MT. But in these cases, the interventions also have had usually only modest ambitions—as well as a hardware orientation, little serious or creative social marketing messaging, and often only the halfhearted sponsorship of utilities who have been quick to shed their EC programs when faced with the prospects of energy system deregulation.[11]

Outside of the utility system, there is an emerging environmental critique of consumerism (Durning, 1992), some environmental organization focus on household conservation (e.g., Fickeisen, 1990), and some evidence of nascent green consumer movements and green market responses (Hawken et al., 1999). But there has been little supporting interest in EC by governments or corporations until the recent energy shortages, blackouts, political turmoil, and (in California's case) renewed demands for social marketing and conservation. There is little reason to believe that the social marketing of EC will be sustained in the absence of crisis. Long-term impetus for serious changes in energy use patterns can only come from a serious U.S. greenhouse gas reduction policy—something that has yet to receive needed support in public discourse or policy debate.

RESEARCH DIRECTIONS

Currently, very little research of any sort related to people and energy—let alone work on EC promotion—is taking place.[12] The recent energy supply problems may offer opportunities for evaluation researchers to examine the effectiveness of different social marketing approaches and incentives schemes under crisis conditions. But the need for basic research will remain.

Stern (2000) argues that, because environmentally significant behavior such as energy consumption is a product of complex interactions among a broad range of factors, effective interventions likely require a combination of intervention approaches (e.g., incentives + community-based programs + information + removal of barriers and constraints). But our knowledge of how these approaches work singly, let alone in combination, is quite limited. So some key areas of inquiry include:

- How might we improve the flow of EC information through the dominant existing channels such as the utility bill? An example might be the provision of comparative information to consumers, allowing them to see how their energy use measures up to neighborhood norms (Brant and Kempton, 1998; also see Schultz, this volume, Chapter 4, on recycling norms). Another prominent source of energy information is the appliance label. U.S. research in this area (e.g., Egan et al., 2000) has produced some provocative preliminary findings about how to make the label more effective through the use of better graphical presentation of energy data, and a recent comprehensive label redesign project in India shows that a good deal more effective social marketing can take place through that vehicle (Dethman et al., 2000). A hybrid experiment might also test the labeling of energy sources (such as fossil, renewables, conservation) on utility bills (see Thøgersen, Chapter 5, on "eco-labeling").

- The increasingly deregulated energy system will offer new opportunities for households to monitor their own consumption and communicate with suppliers and vendors of EC products and services via "smart meters" and the Internet. Not everyone will benefit from these developments, however, and the questions of how the new technologies will be shaped, how they might be used, and who might be excluded from their benefits all warrant study.

- Some fundamental accounting work is needed to compare the consumptiveness of different lifestyles and their relative social and environmental costs (e.g., Lutzenhiser and Hackett, 1993; Lutzenhiser, 1997). Large differences in energy flows, environmentally significant behaviors, target consumer groups, and equity problems in vulnerable populations can be identified in this way.

- We need to better understand the "chicken and egg" nature of demand—the ways in which consumer culture and choices in the marketplace interact with producer/retailer decisions and efforts to create needs, shaping consumption and driving the expansion of environmental impacts (Wilhite and Lutzenhiser, 1999; Wilhite et al., 2001). A hard look at emerging green consumer countermovements also is warranted.

- The continuing expansion of societal energy consumption is evidence of growth in the development and diffusion of energy-using devices and technologies. It is also evidence of the unintended effects of policies not specifically directed toward energy or the environment, such as zoning and land use regulations, fuel subsidies, transportation planning, building codes, industry protection arrangements, and so on. The effects of these policies on consumer choice and the escalation of consumption also should be more carefully examined (Wilhite et al., 2001).

Other research efforts can be imagined as well. However, the listed topics would help us to achieve a better understanding of how consumers can be enabled in their EC efforts through information, social support, removal of constraints, and the disembedding of various forms of consumption.

NOTES

1 Although "conservation" often may be equated with "curtailment" (and attendant suffering), we use the term in a technical sense, denoting reductions in rates of energy use that result from a wide array of choices and actions. Some of these involve simple behaviors such as turning off the lights when not using certain spaces. Others involve conservative strategies for the use of heating and air conditioning that may produce significant energy savings while having little or no noticeable effect on comfort. Still others involve the purchase of new high-performance appliances and equipment. Although these latter hardware choices frequently are identified in policy analysis as "efficiency measures," we treat them here as also falling under the broad rubric of "conservation."

2 The literature on energy conservation is spread across a variety of academic and applied literatures, including energy policy studies, psychology, sociology, anthropology, and applied energy program analysis reported largely in conference proceedings and evaluation reports. As a result, it is not always easily accessible. However, a number of detailed summaries have been undertaken, and these serve as a key set of resources. There are also a number of good literature reviews and metatheoretical articles synthesizing empirical work on human energy use and conservation behavior and critically evaluating various approaches to understanding and influencing that behavior.

Key sources include previous National Academy of Sciences/National Research Council studies on energy and society (National Research Council, 1984), the "human dimensions" of global environmental change (National Research Council, 1992), and the linkages between consumption and the environment (National Research Council, 1997). For this chapter, I have also relied heavily on several other broadly focused critical reviews of the literature (Lutzenhiser, 1993; Lutzenhiser et al., 2001; Shove et al., 1998; Wilhite et al., 2001); specialized reviews of research on conservation behavior change (Katzev and Johnson, 1987; Black et al., 1985; Stern, 1992); and several reviews more narrowly focused on particular subtopics in the area, including work on public opinion on energy and conservation (Farhar, 1993), the effectiveness of targeted incentives (Stern et al., 1986), research on feedback to consumers about their own behavior and consumption (Farhar and Fitzpatrick, 1989), the measurement of energy savings from education/information interventions (Green and Skumatz, 2000), and the effectiveness of various billing formats (Brant and Kempton, 1998; Lutzenhiser et al., 2001).

3   The reasons for this are fairly straightforward. They include: low energy prices, interest by the utilities in sales growth, the marginal status of climate change issues in U.S. policy discourse, and lack of linkage between personal consumption and environmental impacts in consumer consciousness.

4   There is a well-documented "efficiency gap" (Lovins, 1977; Nadel et al., 1998) between what engineers and energy policy analysts know to be technically possible and what's available in the marketplace (e.g., in terms of appliance efficiency, housing design, lighting systems). Also, so much variability in consumption can be observed in the population—with vastly different amounts of energy being used to "power" different households in different social and geographic locations (Socolow and Sonderegger, 1976; Lutzenhiser and Hackett, 1993; Lutzenhiser, 1997)—that what's socially imaginable seems to be potentially quite elastic.

5   See Katzev and Johnson (1987) for a detailed review of the feedback literature, as well as Farhar and Fitzpatrick (1989).

6   This is shown by both the work reported in Lutzenhiser (1992) and a review of the literature from 1990 to 2001 conducted for this chapter. Both reveal a marked decline in energy-related social science research since the mid-1980s.

7   For a more fully developed discussion of these forces, see Lutzenhiser (2001c).

8   See Granovetter (1985) for a discussion of the embeddedness of economic activity—a partial model for the usage proposed here. He noted that economic activity (buying, selling, renting, investing), rather than being part of a separate purely economic sphere, is found deeply rooted in social networks, peer relations, families, communities, institutions, tribes and so on. In fact, it is only through these networks of social relations that economic activities can take place.

9   For example, the information economy and high-priced branded goods can absorb a good deal of money without dramatic increases in environmental withdrawals or deposits.

10   For an expanded discussion of the effects of the "3-Ms" (marketing, media, and merchandising) on consumption, see Lutzenhiser (2001c).

11   Utilities often have been halfhearted participants in efficiency efforts—except for a few large public utilities with environmental constituencies (e.g., Seattle, Tacoma, Eugene, Sacramento) and some privately owned utilities with expanding demand and contracting sources of supply. Most utilities, however, have matched their growing loads with a continuous growth in supply. Supply expansion is part of their organizational culture and business strategy, and unabashed enthusiasm for the supply side has returned, as the movement to deregulate electric utilities has gained momentum and as efforts have grown by the federal government to expand energy production.

12   As DSM programs continue on a limited basis in some locales, and MT efforts are mounted in others, a bit of evaluation research continues. Little of this is household focused, however. Utilities frequently survey their consumers about satisfaction and public opinion on deregulation, but until recently have had little interest in EC or the environment. Not surprisingly, these polls are finding significant popular support for green power, even at premium prices. But utilities can be reluctant to act on these sorts of consumer "demands" when they find them (Farhar and Coburn 2000).

The EPA's EnergyStar™ efforts are supporting some research on billing systems (W. Kempton, University of Delaware, 2000) and a study of possible improvements to the U.S. standard appliance label (Egan et al., 2000). Scattered researchers in the energy system and academic communities continue to offer interesting findings at the biennial American Council for an Energy Efficient Economy conferences (e.g., Lutzenhiser and Goldstone, 2000). But with few exceptions, the research that is being done is fairly ad hoc. It is also interested largely in local program improvements and is disconnected from social science work in this area. For their parts, many social researchers have left the energy and behavior area to work on the more general topic of human systems and environmental change—where analysis seems to be taking a more macro turn (e.g., see Shove et al., 1998; Wilhite et al., 2001).

REFERENCES

Andreasen, A.A.
 1995 *Marketing Social Change: Changing the Behavior to Promote Health, Social Development, and the Environment.* San Francisco: Jossey-Bass.
Archer, D., M. Costanzo, B. Iritani, T.F. Pettigrew, I. Walker, and L. White
 1986 Energy conservation and public policy: The mediation of individual behavior. In *Energy Efficiency: Perspectives on Individual Behavior,* W. Kempton and M. Neiman, eds. Washington, DC: ACEEE Press.
Becker, L.J., C. Seligman, R.H. Fazio, and J.M. Darley
 1981 Relating attitudes to residential energy use. *Environment and Behavior* 13:590-609.
Black, S.J., P.C. Stern, and J.T. Elworth
 1985 Personal and contextual influences on household energy adaptations. *Journal of Applied Psychology* 70:3-21.
Brant, T., and W. Kempton
 1998 Billing System Research Program. Prepared for the California Institute for Energy Efficiency, Market Transformation Research Scoping Project. University of Delaware.
California Energy Commission
 2002 *Summer 2001 Conservation Report.* Sacramento, CA: California Energy Commission.
Coltrane, S., D. Archer, and E. Aronson
 1986 The social-psychological foundations of successful energy conservation programmes. *Energy Policy* 14:133-148.
Costanzo, M., D. Archer, E. Aronson, and T. Pettigrew
 1986 Energy conservation behavior: The difficult path from information to action. *American Psychologist* 41:521-528.
Cottrell, F.W.
 1955 *Energy and Society: The Relation Between Energy, Social Change and Economic Development.* New York: McGraw-Hill.
Dethman, L., I. Unninayar, and M. Tribble
 2000 Transforming appliance markets in India: Consumer research leads the way. *Proceedings, American Council for an Energy Efficient Economy.* 8.51-8.64. Washington, DC: ACEEE Press.
Dillman, D.A., E. Rosa, and J.J. Dillman
 1983 Lifestyle and home energy conservation in the U.S. *Journal of Economic Psychology* 3:299-315.
Durning, A.T.
 1992 *How Much Is Enough? The Consumer Society and the Future of the Earth.* New York: W.W. Norton.
Dyer, R.F., and T.J. Maronick
 1988 An evaluation of consumer awareness and use of energy labels in the purchase of major appliances—A longitudinal analysis. *Journal of Public Policy and Marketing* 7:83-97.

Egan, C., C.T. Payne, and J. Thorne
    2000   Interim findings of an evaluation of the U.S. EnergyGuide label. Pp. 8.77-8.91 in *Proceedings, American Council for an Energy Efficient Economy*. Washington, DC: ACEEE Press.
Electric Power Research Insitute
    1990   *Residential Customer Preference and Behavior: Market Segmentation Using CLASSIFY*. Report No. EM-5908. Palo Alto, CA: Electric Power Research Institute.
Energy Information Administration
    2002   *Energy Perspectives: Trends and Milestones, 1949-2000.* Washington, DC: U.S. Department of Energy.
Ester, P.
    1985   *Consumer Behavior and Energy Conservation.* Dordrecht, Neth.: Martinus Nijhoff.
Ester, P., and R.A. Winett
    1982   Toward more effective antecedent strategies for environmental programs. *Journal of Environmental Systems* 11:201-221.
Ewen, S.
    1976   *Captains of Consciousness: Advertising and the Social Roots of the Consumer Culture.* New York: McGraw Hill.
Farhar, B.
    1993   *Trends in Public Perceptions and Preferences on Energy and Environmental Policy.* Report No. NREL/TP-461-4857. Washington, DC: National Renewable Energy Laboratory.
Farhar, B., and T. Coburn
    2000   Some recent research on the markets for residential renewable energy. *Proceedings, American Council for an Energy Efficient Economy.* 8.93-8.108. Washington, DC: ACEEE Press.
Farhar, B., and C. Fitzpatrick
    1989   *Effects of Feedback on Residential Electricity Consumption: A Literature Review.* Report No. SERI/TR-254-3386. Golden, CO: Solar Energy Research Institute.
Farhar, B., C.T. Unseld, R. Vories, and R. Crews
    1980   Public opinion about energy. *Annual Review of Energy* 5:141-172.
Fickeisen, D.
    1990   The global action plan: Joining a household ecoteam will give you support for getting your life in balance with the earth. *In Context* (Summer):60.
Fishbein, M., and I. Ajzen
    1975   *Belief, Attitude, Intention and Behavior: An Introduction to Theory and Research.* Reading, MA: Addison-Wesley.
Gallup Poll News Service
    2001   *Public Blames Oil and Electric Companies Most for Current Energy Problems.* May 29, 2001. [Online]. Available: http://www.gallup.com/poll/releases/pr010529.asp [Accessed: October 20, 2001].
Gardner, G., and P.C. Stern
    1996   *Environmental Problems and Human Behavior.* New York: Allyn and Bacon.
Geller, H., and S. Nadel
    1994   Market transformation strategies to promote end-use efficiency. *Annual Review of Energy and the Environment* 19:301-346.
Geller, S.
    2001   The challenge of increasing pro-environment behavior. In *Handbook of Environmental Psychology.* R.B. Bechtel and A. Churchman, eds. New York: John Wiley and Sons.

Granovetter, M.

1985 Economic action and social structure: The problem of embeddedness. *American Journal of Sociology* 91:481-510.

Green, J., and L. Skumatz

2000 Evaluating the impacts of education/outreach programs—Lessons on impacts, methods, and optimal education. 8.123-8.136. *Proceedings, American Council for an Energy Efficient Economy.* Washington, DC: ACEEE Press.

Hackett, B., and L. Lutzenhiser

1985 The unity of self and object. *Western Folklore* 4:317-324.

Hawken, P., H. Lovins, and A. Lovins

1999 *Natural Capitalism: Creating the Next Industrial Revolution.* New York: Little, Brown and Co.

Heberlein, T.A., and K. Warriner

1982 The influence of price and attitude on shifting residential electricity consumption from on- to off-peak periods. *Journal of Economic Psychology* 4:107-130.

Heslop, L.A., L. Moran, and A. Cousineau

1981 'Consciousness' in energy conservation behavior: An exploratory study. *Journal of Consumer Research* 8:299-305.

Hirst, E.

1987 *Cooperation and Community Conservation.* Report No. 1300Y 1987. Portland, OR: Bonneville Power Administration..

Katzev, R.D., and T.R. Johnson

1987 *Promoting Energy Conservation: An Analysis of Behavioral Research.* Boulder, CO: Westview.

Keating, K., and C.B. Flynn

1984 Researching the human factor in Hood River: Buildings don't use energy, people do. *Proceedings of the American Council for an Energy Efficient Economy.* Washington, DC: ACEEE Press.

Kempton, W.

1984 Residential hot water: A behaviorally driven system. Pp. 229-244 in *Energy Efficiency: Perspectives on Individual Behavior,* W. Kempton and M. Neiman, eds.. Washington, DC: ACEEE Press.

Kempton, W., J. Boster, and J. Hartley

1995 *Environmental Values in American Culture.* Cambridge, MA: MIT Press.

Kempton, W., and S. Krabacher

1984 Thermostat management: Intensive interviewing used to interpret instrumentation data. Pp. 245-262 in *Energy Efficiency: Perspectives on Individual Behavior,* W. Kempton and M. Neiman, eds. Washington, DC: ACEEE Press.

Kempton, W., and L. Layne

1988 The consumer's energy information environment. *Proceedings of the American Council for an Energy Efficient Economy.* 11.50-11.66. Washington, DC: ACEEE Press.

Kempton, W., and L. Montgomery

1982 Folk quantification of energy. *Energy* 7:817-827.

Klausner, S.

1979 Social order and energy consumption in matrifocal households. *Human Ecology* 7:21-39.

Lovins, A.B.

1977 *Soft-Energy Paths: Toward a Durable Peace.* Cambridge, MA: Ballinger.

Lutzenhiser, L.

1988 Embodied Energy: A Pragmatic Theory of Energy Use and Culture. Ph.D. dissertation, Department of Sociology, University of California-Davis.

1992   A cultural model of household energy consumption. *Energy-The International Journal* 17:47-60.

1993   Social and behavioral aspects of energy use. *Annual Review of Energy and the Environment* 18:247-289.

1997   Social structure, culture and technology: Modeling the driving forces of household consumption. Pp. 77-91 in *Environmentally Significant Consumption: Research Directions,* National Research Council, Committee on the Human Dimensions of Global Change, P.C. Stern, T. Dietz, V.W. Ruttan, R.H. Socolow, and J. Sweeney, eds. Washington, DC: National Academy Press.

2001a   The contours of U.S. climate non-policy. *Society and Natural Resources* 14:511-523.

2001b   Findings from research in progress.

2001c   Greening the economy from the bottom-up? Lessons in consumption from the energy case. Pp. 345-356 in *Readings in Economic Sociology,* N.W. Biggart, ed. Oxford, Eng.: Blackwell.

Lutzenhiser, L., and S. Goldstone

2000   Introduction: Panel 8, consumer behavior and non-energy effects. *Proceedings, American Council for an Energy Efficient Economy.* 8.xiii-8.xvi. Washington, DC: ACEEE Press.

Lutzenhiser, L., and M.H. Gossard

2000   Lifestyle, status and energy consumption. *Proceedings, American Council for an Energy Efficient Economy.* 8.207–8.222. Washington, DC: ACEEE Press.

Lutzenhiser, L., and B. Hackett

1993   Social stratification and environmental degradation: Understanding household CO_2 production. *Social Problems* 40:50-73.

Nadel, S., L. Rainer, M. Shepard, M. Suozzo, and J. Thorne

1998   *Emerging Energy-Saving Technologies and Practices for the Buildings Sector.* Report A985. Washington, DC: ACEEE Press.

National Research Council

1984   *Energy Use: The Human Dimension.* Committee on the Behavioral and Social Aspects of Energy Consumption and Production. P.C. Stern and E. Aronson, eds. Washington, DC: National Academy Press.

1992   *Global Environmental Change: Understanding the Human Dimensions.* Committee on the Human Dimensions of Global Change. P.C. Stern, O. Young, and D. Druckman, eds. Washington, DC: National Academy Press.

1997   *Environmentally Significant Consumption: Research Directions.* Committee on the Human Dimensions of Global Change. P.C. Stern, T. Dietz, V. Ruttan, R.H. Socolow, and J. Sweeney, eds. Washington, DC: National Academy Press.

Olsen, M.E.

1981   Consumers attitudes toward energy-conservation. *Journal of Social Issues* 37:108-131.

Ritchie, B., and G. McDougall

1985   Designing and marketing consumer energy conservation policies and programs: Implications from a decade of research. *Journal of Public Policy and Marketing* 4:14-32.

Rosa, E.A., G.E. Machlis, and K.M. Keating

1988   Energy. *Annual Review of Sociology* 14:149-172.

Schipper, L., S. Bartlett, D. Hawk, and E. Vine

1989   Linking lifestyles to energy use: A matter of time? *Annual Review of Energy* 14:273-318.

Schwartz, D., and B. True

1990   What households do when electricity prices go up: An econometric analysis with policy implications. *Proceedings, American Council for an Energy Efficient Economy.* 2.121-2.130. Washington, DC: ACEEE Press.

Seligman, C., M. Kriss, J. Darley, R.H. Fazio, L.J. Becker, and J.B. Pryor
 1979   Predicting summer energy consumption from homeowners' attitudes. *Journal of Applied Social Psychology* 9:70-90.
Shove, E., L. Lutzenhiser, S. Guy, B. Hackett, and H. Wilhite
 1998   Energy and social systems. Pp. 201-234 in *Human Choice and Climate Change*, S. Rayner and E. Malon, eds. Columbus, OH: Battelle Press.
Singh, N., P. Ong, and S. Holt
 1989   Evaluating Biases in Conditional Demand Models. Report to the Universitywide Energy Research Group and the California Energy Commission, University of California, Santa Cruz, CA.
Socolow, R.H., and R.C. Sonderegger
 1976   The Twin Rivers Program on Energy Conservation in Housing: Four Year Summary Report. Report No. 32. Center for Energy and Environmental Studies, Princeton University, Princeton, NJ.
Stein, B.
 1979   *The View from Sunset Boulevard: America as Brought to You by the People Who Make Television.* New York: Basic Books.
Stern, P.C.
 1986   Blind spots in policy analysis: What economics doesn't say about energy use. *Journal of Policy Analysis and Management* 5:200-227.
 1992   What psychology knows about energy conservation. *American Psychologist* 47:1224-1232.
 2000   Toward a coherent theory of environmentally significant behavior. *Journal of Social Issues* 56:407-424.
Stern, P.C., E. Aronson, J. Darley, D.H. Hill, E. Hirst, W. Kempton, and T.J. Wilbanks
 1986   The effectiveness of incentives for residential energy-conservation. *Evaluation Review* 10:147-176.
Stern, P.C., and S. Oskamp
 1987   Managing scarce environmental resources. Pp. 1043-1088 in *Handbook of Environmental Psychology,* D. Stokols and I. Altman, eds. New York: John Wiley and Sons.
Stutzman, T., and S. Green
 1982   Factors affecting energy consumption: Two field tests of the Fishbein-Ajzen model. *Journal of Social Psychology* 117:183-201.
Underhill, P.
 1999   *Why We Buy: The Science of Shopping.* New York: Simon and Schuster.
Wilhelm, M., and D. Iams
 1984   Attitudes and energy conservation behaviors of desert and non-desert residents of Arizona. Pp. 405-414 in *Families and Energy: Coping With Uncertainty,* B.M. Morrison and W. Kempton, eds. East Lansing, MI: Michigan State University.
Wilhite, H., and L. Lutzenhiser
 1999   Social loading and sustainable consumption. *Advances in Consumer Research* 26:281-287.
Wilhite, H., E. Shove, L. Lutzenhiser, and W. Kempton
 2001   Twenty years of energy demand management: We know more about individual behavior but next to nothing about demand. In *Society, Behaviour and Climate Change Mitigation,* J. Ebarhard, ed. Berlin: Kluwer Academic Publishers.
Wilhite, H., and R. Wilk
 1987   A method for self-recording household energy-use behavior. *Energy and Buildings* 9:73-79.
York, D., ed.
 1999   *A Discussion and Critique of Market Transformation.* Madison, WI: The Energy Center of Wisconsin.

4

Knowledge, Information, and Household Recycling: Examining the Knowledge-Deficit Model of Behavior Change

P. Wesley Schultz

E ducation is often seen as the key to changing behavior. Indeed, how can people engage in environmentally significant behaviors if they do not know about the impacts of their actions, or about the details of how to engage in a specific behavior? Recycling and other conservation behaviors are becoming increasingly important as the harmful effects of human behavior on the natural environment become more evident. Each year, reports are presented about the increasing damage that human behavior is having on the natural environment—ozone holes, deforestation, overpumping of groundwater, and an over-reliance on oil as an energy source. But is education sufficient to change behavior? This chapter examines the research on the effects of one educational approach—knowledge-based interventions designed to increase residential recycling rates. The knowledge-deficit model for information campaigns is presented, and research on three aspects of the model is summarized. Although the chapter focuses on a specific behavior (recycling), the basic principles discussed are believed to generalize to a range of environmentally related activities. Finally, an alternative educational approach focusing on social norms is presented, and some recommendations for implementing normative education programs are provided.

Before examining the knowledge-deficit model, it is important to clarify what I mean by "education." In the context of a social marketing approach, "educate" is often synonymous with "provide information." In working with recycling companies and with city and county recycling coordinators, I have frequently heard the phrase "We need to educate people about ___." Indeed, this same phrase can be found across a range of social marketing programs, and with regard to a range of behaviors. In essence, the educational activities involve disseminating information about the topic or about the behavior, with the goal of motivating people to act. It is, however, important to point out that this is just one narrowly conceived approach, and that not all educational efforts are information-based (cf. Andrews et al., this volume, Chapter 10; Ramsey and Hungerford, this volume, Chapter 9). There is a rich literature on environmental education, much of which is experiential or affect-based. What follows is an analysis of information-based education interventions used to promote recycling.

SOLID WASTE AND RESIDENTIAL RECYCLING PROGRAMS

The disposal of solid waste is becoming both an environmental and economic burden. In 1999, the average person in the United States generated 4.6 pounds of trash each day—a figure that was up dramatically from 2.7 pounds per day in 1960 (U.S. Environmental Protection Agency [EPA], 2000). Combined, people in the United States generated 230 million tons of trash in 1999. Of this total, residential solid waste accounted for an estimated 60 percent, with the remaining 40 percent coming from commercial sources (EPA, 2000).[1]

By far, the bulk of the trash generated in the United States is buried in landfills. Of the 230 million tons generated in 1999, approximately 72 percent was buried in landfills, at an estimated disposal cost of around $30 billion each year—a cost figure that is projected to grow substantially in the next few years (EPA, 1998a, 1998b, 2000). Lowering the amount of trash buried in landfills has important economic as well as environmental consequences. Less trash means lower disposal fees, less strain on the diminishing number of landfills open to accept waste, and less consumption of raw materials.

The approaches to reducing the amount of solid waste generated by households in the United States can be classified as either Reduce, Reuse, or Recycle. The *Reduce* approach focuses primarily on purchasing—for example, purchasing items with minimal packaging or items that can be composted. An article summarizing the research on "green buying" by Thøgersen can be found in Chapter 5 of this volume (see also Hormuth, 1999; Mainieri et al., 1997). *Reuse* focuses on repeated uses of purchased items—for example, using canvas shopping bags or purchasing beverages in refillable containers. *Recycle,* the focus of this chapter, refers to the collection of used items for use in the manufacturing of new items. Nationally, more than 9,000 curbside recycling programs in the United States serve more than 134 million people, and both numbers are growing rapidly (EPA, 1998a).

At the city and county levels, when people talk about recycling it is often in the context of technical issues like the implementation of a new program, changing to an automated collection system, the distribution of recycling bins, zoning or siting landfills or transfer stations, and different types of recycling programs (e.g., commingled, pay-to-throw, or source separated). However, it is important to point out that recycling is a behavior, and like all human behaviors, recycling is motivated and constrained by the context in which it occurs. The success or failure of a recycling program hinges on participation by community residents.

Recent reviews of the research on recycling and other environmentally significant behaviors have distinguished between personal and situational determinants (Hornik et al., 1995; Schultz et al., 1995). A *personal* behavioral predictor refers to a characteristic that exists within an individual. Examples of personal predictors include knowledge, attitudes, beliefs, personality, perceived control, and level of ascribed personal responsibility. *Situational* predictors are characteristics of the context that are related to the behavior. Examples of situational predictors include the types of materials collected, the location of collection bins, and the qualities associated with collection bins (color, shape, labeling).

Recycling programs have become common in communities throughout the United States and Canada. Within the past 10 years, all 50 U.S. states have passed laws requiring reductions in the amount of trash sent to landfills. California, like many states, has set a goal of a 50 percent diversion rate—that is, 50 percent less trash sent to landfills. In response to these laws, cities and counties have implemented many types of programs, one of which is curbside recycling. Although laudable, these diversion goals are difficult to reach, and many communities are struggling to meet diversion mandates. In an effort to encourage people to recycle, a number of intervention programs have been developed and implemented. These programs target both personal and situational variables, but the effectiveness is often questionable. Of the interventions used to promote recycling, the most common approach is based on knowledge.

KNOWLEDGE, INFORMATION, AND BEHAVIOR

Recycling coordinators for cities, advisors for technical councils at the county level, and directors and other administrators of recycling companies often believe that low recycling rates (or many other behaviors, for that matter) result from a lack of knowledge. From this basic assumption, the solution for increasing recycling rates is the distribution of educational materials about recycling. The basic assumption of this *knowledge-deficit theory* is that increasing knowledge will translate into a change in behavior. Three testable hypotheses can be derived from this theory. First, knowledge about recycling will be correlated with recycling behavior. Second, distributing educational materials containing information about recycling will lead to an increase in knowledge about recycling. Third, an increase in knowledge about recycling will lead to an increase in recycling behavior.

Before summarizing the research on these three topics, it is important to define some terms and to distinguish between different types of knowledge. Most of the research on knowledge as a predictor of recycling behavior has focused on *procedural knowledge*—that is, knowledge about the where, when, and how of recycling. For example, a resident may know that recycling is collected on Tuesdays by placing three bins at the curb: one for newspapers, a second for glass and cans, and a third for plastics. This can be distinguished from *impact knowledge,* which refers to an individual's beliefs about the consequences of recycling. For example, making aluminum from recycled cans requires 95 percent less energy and generates 95 percent less pollution than mining and processing raw aluminum. This type of knowledge is especially important in the value-belief-norm theory, where beliefs act with values and norms as joint determinants of behavior (cf. Stern, 2000; Stern et al., 1999; Stern et al., 1993). A third type of knowledge is *normative knowledge*—beliefs about the behaviors of others. We use the term *belief* to refer to an individual's subjective understanding of the procedure, impact, or normativeness of recycling; knowledge refers to accurate beliefs. Beliefs may or may not be accurate, and yet, may still predict recycling behavior.

Does Knowledge Predict Recycling Behavior?

The short answer to this question is "Yes." The research addressing this issue has focused almost exclusively on procedural knowledge. For example, knowledge is often measured by asking participants to identify which materials are or are not recycled in their recycling program and coding the percentage of items correctly classified. The overwhelming finding from the research is that knowledge is a strong and consistent predictor of recycling behavior. In general, the more knowledgeable a person is about which materials are recyclable, and when and where materials are collected, the more likely that person is to recycle (De Young, 1989; Gamba and Oskamp, 1994; Lindsay and Strathman, 1997; Vining and Ebreo, 1990). In a meta-analysis of the correlates of recycling behavior, Hornik et al. (1995) identified 17 studies that examined the relationship between knowledge about recycling (i.e., procedural knowledge) and recycling behavior. The aggregate relationship across these studies was $r=.54$ ($N=5,376$). Among the variables examined in their review, knowledge was the strongest correlate of recycling.

This finding should not be surprising. Indeed, research on a variety of other behaviors (e.g., condom use, cigarette smoking, substance use among adolescents, energy conservation) has consistently found knowledge to be a strong correlate. Illustrative research on the relationship between knowledge and behavior can be found in the literature examining Fisher's Information-Motivation-Behavioral skills (IMB) model (cf. Bryan et al., 2000; Fisher et al., 1994; Fisher and Fisher, 1996).

Although encouraging, the strong relationship between knowledge and behavior may not be causal. There are three possible causal relationships. The first, and the one implicitly assumed in the knowledge-deficit model, is that knowledge causes action. Knowing more about recycling causes a person to recycle more often. A second possible causal relationship is that action causes knowledge. That is, when a person recycles, he or she learns more about the behavior. Finally, an unspecified third variable may cause both knowledge and action. For example, a general interest in community activities may cause the individual to seek out information about recycling, and also to participate in a community-sponsored recycling program. Without additional data, the causal link is unclear.

Does Distributing Information Increase Knowledge About Recycling?

The second assumed connection in the knowledge-deficit model is that distributing information materials will cause an increase in knowledge. There is a substantial body of research on the development of persuasive educational materials, and a thorough review of this literature is beyond the scope of this chapter. However, chapters 6 and 7 in this volume address this issue (see also Petty and Wegener, 1998). Some of the key issues in developing an effective educational program are the complexity of the information presented, the medium through which the information is presented (e.g., newspapers, television or radio, printed brochures, posted signs or prompts), the framing of the message, and the credibility of the source. With these considerations in mind, researchers have been successful at creating information materials that increase knowledge about recycling (Littlejohn, 1997; Werner et al., 1997).

Does Changing Knowledge About Recycling Lead to a Change in Recycling Behavior?

This is the linchpin of the knowledge-deficit model. We have seen that knowledge correlates with behavior and that a well-designed education campaign can change beliefs and increase knowledge. But does this change in knowledge cause a change in behavior?

To address this question, an experimental study is needed in which households are randomly assigned to either receive or not to receive educational materials and subsequent changes in behavior are monitored. In the area of recycling, several studies have used this approach (cf. Schultz, 1999; Schultz and Tyra, 2000; Werner et al., 1997). The basic finding from these studies is that although distributing information materials can increase knowledge, this change in knowledge is associated with only a small, short-term change in behavior. For example, Schultz (1999) reported the findings from an experiment that disseminated information about the specifics of a local curbside recycling program to community residents. Residents were given information about the types of materials that were recyclable, along with information about collection procedures. Results showed only a small increase in recycling rates and the amount of material recycled, and no significant change relative to a control condition. In essence, information was not sufficient to produce a change in behavior.

Information Campaigns Are Cheap, but the Effects Are Short Lived

The results from the studies summarized suggest that the effectiveness of education campaigns to produce durable changes in behavior is dubious. In addition, changes that have been observed following information interventions to promote recycling are typically short lived.

So why, in the absence of evidence regarding the effectiveness of information-based campaigns, are they still widely used? First, they are cheap. Relative to other types of interventions, or to altering the recycling program itself, creating and disseminating educational materials is inexpensive. Schultz (1999) estimated a cost of approximately 3 cents per household to create and disseminate the materials for an information-based campaign. Second, creating informational materials is believed to require no special training in psychology or marketing, and it is a task that can be done by staff members already involved with the program. If residents simply need to be educated about recycling, then all that is needed is a list with details about recycling—most commonly to be included as an insert in the trash bill.

The reason that information campaigns often are ineffective is that they ignore the motives behind behavior. People recycle (or don't recycle) for reasons. A sizable number of studies have examined the reasons that people give for recycling, with some consistent findings (Gamba and Oskamp, 1994; McCarty and Shrum, 1994; Vining and Ebreo, 1990; Werner and Makela, 1999). Oskamp et al. (1998) identified four motivational factors associated with the level of recycling behavior:

- The benefits of recycling (e.g., satisfaction of saving natural resources, decreasing landfill use, saving energy),
- Personal inconvenience (e.g., no space for bins, no time to prepare materials, hard to move recycling bins to the curb),
- External pressure (e.g., friends and neighbors are doing it, pressure from friends, pressure from family), and
- Financial motives (earn money, decrease garbage costs).

Knowledge is not a motive for recycling. However, lack of knowledge can be a barrier to recycling. Several recent articles have suggested that in developing or modifying a social program, researchers must consider the barriers for the desired behavior (see also Gardner and Stern, 1996; McKenzie-Mohr and Smith, 1999). For example, McKenzie-Mohr (2000) argued that the first step in effective community-based social marketing is to uncover the barriers to the targeted behavior. These barriers can be external to the individual (e.g., lack of storage space for recycling bins) or internal (lack of knowledge about which materials are recyclable). Thus, lack of knowledge can be a barrier to recycling, and we would predict that an individual who knows what, when, and how to recycle would be more likely to do it.

When to Use Information

The research reviewed to test the knowledge-deficit model was, without exception, conducted within existing recycling programs. The findings clearly indicate that in such programs, disseminating information will not lead to a change in recycling behavior. However, given McKenzie-Mohr's work on community-based social marketing, it does seem that disseminating information can lead to a change in behavior in situations where lack of knowledge is a barrier to action. Three specific instances emerge in which knowledge may be a barrier to action. It is important to note that in each of these instances, people are motivated to act, but fail to do so because they do not know how.

1. *New program.* At the start of a new program, it is safe to assume that most people will not know the procedures for recycling. Disseminating information about the new program is likely to produce substantially more recycling behavior.
2. *Changing an existing program.* When an established program is changed, the change should be accompanied by information. For example, changes in the days of collection or the type of materials that are recyclable should be accompanied by information. To minimize the knowledge barrier, changes should be made sparingly.
3. *Complexity of procedures.* Programs that require procedures that are complex or difficult to remember should regularly disseminate information. For example, recycling programs with a long of list of materials that are, and are not, recyclable should disseminate this information on a regular basis.

In each of these instances, a lack of knowledge can be a barrier to action, and disseminating information is likely to produce an increase in recycling behavior. However, in existing programs where people have a basic understanding of the program, increasing knowledge will not lead to a change in behavior.

NORMATIVE EDUCATION: AN ALTERNATIVE APPROACH

The bulk of the research on knowledge about recycling and educational interventions to promote recycling has focused on increasing *procedural knowledge*. Given the limited effectiveness of education aimed at increasing procedural knowledge, it is useful to examine the research on *normative knowledge*—an understanding of the behaviors of others. In essence, these beliefs are perceived social norms. Cialdini and colleagues (1990) have distinguished between descriptive and injunctive social norms. *Descriptive* social norms are beliefs about what other people are doing—what Kallgren and colleagues (2000) refer to as norms of *is*. *Injunctive* social norms, in contrast, are beliefs about what other people think *should* be done—norms of *ought.* Social norms can be distinguished from personal norms, which are feelings of obligation to act in a particular manner in specific situations. Schwartz and Fleishman (1978:307) define personal norms as "self-expectations for behavior backed by the anticipation of self-enhancement or [self]-deprecation." Personal norms differ from social norms in that they refer to internalized self-expectations, whereas social norms refer to external perceptions about the appropriateness of behaviors. The focus here is on normative beliefs that an individual holds about the behavior of others. Our interest is in normative beliefs, regardless of the accuracy of these beliefs.

Normative Beliefs Predict Behavior

There is considerable evidence, from a number of lines of psychological research, that normative beliefs (both descriptive and injunctive) are good predictors of behavior. A number of studies focused on recycling have reported a strong, positive relationship between normative beliefs and recycling behavior. In these studies, normative beliefs often are measured by asking about perceptions of social pressure to recycle—for example, from friends, family, or neighbors. These are perceptions of injunctive social norms—that is, they are an individual's belief that others think he or she should be recycling. In addition, several studies have asked residents about their perceptions of the frequency with which other people recycle. For example, studies may ask questions like, "How often do your neighbors put recyclables at the curb to be collected?" In their meta-analysis of recycling studies, Hornik et al. (1995) found an aggregate correlation of $r=.43$ ($N=2,828$) between perceptions of social influence and recycling behavior.

Research showing a positive association between normative beliefs and many different behaviors also can be found among studies utilizing the Theory of Reasoned Action and the Theory of Planned Behavior. The Theory of Planned Behavior proposes that attitudes, subjective norms, and perceived control predict behavioral intentions, which in turn lead to behavior (Ajzen, 1991; Fishbein and Ajzen, 1975; Fishbein et al., 1994). *Subjective norms* refer to a person's perceptions of the social pressure to perform a behavior; they are an individual's perceptions of how other people or groups think he or she should act. Subjective norms have been found to be strong predictors of a variety of behaviors (Ajzen, 1991). A sampling of some of these research areas includes studies using subjective norms as predictors of condom use (Baker et al., 1996; Richardson et al., 1997), substance use among adolescents (Morrison et al., 1996), intentions to commit driving violations (Parker et al., 1992), compliance with lithium treatment among people with bipolar affective disorder (Cochran and Gitlin, 1988), intentions to wear seatbelts (Thuen and Rise, 1994), and occupational choice among women (Greenstein et al., 1979).

Schultz and Tyra (2000) found that descriptive normative beliefs were strong predictors of recycling behavior, and that normative beliefs about people closer to self were stronger predictors than beliefs about those who were more socially distant. For example, beliefs about the frequency of recycling by neighbors correlated *r*=.44 with recycling behavior; beliefs about recycling by "people in your neighborhood" correlated *r*=.31 with recycling behavior; and beliefs about recycling rates across the city correlated *r*=.17 with behavior.

The research just summarized clearly indicates that normative beliefs, both descriptive and injunctive, are predictive of a variety of behaviors. But do normative beliefs cause behavior? Guerra et al. (1995) suggest that commonly used rules may become injunctive simply because they are shared by many people. Although this assertion may hold for some behaviors, there are quite a few instances in which this would not apply—for instance, where an individual perceives that a behavior is desired but does not perceive that others are doing it, or in situations where the behavior is not directly observable by other community members. In these situations, normative beliefs would not predict behavior. Some of these situations can be characterized as commons dilemmas (Hardin, 1968), where a behavior is prescribed (both individually and collectively), but not commonly observed.

Recycling is a behavior that benefits the collective, with few direct rewards for the individual (excluding cash redemption centers), and recycling has a cost to the individual in terms of convenience, sorting, and storage. The behavior is socially accepted, and people generally believe they should recycle. But if no one else is doing it, why should I? One of the consistent findings from research on the commons dilemma is that communication can lead individuals to act in the interest of the group (Dawes et al., 1990; Schelling, 1966). Individuals are considerably more likely to reduce their use of the common when they believe that others who share access to the common also will limit their use. For example, I am much more likely to conserve energy if I believe my neighbor is also making an effort to conserve energy. Thus, disseminating information about the behavior of others (i.e., descriptive norms) is a mechanism for communication and an important way to overcome the commons dilemma.

Does Changing Normative Beliefs Cause a Change in Behavior?

The data just described suggest that normative beliefs can predict behavior. But what about instances in which an individual does not already possess injunctive or descriptive normative beliefs regarding a behavior, or instances where a descriptive belief is too low (i.e., hardly anyone in my community recycles) or too high (nearly everyone in my school smokes marijuana)? To what extent does changing a normative belief lead to a change in behavior? The opposite sequence seems possible, if not likely—that is, that many descriptive social norms are created to justify or validate behavior. For example, an individual may come to believe that a behavior (e.g., smoking, recycling, drug use, wearing seatbelts) is more common than it really is only after engaging in the behavior. Likewise, one may come to believe that a behavior (e.g., cheating on an exam, littering, running a red light) is more reprehensible after not engaging in it. Such an effect may result from a "false consensus," wherein we tend to believe that others share our views (Fabrigar and Krosnick, 1995; Suls et al., 1988). One way to assess the causal link is to change normative beliefs experimentally and then to observe any subsequent change in behavior.

Only a few studies have attempted to manipulate normative beliefs experimentally. One such study came from a program to reduce and prevent adolescent drug use. Donaldson and his colleagues (Donaldson et al., 1994; Donaldson et al., 1995) reported a series of studies on a normative education intervention designed to change adolescents' beliefs about the prevalence of substance use by their peers. Over a period of five sessions, the program presented information about alcohol and drug use that established a conservative normative atmosphere in the school regarding substance use. Results from the longitudinal study showed that the effectiveness of an adolescent drug use prevention program was mediated largely by changes in beliefs about the prevalence and acceptability of substance use among peers.

Schultz (1999) reported a study on the effects of a normative intervention within a community curbside recycling program. Study participants were community residents in a large metropolitan suburb. Approximately 120 houses were systematically assigned to each of five experimental conditions: individual normative feedback (targeting injunctive social norms), group normative feedback (targeting descriptive social norms), information, plea only, and control. The results showed that, overall, households in the injunctive norm condition recycled significantly more often and more material per week during a 4-week followup period than they did during the baseline period. For the descriptive norm condition, results showed a similar significant increase in the frequency of participation and in the amount of material recycled. The information, plea-only, and control conditions showed no significant change across time.

Practical Approaches for Making Recycling Normative

The research summarized shows that normative beliefs are causally linked with behavior. That is, normative beliefs predict behavior, and changing normative beliefs can cause a change in behavior. A remaining question is: how do we change normative beliefs? I offer two suggestions here.

Block leaders. A number of studies have examined the effectiveness of neighborhood leaders at promoting recycling. Within this approach, communities are divided into small residential areas, and volunteers are recruited from each area to serve as a block leader. These leaders are asked to take responsibility for the recycling within their neighborhood, to recycle diligently, and to encourage neighbors to recycle. Studies on the effects of block leader programs indicate that they have a direct effect on normative beliefs (particularly injunctive normative beliefs, but descriptive norms as well). In addition, block leader programs have been very successful at producing sustained increases in recycling behavior (Burn, 1991; Hopper and Nielsen, 1991; Shrum et al., 1994).

Disseminating data on community recycling rates. A second approach to making recycling normative is through the dissemination of recycling information to residents. This can occur through community newsletters, newspaper articles, public service announcements, or inserts in the recycling or trash bill. Note that this approach is most effective in areas where there is a low descriptive norm, but many people are actually recycling. That is, the disseminated normative information must be higher than the overall normative belief among residents. Publishing a statistic that "50 percent of residents in San Marcos recycle regularly" will only lead to an increase in recycling if the existing normative belief among residents is that fewer than 50 percent recycle regularly. Some types of normative information that can be distributed include percentages of people who recycle each week, the percentage of solid waste that is recycled by residents, or the number of recycling bins placed at the curb by residents each week.

In all cases, it is important to keep the normative information specific to the level of the individual, providing a standard against which an individual can compare his or her behavior. For example, providing information about the recycling rates across the city, or about the citywide diversion rate, will be unlikely to change behavior. This type of information is not connected with a specific behavior, and does not provide a clear standard against which a person can compare his or her behavior. Instead, we advocate targeting specific behaviors like "place recyclables at the curb to be collected" or "use designated bins for greenwaste." Likewise, we advocate using comparison groups that are closer to the individual with statements like "people in your community" or "your neighbors" rather than broader comparisons at the city, county, or even state levels.

The Limits of Normative Intervention

The previous discussion suggests that, unlike knowledge, normative beliefs can be a powerful motive for action. However, it is important to point out that normative beliefs are more likely to lead to behavior under a specified set of conditions.

A large body of social psychological research on conformity suggests that beliefs about the behavior of others are more likely to influence our own actions under a specific set of conditions. One of the most important considerations is whether the behavior is publicly observable. Behaviors that are more observable are more likely to be affected by changes in normative beliefs. Classic social psychological studies of conformity have shown that people conform more when they respond publicly (in front of others) than when they respond privately (Asch, 1946, 1955). The observability of a behavior interacts with normative beliefs in two ways. First, behaviors that are observable can be monitored by others. In situations where there is an injunctive normative belief for the behavior (e.g., people in my community think I should recycle), this monitoring function is likely to lead to an increased compliance with the norm. If, on the other hand, the behavior is not publicly observable (for example, household energy conservation, backyard composting, or proper disposal of hazardous household waste), then promoting a normative belief is less likely to change behavior.

A second aspect of the observability of a behavior has to do with the development of descriptive normative beliefs. Behaviors that are publicly observable reinforce (or undermine) existing descriptive normative beliefs. When we can monitor the behavior of others, their actions will directly affect our normative beliefs. For example, observing that my neighbors rarely put recyclables at the curb to be collected is likely to produce a low descriptive norm. Even if presented with information that a higher percentage of people in my community recycle, unless I observe my neighbors doing it, I am unlikely to believe the message or change my normative belief. On the other hand, if my neighbors regularly put a great deal of recyclables at the curb each week, my observations of their behavior will lead to a high descriptive norm.

Although the observability of a behavior is one of the more powerful conditions under which normative beliefs will affect behavior, other aspects of the situation can also play important roles. Variables like perceived similarity with others in the community, status of people who are engaging in the behavior, prior commitment to act in a particular manner, size of the group, and cohesion of the group are all variables that can affect the effectiveness of a normative intervention.

Overall, social psychological research on conformity suggests that normative social influence works best with behaviors that are publicly observable—like curbside recycling. Other behaviors that are less observable, like energy consumption or proper disposal of hazardous household waste, may be less affected by normative social influence.

CONCLUSIONS

This chapter has synthesized the research findings regarding knowledge and the effectiveness of certain educational interventions intended to promote recycling. We distinguished between procedural knowledge, impact knowledge, and normative knowledge. The results from a variety of studies suggest that knowledge about recycling is a strong correlate of recycling behavior. This conclusion is qualified by the concept that knowledge does not provide a motive for behavior, but instead it is a lack of knowledge that is a barrier to behavior. Research also demonstrates that it is possible to increase knowledge about a behavior (procedural or normative) by disseminating information. However, the findings show that although information can lead to an increase in knowledge, its effect on behavior tends to be small and short term. An alternative to procedural information is to distribute normative information to residents. Like procedural knowledge, normative beliefs are strong predictors of behavior, and they can be changed through the use of education. However, unlike procedural knowledge, normative beliefs provide a motive for behavior, and changing normative beliefs can cause a change in behavior.

The data used to support these conclusions were drawn primarily from the literature on recycling. However, I believe that the findings apply to a range of other human behaviors. Previous research in other areas of applied psychology (particularly health psychology) has found similar results, and the findings would seem to generalize to many of the behaviors addressed in this volume: energy conservation, "green" buying, public health communication, household disaster preparedness, pollution prevention, and more general environmental education. Across these areas, the basic argument outlined in this chapter would apply: Increasing knowledge does not translate into a change in behavior.

NOTE

1   Industrial and transportation-related wastes are not included in these statistics.

REFERENCES

Ajzen, I.
  1991   The theory of planned behavior. *Organizational Behavior and Human Decision Processes* 50:179-211.
Asch, S.E.
  1946   Forming impressions of personality. *Journal of Abnormal and Social Psychology* 41:258-290.
  1955   Opinions and social pressure. *Scientific American* (November):31-55.
Bryan, A., J. Fisher, W. Fisher, and D. Murray
  2000   Understanding condom use among heroin addicts in methadone maintenance using the Information-Motivation-Behavioral Skills Model. *Substance Use and Misuse* 35:451-471.
Burn, S.
  1991   Social psychology and the stimulation of recycling behaviors: The block leader approach. *Journal of Applied Social Psychology* 21:611-629.
Cialdini, R.B., R.R. Reno, and C.A. Kallgren
  1990   A focus theory of normative conduct: Recycling the concept of norms to reduce littering in public places. *Journal of Personality and Social Psychology* 58:1015-1026.
Cochran, S.D., and M.J. Gitlin
  1988   Attitudinal correlates of lithium compliance in bipolar affective disorders. *Journal of Nervous and Mental Disease* 176:457-464.
Dawes, R.M., A.J.C. van de Kragt, and J.M. Orbell
  1990   Cooperation for the benefit of us—Not me, or my conscience. In *Beyond Self-interest,* J.J. Mansbridge, ed., Chicago: University of Chicago Press.

De Young, R.
  1989   Exploring the differences between recyclers and nonrecyclers: The role of information. *Journal of Environmental Systems* 18:341-351.
Donaldson, S., J. Graham, and W.B. Hansen
  1994   Testing the generalizability of intervening mechanism theories: Understanding the effects of adolescent drug use prevention interventions. *Journal of Behavioral Medicine* 17:195-216.
Donaldson, S., J. Graham, A. Piccinin, and W. Hansen
  1995   Resistance-skills training and onset of alcohol use: Evidence for beneficial and potentially harmful effects in public schools and in private Catholic schools. *Health Psychology* 14:291-300.
Fabrigar, L.R., and J. Krosnick
  1995   Attitude importance and the false consensus effect. *Personality and Social Psychology Bulletin* 21:468-479.
Fishbein, M., and I. Ajzen
  1975   *Belief, Attitude, Intention, and Behavior: An Introduction to Theory and Research.* Reading, MA: Addison-Wesley.
Fishbein, M., S.E. Middlestadt, and P.J. Hitchcock
  1994   Using information to change sexually transmitted disease-related behaviors: An analysis based on the theory of reasoned action. In *Preventing AIDS: Theories and Methods of Behavior Interventions,* R.J. DiClemente and J. Peterson, eds. New York: Plenum.
Fisher, J.D., and W.A. Fisher
  1996   The Information-Motivation-Behavioral skills model of AIDS risk behavior change: Empirical support and application. In *Understanding and Preventing HIV Risk Behavior: Safer Sex and Drug Use,* S. Oskamp and S. Thompson, eds. Thousand Oaks, CA: Sage.
Fisher, J.D., W.A. Fisher, W. Sunyna, and M. Thomas
  1994   Empirical tests of an information-motivation-behavioral skills model of AIDS-preventive behavior with gay men and heterosexual university students. *Health Psychology* 13:238-250.
Gamba, R., and S. Oskamp
  1994   Factors influencing community residents' participation in commingled curbside recycling programs. *Environment and Behavior* 26:587-612.
Gardner, G.T., and P.C. Stern
  1996   *Environmental Problems and Human Behavior.* Needham Heights, MA: Allyn and Bacon.
Greenstein, M., R.H. Miller, and D.E. Weldon
  1979   Attitudinal and normative beliefs as antecedents of female occupational choice. *Personality and Social Psychology* 5:356-362.
Guerra, N.G., L.R. Huesmann, and L. Hanish
  1995   The role of normative beliefs in children's social behavior. In *Social Development: Review of Personality and Social Psychology,* N. Eisenberg, ed. Thousand Oaks, CA: Sage.
Hardin, G.
  1968   The tragedy of the commons. *Science* 162:1243-1248.
Hopper, J., and J. Nielsen
  1991   Recycling as altruistic behavior: Normative and behavioral strategies to expand participation in a community recycling program. *Environment and Behavior* 23:195-220.
Hormuth, S.
  1999   Social meaning and social context of environmentally-relevant behaviour: Shopping, wrapping, and disposing. *Journal of Environmental Psychology* 19:277-286.

Hornik, J., J. Cherian, M. Madansky, and C. Narayana
1995   Determinants of recycling behavior: A synthesis of research results. *Journal of Socio-Economics* 24:105-127.
Kallgren, C.A., R.R. Reno, and R.B. Cialdini
2000   A focus theory of normative conduct: When norms do and do not affect behavior. *Personality and Social Psychology Bulletin* 26:1002-1012.
Lindsay, J.J., and A. Strathman
1997   Predictors of recycling behavior: An application of a modified health belief model. *Journal of Applied Social Psychology* 27:1799-1823.
Littlejohn, C.R.
1997   Measuring the Effects of Message Framing on the Behavior of Recycling in a Residential Recycling Program. Unpublished doctoral dissertation, Florida State University.
Mainieri, T., E.G. Barnett, T.R. Valdero, J.B. Unipan, and S. Oskamp
1997   Green buying: The influence of environmental concern on consumer behavior. *Journal of Social Psychology* 137:189-204.
McCarty, J.A., and L.J. Shrum
1994   The recycling of solid wastes: Personal values, value orientations, and attitudes about recycling as antecedents of recycling behavior. *Journal of Business Research* 30:53-62.
McKenzie-Mohr, D.
2000   Promoting sustainable behavior: An introduction to community-based social marketing. *Journal of Social Issues* 56:543-554.
McKenzie-Mohr, D., and W. Smith
1999   *Fostering Sustainable Behavior: An Introduction to Community-based Social Marketing,* 2nd ed. Gabriola Island, British Columbia, Can.: New Society.
Morrison, D.M., M.R. Gillmore, E.E. Simpson, and E.A. Wells
1996   Children's decisions about substance use: An application and extension of the theory of reasoned action. *Journal of Applied Social Psychology* 26:1658-1679.
Oskamp, S., R. Burkhardt, P.W. Schultz, S. Hurin, and L. Zelezny
1998   Predicting three dimensions of residential curbside recycling: An observational study. *Journal of Environmental Education* 29:37-42.
Parker, D., A.S. Manstead, S.G. Stradling, and J. Reason
1992   Intention to commit driving violations: An application of the theory of planned behavior. *Journal of Applied Psychology* 77:94-101.
Petty, R.E., and D.T. Wegener
1998   Attitude change: Multiple roles for persuasion variables. In *Handbook of Social Psychology,* 4th ed., D. Gilbert, S. Fiske, and G. Lindzey, eds. Boston: McGraw-Hill.
Richardson, H.R., R.P. Beazley, M.E. Delaney, and D.B. Langille
1997   Factors influencing condom use among students attending high school in Nova Scotia. *Canadian Journal of Human Sexuality* 6:185-196.
Schelling, T.C.
1966   The life you save may be your own. In *Problems in Public Expenditure Analysis,* S. Chase, ed. Washington, DC: Brookings Institute.
Schultz, P.W.
1999   Changing behavior with normative feedback interventions: A field experiment of curbside recycling. *Basic and Applied Social Psychology* 21:25-36.
Schultz, P.W., S. Oskamp, and T. Mainieri
1995   Who recycles and when? A review of personal and situational factors. *Journal of Environmental Psychology* 15:105-121.
Schultz, P.W., and A. Tyra
2000   *A Field Study of Normative Beliefs and Environmental Behavior.* Poster presented at the meeting of the Western Psychological Association, Portland, OR, April 13-16.

Schwartz, S., and J.A. Fleishman
    1978   Personal norms and the mediation of legitimacy effects on helping. *Social Psychology*
            41:306-315.
Shrum, L.J., T.M. Lowrey, and J.A. McCarty
    1994   Recycling as a marketing problem: A framework for strategy development. *Psychology*
            and Marketing 11:393-416.
Stern, P.C.
    2000   Toward a coherent theory of environmentally significant behavior. *Journal of Social*
            Issues 56:407-424.
Stern, P.C., T. Dietz, T. Abel, G.A. Guagnano, and L. Kalof
    1999   A social psychological theory of support for social movements: The case of environ-
            mentalism. *Human Ecology Review* 6:81-97.
Stern, P.C., T. Dietz, and L. Kalof
    1993   Value orientations, gender, and environmental concern. *Environment and Behavior*
            25:322-348.
Suls, J., C. Wan, and G. Sanders
    1988   False consensus and false uniqueness in estimating the prevalence of health-protective
            behaviors. *Journal of Applied Social Psychology* 18:66-79.
Thuen, F., and J. Rise
    1994   Young adolescents' intention to use seat belts: The role of attitudinal and normative
            beliefs. *Health Education Quarterly* 9:215-223.
U.S. Environmental Protection Agency
    1998a   *Municipal Solid Waste Factbook.* Washington, DC: U.S. Environmental Protection
            Agency.
    1998b   *Characterization of Municipal Solid Waste in the United States: 1997 Update.* Wash-
            ington, DC: U.S. Environmental Protection Agency.
    2000   *Municipal Solid Waste in the United States: 1999 Facts and Figures.* Washington, DC:
            U.S. Environmental Protection Agency.
Vining, J., and A. Ebreo
    1990   What makes a recycler? A comparison of recyclers and nonrecyclers. *Environment and*
            Behavior 22:55-73.
Werner, C., and E. Makela
    1999   Motivations and behaviors that support recycling. *Journal of Environmental Psycholo-*
            gy 18:373-386.
Werner, C., M. Rhodes, and K. Partain
    1997   Designing effective instructional signs with schema theory: Case studies of polystyrene
            recycling. *Environment and Behavior* 30:709-735.

5

Promoting "Green" Consumer Behavior with Eco-Labels

John Thøgersen*

E co-labeling is one among a number of policy tools that are used in what has been termed an Integrated Product Policy (Nordic Council of Ministers, 2001). The increasing popularity of product-oriented environmental policy in Europe and elsewhere is based on the perception that the abatement of pollution from industrial and other large sources is now within reach. Hence, the relative importance of pollution from "nonpoint" sources (Miljøstyrelsen, 1996), particularly pollution (and resource use) associated with private consumption (Geyer-Allély and Eppel, 1997; Norwegian Ministry of Environment, 1994; Organization for Economic Co-operation [OECD], 1997b; Sitarz, 1994), has increased. However, not only the composition, but also the volume of consumption in the industrialized countries is increasingly acknowledged to be unsustainable. If widely accepted prognoses for the growth in global consumption are realized, a factor 4 or greater reduction in the environmental impact per produced unit is needed in the next 40 to 50 years just to keep the total environmental impact at the current level (Miljøstyrelsen, 1996).

As a means to reduce the pollution and resource use following from consumption, attempts are made to motivate consumers to switch to less environmentally harmful and resource-consuming products. One of the increasingly popular tools is to label the least harmful products in such a way that consumers can distinguish them from others (OECD, 1991, 1997a; U.S. Environmental Protection Agency [EPA], 1998). The hope is that consumers' choices will give producers of (relatively) environment-friendly products a competitive advan-

*The author would like to express gratitude to Doug McKenzie-Mohr and Paul Stern for helpful comments on an earlier draft of this chapter.

tage, allowing them to gradually push less environment-friendly products out of the market (Miljø- og Energiministeriet, 1995; OECD, 1991). In addition, it is hoped that the anticipated competitive advantage gives companies an incentive to develop new products that are more friendly to the environment (Backman et al., 1995; Miljøstyrelsen, 1996; OECD, 1991; EPA, 1998).

Other tools in the Integrated Product Policy toolbox are mandatory standards, taxes and subsidies, and voluntary agreements. These means are not necessarily alternatives to labeling, of course. They may be—and have been—used in combination. An important advantage of voluntary means is that they make it possible to proceed faster than is politically feasible by means of legal restrictions and taxes. Eco-labeling is unique in that it rewards proactive companies and thereby has the capacity to harness their innovative creativity to the environmental policy carriage, instead of directing it toward ways of avoiding the consequences of regulation (e.g., Tenbrunsel et al., 1997). In addition, it is hoped that eco-labeling will help increase consumer attention toward, and knowledge about, the environmental risks associated with consumption (Backman et al., 1995; Miljø- og Energiministeriet, 1999; Nordic Council of Ministers, 2001; OECD, 1991, 1997a; EPA, 1998).

Others have expressed fear that environmental claims on products may legitimize continued consumerism (e.g., Davis, 1992; Durning, 1992) and that the possible environmental gain from a shift to less harmful products may be more than offset by the continued rapid growth in the volume of consumption (e.g., Matthews et al., 2000; United Nations Environment Program, 1994). For example, many serious environmental impacts from traffic are still increasing in spite of more energy-efficient engines and catalytic converters (Mackenzie, 1997; Noorman and Uiterkamp, 1998), and the volume of waste is still growing in spite of increased recycling (Miljø- og Energiministeriet, 1999; Waller-Hunter, 2000). Whether eco-labeling contributes to consumer ignorance concerning such developments or, on the contrary, makes them more attentive to the problems associated with growing consumption is a question still not settled by research, to my knowledge.

The effectiveness of eco-labeling, in a narrow sense, is reflected in the reduction in pollution and resource use that can be attributed to the labeling. To calculate its efficiency, the costs of using this measure also should be included (Morris, 1996). However, the full picture of eco-labeling's effectiveness and efficiency includes positive and negative effects on consumer/citizens' perceptions about, attentiveness toward, and readiness to act to solve environmental problems in general. To complicate the issue further, the effectiveness of eco-labels, both in a narrow and in a wider sense, may depend on the mutual implementation of other policy measures (e.g., Gardner and Stern, 1996), notably environmental education and information about the labels.[1]

ECO-LABELS AND CONSUMER DECISION MAKING

Consumer decision making concerning eco-labeled products involves considerations about the label as well as about the specific product itself. To reduce the analytical complexity, I consider the decision making as consisting of two interwoven, but partly independent decision—and learning—processes: one concerning a specific product and one concerning a specific label.

At least in the eyes of the consumer, a product that suddenly comes with an eco-label is an innovation, that is, a new product that differs more or less from the nonlabeled product that it may have replaced and from other nonlabeled products in the same category. The eco-label documents and communicates that the product has certain characteristics leading to outstanding eco-performance. Innovation adoption theory describes the decision to buy such a product as a learning process, consisting of a number of successive phases, where the consumer obtains, accumulates, and integrates knowledge about the product and evaluates its self-relevance (e.g., Peter et al., 1999; Rogers, 1995). Communicationwise, the process may be conceived as a hierarchy of stages (or effects) that the consumer needs to go through before making a decision to buy the new product. What these stages are, as well as their succession, depends on a number of circumstances, notably how risky the decision is perceived to be (e.g., Hoyer and MacInnis, 1997). Because the decision making process may be lengthy, and can be interrupted anywhere in the process, the evaluation of an eco-labeling scheme's success should be based not only on its eventual environmental outcomes, but also on its influence on the move from one stage in the decision process to the next (Abt Associates Inc., 1994; Nordic Council of Ministers, 2001).

An eco-label is an innovation in itself. Hence, the process through which the consumer learns about and adopts the eco-label also may be described as an innovation adoption process in which the final adoption is reflected in the purchase of products carrying the label.

The purchase of "x-labeled" (an eco-label) products is a behavioral category consisting of many independent actions, rather than just a single action (Ajzen and Fishbein, 1980). An important question, which to my knowledge remains to be answered, is whether consumers form mental categories based on eco-labels, as they have been known to do based on (some) other product characteristics (e.g., Cohen, 1982; Sujan, 1985). Because eco-labels typically are not restricted to one established product category, new mental categories based on eco-labels may cross established boundaries. The formation of such new mental categories is not likely unless consumers perceive environment friendliness as an important product attribute, both in an absolute sense and relative to other salient attributes (Gutman, 1982). Therefore, new cross-boundary eco-categories seem more likely to emerge in traditional low-involvement areas, such as groceries, than in traditional high-involvement areas, such as furniture, white goods, and electronic equipment. For example, it seems more likely that consumers will form a new cross-boundary product category for organic food products carrying a third-party eco-label, such as the Danish Ø-label, than for energy-efficient white goods carrying, say, European Union's (EU's) mandatory energy labeling's A-classification. If consumers form such categories, they may use them as the basis for category-based decision making in future choice situations when encountering labeled products of the same or different kind(s) (Cohen, 1982; Fiske and Pavelchak, 1986; Sujan, 1985). This would increase the likelihood of repeat purchase of eco-labeled products and speed up the adoption process for other new products wearing the same label. There is evidence that mental categories carry affect, which is used when evaluating entities that fit the category (Cohen, 1982; Fiske and Pavelchak, 1986). Because environmental attitudes seem to have acquired a moral basis for many people in modern society (e.g., Harland et al., 1999; Heberlein, 1972; Thøgersen, 1996b, 1999), the affect associated with eco-categories may be more charged than usual product-related attitudes (e.g., Peter et al., 1999). Strong category-based affect further increases the likelihood that eco-categories have behavioral implications (Verplanken et al., 1998).

RESEARCH ON THE EFFECTIVENESS OF ECO-LABELS

Of course, environmental labels are useful from an environmental policy perspective only if consumers use them in their decision making. However, there are still few published studies of the effectiveness of labeling schemes in this respect (OECD, 1997a). Most of the published studies focus on consumers' recognition of or knowledge about labels and/or their trust in them (Bekholm and Sejersen, 1997; Tufte and Lavik, 1997), implicitly or explicitly assuming that these are fundamental prerequisites for the use of a label in decision making. However, practically all studies are purely descriptive, leaving the question of *why* consumers know, notice, and use labels only sporadically answered. With few exceptions (e.g., Verplanken and Weenig, 1993), it is not systematically considered how the decisions that the labels are meant to influence are made and/or the implications of the decision making process for the functioning and effectiveness of labeling. For example, plenty of evidence shows that how and how much consumers attend to information in a buying situation depends on their involvement (e.g., Celsi and Olson, 1988; Herr and Fazio, 1993; Kokkinaki, 1997). In general, one cannot count on information about environmental consequences, in the form of a label or otherwise, producing high involvement in itself. The isolated consequences— environmental as well as personal—of each individual decision are simply too small in most cases (Thøgersen, 1998). If this is the case, and if other self-relevant information competes for the consumer's attention—sometimes to a degree to which the consumer experiences information overload (Jacoby, 1984)—consumers may easily fail to notice relevant labels in the buying situation.

In a recent publication, I have reasoned at length about how and why consumers attend to eco-labels (Thøgersen, 2000b). It is emphasized that "paying attention to eco-labels" is hardly a goal in itself, but rather a means to a goal: buying environment-friendly products, which is a means to a more abstract goal about protecting the environment. Thus, it is unlikely that a consumer will pay attention to an environmental label unless he or she values protecting the environment, perceives that buying (more) environment-friendly products is an effective means to achieve this goal, and finds that the information the label conveys is useful for this purpose. In addition, the availability of eco-labeled products in the shops and the consumer's ability to recognize and understand the labels undoubtedly influence attention toward this type of label.

Empirically, I find that a large majority of the consumers in four analyzed countries pay attention to eco-labels at least sometimes. As predicted, paying attention to eco-labels is strongly influenced by the belief in considerate buying as a means to protect the environment and by the trust in the labels. The personal importance of environmental protection (proenvironmental attitude) and perceived effectiveness regarding the solving of environmental problems also influence paying attention to eco-labels, but this influence is mediated through the former two concepts (belief and trust). In three of the analyzed cases, there is also an interaction effect between proenvironmental attitude and trust, meaning that the influence of proenvironmental attitude on paying attention is higher when the consumer trusts the label (and the influence of trust higher when the consumer holds a proenvironmental attitude).

Environmental Outcomes

Only a few studies have attempted to estimate the environmental impact of eco-labels. The most thoroughly evaluated schemes—the Swedish Society for Nature Conservation's "Good Environmental Choice" label, the Nordic Council of Ministers' Swan label, and the German Blue Angel label—are presumably also the most successful ones. For example, the Blue Angel has been credited for a reduction in emissions of sulphur dioxide, carbon monoxide, and nitrogen oxides from oil and gas heating appliances by more than 30 percent and for a reduction in the amount of solvents emitted from paints and varnishes into the environment by some 40,000 tons (United Nations Conference on Trade and Development, 1995). In Sweden, the Good Environmental Choice and the Nordic Swan labels have been credited for a considerable reduction in (1) chlorinated compounds, acids, and other pollutants from the Swedish forest industry (paper products) (Naturvårdsverket, 1997), and (2) the volume and toxicity of household chemical emissions, particularly laundry detergents, down the drains (Beckerus and Rosander HB, 1999; Scandia Consult Sverige AB, 1999; The Swedish Society for Nature Conservation, 1999). I will elaborate on the latter case.

Laundry detergents represent 70 percent of the annual consumption of household chemicals in Sweden,[2] which makes it a particularly environmentally significant product category. Since the Good Environmental Choice and the Nordic Swan labels were introduced in the late 1980s, Swedish consumers have changed their demand from less to more concentrated products and have rejected the most environmentally harmful chemicals, a development that has been largely attributed to the two labels (Backman et al., 1995). Specifically, the sales volume of household chemicals for cleaning and personal care decreased by 15 percent between 1988 and 1996. Furthermore, in 1996, 60 percent of the chemical ingredients used in soap, shampoo, detergents, and cleaners in 1988 had been removed or replaced by less harmful substances. In 1997, eco-labeled detergents had a market share of more than 90 percent in Sweden.

As already mentioned, these are undoubtedly some of the most successful eco-labeling schemes. But still, they encouragingly demonstrate that under the right circumstances, eco-labeling has the power to produce a substantial reduction in the environmental pressure from serious sources of household pollution. Important prerequisites are consumer receptiveness to information about products' environmental attributes (i.e., environmental concern and belief in responsible consumer behavior as a means to solve the problem), company willingness to adopt eco-labeling schemes, and sufficient effort in promoting the schemes to consumers. Together, these conditions decisively influence the speed by which consumers become aware of eco-labels and of new eco-labeled products and by which they pass through the subsequent stages in the decision making process.

The Eco-Label Hierarchy of Effects

Awareness

Knowing that a label exists is a prerequisite for using it in decision making. This basic type of knowledge is typically measured as (aided and/or unaided) recall in surveys (e.g., Dyer and Maronick, 1988; OECD, 1997a). The results vary widely, reflecting the presence of labels in the stores, the efforts put into promoting a label, the clarity of the label's profile, and its perceived self-relevance for consumers (Van Dam and Reuvekamp, 1995). A 1999 survey in the Nordic countries found that between 61 and 75 percent of random samples in Norway, Sweden, and Finland were able to recall the Swan label unaided when asked about which eco-labels could be found on products in their country (Palm and Jarlbro, 1999). Recurrent surveys show that awareness about the Swan label was built gradually in these countries since its introduction in the early 1990s (Backman et al., 1995). In Denmark the unaided recall in 1999 was a much lower 18 percent. Although the Swan label was introduced in the other Nordic countries in 1989, Denmark only became a full member of this labeling scheme in the beginning of 1998, which undoubtedly explains the difference. Between 1997 and 1999, aided recall of the Swan label in Denmark rose from 37 to 51 percent. During that time, the label was promoted through newspaper and magazine ads, leaflets in shops, and public relations work and the number of Swan-labeled products in the shops rose from 1,000 to 1,300 (Kampmann, 2000). In Denmark, 31 percent of the respondents mentioned the national organic food label (the Ø-label) unaided, which is substantially higher than in the other Nordic countries.[3] The unaided recall of EU's Flower label was below 2 percent in all four countries, and most other environment-related labels also achieved low unaided recall (Palm and Jarlbro, 1999).

An indicator of label awareness with particularly high face validity is the recognition of visual images of the label. A Dutch study found a wide variation in the recognition of 11 environment-related labels—from 11.5 percent recognition of the Society of Plastic Industry Symbol to 92.7 percent recognition of the chasing-arrows recycling symbol (Van Dam and Reuvekamp, 1995). The length of time a label was on the market generally correlated with an increase in recognition. Recognition also depended on the type and amount of promotion backing the label. A similar study in Denmark in 1997 investigated the recognition of five environment-related and five safety and/or health-related labels (Bekholm and Sejersen, 1997). On average, the environment-related labels were better known, but as in the Netherlands, recognition varied widely, from 18 percent recognizing EU's Flower label to 89 percent recognizing the chasing-arrows recycling symbol. This study was conducted a few months before Denmark joined the Nordic Swan labeling scheme. Hence, with few Swan-labeled products in the shops and no official promotion of the label, it is no wonder that the Swan label was recognized by only 29 percent of respondents. The promotion campaign and increased presence of Swan-labeled products boosted the recognition of the label to just over 40 percent in June 1998 and 52 percent in October 1999 (Palm and Jarlbro, 1999). Also reflecting promotion activities and presence in the shops, the Danish Ø-label ("State Controlled Organic" label for organic food products) was recognized by 43 percent of a broad sample of consumers in 1995, 5 years after its introduction (Thøgersen and Andersen, 1996), and by 79 percent in 1997 (Bekholm and Sejersen, 1997).[4]

Even consumers who know a relevant environmental label will not use it if they fail to notice it because of information overload (Jacoby, 1984) or for other reasons. For example, in 1992 it was estimated that 400 to 600 private labels, in addition to 36 labeling schemes issued by public authorities, targeted Danish consumers (Forbrugerstyrelsen, 1993). In 1996, a study found environmental claims on 63 percent of the packaged goods within 16 product categories in the major supermarkets in Oslo (Enger, 1998). A minority of 8 percent of the goods carried a third-party environmental label. The study was a partial replication of a 1994 U.S. study that found environmental claims on 65 percent of the packaged goods in 16 product categories in major supermarkets in five large population centers throughout the United States (Mayer and Gray-Lee, 1995). Only 0.3 percent of the American packages carried an environmental label issued by a third party.

Comprehension

Recognizing a label is not the same as understanding the exact, or even the approximate, meaning of it. It is well known from other areas that consumers often have a hard time understanding labels (e.g., Laric and Sarel, 1981; Parkinson, 1975). Van Dam and Reuvekamp (1995) suggest that eco-labels suffer from a double confusion: the "generic" confusion from the limited meaning of seals and certifications and a remarkable amount of uncertainty and misunderstanding concerning environmental claims and terminology. Confirming this, one study found that only about 5 percent of a representative sample of U.S. consumers exhibited a thorough understanding of the terms "recycled" and "recyclable" (Hastak et al., 1994; see also Morris et al., 1995). Hence, campaigns that effectively target the confusion may lead to a substantial increase in the sale of labeled products, as illustrated by the "Get in the Loop, Buy Recycled" campaign in the state of Washington in 1994-1995.[5] Through a focused effort to increase awareness of products with recycled content and comprehension of the claim, the campaign produced a 58-percent increase in sales of recycled products in participating grocery stores. The campaign included prompts placed below products, which served to highlight product availability and substantiate manufacturer recycled content claims. In addition, posters, employee buttons, and door decals served as reminders for consumers.

Of course, less than a thorough understanding may be sufficient for decision making, particularly under low-risk circumstances. Van Dam and Reuvekamp (1995) classified respondents' understanding of 11 seals found on Dutch packages in three groups: adequate, underestimation, and overestimation of environmental implications. Among those recognizing a label, from 9 to 95 percent, depending on the label, had an adequate understanding of its environmental implications. Misunderstandings more often were in the direction of underestimation than overestimation. The higher the recognition of a label, the more likely it was also understood accurately (see also Bekholm and Sejersen, 1997), attention seemingly shading off into comprehension (e.g., Peter et al., 1999). As with recognition, there was a positive relationship between understanding and the length of time the label had been on the market. Understanding also depended on the type and extent of promotion, on the label's self-relevance, and on the clarity of its environmental profile. For example, two labels that particularly few understood were the German "Green Dot" and the Dutch Union of Housewives' seal. The former appears on many Dutch packages, but it has no relevance outside Germany. With regard to the latter, the environmental assessment is drowning in the long range of criteria influencing whether the Union endorses the product.

Uncertainty about what a label means often is accompanied by mistrust. A consumer only will use a label (as intended) in decision making if he or she trusts the message it conveys (Hansen and Kull, 1994). A large number of studies have found that consumers tend to be skeptical towards "green" product claims (see Peattie, 1995). One study cited by Peattie (1995) found that 71 percent of British consumers thought that companies were using green issues as an excuse to charge higher prices. However, many studies find that third-party labels and environmental information are trusted more than information provided by producers or retailers (e.g., Bekholm and Sejersen, 1997; Eden, 1994/95; Enger and Lavik, 1995; Tufte and Lavik, 1997). Unfortunately, and perhaps because they are outnumbered so many times by private labels and other types of environmental information, consumers often are uncertain or hold outright erroneous beliefs about who issues third-party labels (e.g., Bekholm and Sejersen, 1997; Tufte and Lavik, 1997). A Norwegian study found that such mistakes reduce the trust in the Nordic Swan label (Tufte and Lavik, 1997).

Attitude

Consumers generally welcome informative product labeling (Bekholm and Sejersen, 1997; Forbrugerstyrelsen, 1993). Specifically regarding eco-labels, a previously mentioned study found that from 64 to 91 percent of representative samples in Denmark, Norway, Sweden, and Finland agreed that eco-labels are needed (Palm and Jarlbro, 1999). A positive attitude toward eco-labels depends on the consumer believing that he or she can help attain a valued goal (e.g., Forbrugerstyrelsen, 1993; Nilsson et al., 1999; Palm and Windahl, 1998). Just as unit pricing helps the consumer obtain the goal of value for money and nutrition declarations facilitate health-related goals, environmental labeling helps consumers obtain environmental goals. Hence, a positive attitude toward eco-labels is only likely if consumers desire environment-friendly products.[6]

Intention and Behavior

The intention to buy eco-labeled products is reflected most clearly in consumers' search for and attention to this kind of information. Based on survey data collected by the European Consortium for Comparative Social Surveys (COMPASS) in 1993, I analyzed the frequency of paying attention to eco-labels in Britain, Ireland, Italy, and (two samples from) Germany (Thøgersen, 2000b). A large majority of consumers in these countries seem to pay attention to eco-labels when they shop, at least sometimes. Only from 8 percent (Great Britain) to 15 percent (Ireland) never do that. Other more recent studies find a similar attentiveness to environmental information. For example, a survey in 1997 found that 61 to 71 percent of random samples of consumers in the Nordic countries claimed that they "sometimes" or "always" check out the environment friendliness of the products they buy (Lindberg, 1998).

The Swedish Consumer Agency monitored the self-reported purchase of eco-labeled products yearly between 1993 and 1997. In this period, the share of respondents claiming that they bought eco-labeled products regularly rose from 37 to 51 percent (Konsumentverket, 1993, 1995/96, 1998). These numbers are supported by market data. For example, in 1994 eco-labeled products already had captured more than 60 percent of the detergent market and more than 80 percent of the copying and printing paper market in Sweden (Backman et al., 1995).

Repeat Purchase

There is a lack of studies of repeat purchase of eco-labeled products. It seems that most researchers implicitly assume that all decisions to purchase such products are the same, independent of the consumer's buying history. That this is hardly true is indicated by some of my own research (Thøgersen, 1998). Not unexpectedly, I found that a person's beliefs about product attributes and consequences of buying Ø-labeled products depend on the length of his or her experience with buying such products. Beliefs are changed or strengthened based on experience. I also found that experience has a direct and positive influence on the attitude toward buying organic products (after controlling for salient beliefs). Therefore, it seems that the longer a person has bought (labeled) organic products, the more positive the person's attitude is toward buying such products and the less it is based on thorough consideration of the pros and cons of doing so.

A followup study by two of my master students[7] investigated consumer purchase of 16 different food products (Andersen and Vestergaard, 1998). Based on their data set, I have made the calculations presented in Table 5-1.

Table 5-1 indicates that once consumers have started to buy Ø-labeled products, they tend to do so increasingly over time, and their propensity to choose labeled products is extended to an increasing number of product categories. Both tendencies are highly significant. In the beginning of this chapter, it was suggested that eco-labels may lead consumers to form new mental categories and that affect related to such a category can have a strong influence on their subsequent behavior. The results presented in Table 5-1 are consistent with this suggestion.

TABLE 5-1   Breadth and Depth of Organic Buying Within 16 Product Categories and Length of Buying Experience, Aarhus, Denmark, 1998 (N=232)

	< 1 Year	1-3 Years	3-5 Years	> 5 Years	F test
Pct. of food products organic*	2.3	2.6	2.8	3.5	13.0
Number of organic foods bought	6	8	9	11	13.8

*1=0%, 2=10%, 3=25%, 4=50%, 5=75%, 6=100%.

The Environment-Friendly Product Hierarchy of Effects

Studies have found that large segments of Western European and North American consumers demand environment-friendly products in diverse areas such as packaging (Bech-Larsen, 1996; Thøgersen, 1996a), food products (Biel and Dahlstrand, 1997; Grunert and Juhl, 1995; Sparks and Shepherd, 1992; Thøgersen, 1998), paint (Buchtele and Holzmuller, 1990), and heating systems (Berger et al., 1994). Few products are acquired with the sole (or main) purpose of protecting the environment, however. Typically, consumers buy goods for the private utility they provide. Still, many consumers are willing to make an effort to diminish the negative environmental impact of their consumption, and environmental labels are welcomed as a tool for this purpose. Given that environmental attributes—as long as they do not represent any personal threat—are peripheral to what the consumer wants to achieve through their purchase, the issue usually should not be expected to be a high-involvement one. It is well documented in the cited studies that proenvironmental attitudes increase consumers' propensity to buy environment-friendly products. Less researched in this connection is Fazio's (1986; Roskos-Ewoldsen and Fazio, 1992) proposition that attitudes also influence which information about a product a consumer pays attention to, including information about the product's environmentally relevant characteristics (but see Thøgersen, 1999).

The limited space available here makes it impossible to thoroughly review the huge literature on environment-friendly consumer behavior. Thus, I concentrate on the two areas where I believe that eco-labels have the greatest potential impact: (1) increasing consumer confidence in green claims, and (2) helping consumers carry out intentions to choose environment-friendly products.

Confidence in Green Claims

Basically green purchase behavior depends on the compromise consumers have to make in the form of higher price and/or lower quality and on the confidence they have in their choice leading to desirable environmental consequences (Peattie, 1999). The toughest green products to sell are those that require a large compromise and where consumers' confidence in it making any environmental difference is low. Successful green products typically enjoy high confidence and demand no or low compromise from consumers. Thus, by increasing consumer confidence in the credibility and the significance of green claims, third-party eco-labels may greatly improve the market prospects of environment-friendly products. Calculations based on data collected for a master thesis that I supervised may serve as an illustration (Andersen, 1995).

Respondents were a broad sample of individuals[8] responsible for their household's food shopping. One sample was interviewed about their purchase of organic milk, another about organic carrots. The most important environmen-

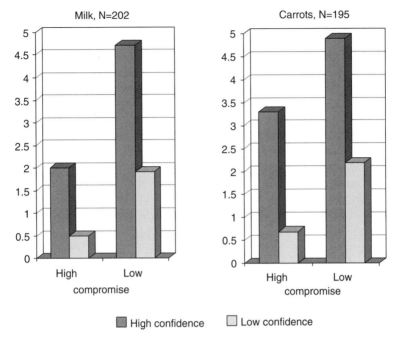

FIGURE 5-1 The purchase of organic milk and carrots in groups differing in confidence in environmental consequences and perceived compromise, Denmark 1995.

tal benefit from organic production is that it leads to less groundwater pollution than chemically based agriculture. Hence, agreement with the statement that by buying the organic product in question the consumer contributes to groundwater protection is used as an indicator of confidence in it making an environmental difference. In Denmark, the only real compromise when buying organic food products is higher price. Thus, agreement with the statement that the organic product in question is expensive is used as an indicator of perceived compromise. The average number of times the respondent reportedly chose organic out of the last 10 purchases of the product is shown graphically in Figure 5-1, using the confidence and perceived compromise indicators (split at the scale's midpoint) as grouping variables. In both cases, both grouping variables make a highly significant difference (F-test), the lowest F-value being 5.664. There are no significant interactions.

It is obvious from Figure 5-1 that consumers with a high confidence and who perceive the compromise as low are also most likely to buy organic products, and that the reverse combination of beliefs is much less facilitating. It is also apparent that even consumers who perceive the compromise to be high are much more likely to buy organic products if they also have a high confidence in the contribution's environmental implications. Therefore, if an eco-label increases consumer confidence in the implied green claim, the impact on the purchase of an environment-friendly product may be substantial. In fact, in the present case an eco-label had exactly this effect. In Denmark, organic food products carry the Ø-label (with the text "State Controlled Organic"). Respondents in this study were asked to point out the correct Ø-label among three alternative designs. Forty-three percent of both samples were able to do that. Those who knew the label (i.e., who picked the right one) had a significantly higher confidence in the choice making an environmental difference than those who did not (t_{milk} = 3.467, p < .001; $t_{carrots}$ = 3.488, p < .001).

The Implementation of Decisions to Buy Green

Several studies have demonstrated that environment-friendly behavior often depends on specific, task-related information (e.g., Bell et al., 1996; Kearney and De Young, 1995; Pieters, 1991; Thøgersen, 2000a). Consumers need specific and reliable information in order to be able to choose the most environmentally friendly alternative when competing options are offered or to do the right thing when asked to change a behavioral routine.

Figure 5-2 illustrates the importance of (knowing) an environmental label, the Danish Ø-label for organic products, for transforming environment-friendly

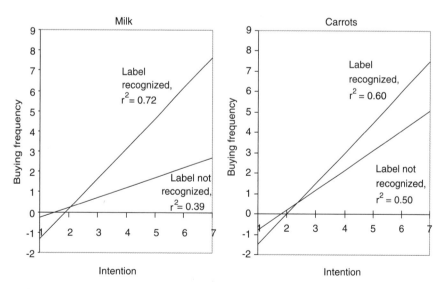

FIGURE 5-2 The influence of knowing the Ø-label on the relationship between buying intention and buying frequency regarding organic milk and organic carrots, Denmark, 1995.

buying intentions into action. The data set is the same as that used in Figure 5-1, but in this case I analyze whether the respondent's ability to point out the true Ø-label among three alternative designs influences the relationship between buying intentions and buying frequency (number of organic out of the last 10 liter/kilo). Separate regression analyses were made for split samples concerning each product: those choosing the correct design (43 percent of the sample in each case) and the rest. The lines are regression lines.

The results are in principle identical in the two cases, but the effect of knowing the Ø-label is somewhat stronger for milk than for carrots.[9] The difference may be due to some organic carrots being grown in one's own garden or bought at open markets, where there are other means to identify an organic product than the Ø-label, while organic milk can only be bought only from retail outlets.

The regression analyses illustrate that knowing the Ø-label has a substantial effect on buying frequency among those with a high buying intention, but no effect among those with a low buying intention. They also show that the relationship between buying intentions and buying frequency is stronger among consumers who are able to identify the correct Ø-label than among consumers who are not.[10] Hence, the study shows that by increasing consumers' ability to distinguish environment-friendly products, eco-labels can facilitate the implementation of environment-friendly intentions.

SUMMARY, CONCLUSIONS, AND IMPLICATIONS

Eco-labeling is aimed at reducing pollution and resource use associated with consumption by influencing consumer choices and, through these, companies' product policies. In the past couple of decades, eco-labeling has become a popular environment policy instrument in countries all over the world. Few schemes have been sufficiently thoroughly evaluated to be able to draw conclusions about their success. From those that have, it seems that, under the right conditions, eco-labeling can indeed lead to a substantial reduction in pollution and resource use. However, it takes time and a committed effort to build eco-labeling success. In particular, consumers have to go through an often time-consuming decision making process through which they first become aware of the label, and of labeled products, and then acquire sufficient knowledge to use it as a guide in decision making and to trust the message it conveys. A positive attitude toward eco-labels probably follows more or less automatically from knowledge and trust, but forming a positive attitude toward buying a specific eco-labeled product may take longer because time-consuming tradeoffs need to be made. Therefore, decision making about eco-labels is a gradual process and one that consumers go through at an uneven pace. Among other things, consumer receptiveness toward this kind of information and, hence, the pace depends on their environmental concern. The speed of diffusion of eco-labels also depends on the clarity of the label's profile, the intensity of its promotion, and its presence in the shops. The latter is particularly crucial for the outcome of the decision making.

There are, of course, a variety of other means that governments can use in their efforts to reduce the environmental impact of consumption. Labeling is obviously no substitute for legal restrictions and standards regulating, for example, the flow of harmful substances through the household, and taxes and subsidies— attempting to secure that nonmarket environmental impacts are reflected in the relative prices—may effectively influence consumer choices (e.g., Andersen and Sprenger, 2000; Von Weizsacker and Jesinghaus, 1992). There is no reason to believe that eco-labeling renders any of these means obsolete— on the contrary. Just keeping the environmental impact of consumption from increasing is an ambitious goal that will demand the effective use of all available means. In addition, there may be important synergies to be obtained from the coordinated implementation of several means (see, e.g., Gardner and Stern, 1996; Stern, 1999).

The fact that eco-labels compete with many other types of information in the shopping situation, including other informative labels and producers' noncertified green claims, acts as a noise wall that third-party eco-labels need to break through. Studies have shown that many consumers are uncertain about who issues third-party eco-labels and that this uncertainty reduces the trust in such labels. On the other hand, it has been shown that third-party eco-labels can increase the confidence in green claims and help distinguish environment-friendly products, thus increasing the likelihood of such products being bought. There is also evidence that experience with buying a product with an eco-label facilitates future decisions about buying this product, as well as other products wearing the same label. I suggested that the latter effect might be due to consumers forming new mental categories based on eco-labels and that such categories may carry affect. Research on the commitment approach to behavior change is informative regarding the type of affect in question (see, e.g., McKenzie-Mohr and Smith, 1999). According to this line of research, the purchase of an eco-labeled product can alter a person's self-perception to that as the type of person who buys eco-labeled products (cf., e.g., Hutton, 1982). There is plenty of evidence indicating that many people perceive a strong internal pressure to behave consistently with such a self-perception. Expressions of commitment seem to have stronger impacts on future behavior when they are voluntary (e.g., Shippee and Gregory, 1982) and public (e.g., Pallak et al., 1980), both of which typically characterize individual purchase acts.

The mentioned conclusions are based on scattered evidence and the evaluation of few schemes. There is a need for more, and more thorough and systematic, evaluations of eco-labeling schemes, particularly with a view to better identify manageable conditions for success. Special attention should be directed toward design characteristics that influence how consumers use labeling schemes in their decision making, including characteristics that facilitate and amplify the use of eco-labels as a basis for category-based decision making. Other more basic questions about eco-labeling remain unanswered, such as whether it contributes to consumer ignorance or makes them more attentive toward the problems associated with the continued rapid growth in private consumption. Hopefully this chapter will inspire future research on these topics.

NOTES

1   In this volume, Chapter 6, Valente and Schuster, make a similar point with regard to public health communication.

2   Dry weight.

3   The national organic food label was mentioned by 16 percent in Sweden, 5 percent in Finland, and 1.4 percent in Norway.

4   The two studies used different ways to measure recognition, meaning they are not strictly comparable.

5   More information about this case can be found at http://www.toolsofchange.com/English/CaseStudies/default.asp?ID=8 and at McKenzie-Mohr's Web site, http://www.cbsm.com. I am grateful to him for bringing the case to my attention.

6   Unless they believe other advantages are associated with environmental friendliness (Thøgersen, 1998). A recent Danish study found that "quality conscious" consumers use the Danish Ø-label as one among several cues indicating high product quality (Juhl et al., 2000).

7   A mall-intercept survey carried out in three shopping centers in Aarhus, Denmark, in 1998.

8   Fourteen acquaintances of the master students all over Denmark distributed questionnaires to some of their acquaintances, with the instruction to cover age groups (above 20 years) as broadly as possible. The data were collected in 1995.

9   The interaction between buying intention and knowing the Ø-label is statistically significant ($p <$ than 0.05) in both cases. The hierarchical regression analysis used to test for the interaction effect is reported in Thøgersen and Andersen (1996).

10   The somewhat surprising positive correlation between intention and behavior among those who are not able to identify the correct design may be because only one label design is available in the supermarket or because consumers in some cases—correct or mistakenly—use other cues to identify organic products. Of course, it also may be caused by a tendency to exaggerate organic buying that is correlated with stated buying intentions.

REFERENCES

Abt Associates Inc.
    1994   *Determinants of Effectiveness for Environmental Certification and Labeling Programs.* Cambridge, MA: U.S. Environmental Protection Agency.
Ajzen, I., and M. Fishbein
    1980   *Understanding Attitudes and Predicting Social Behavior.* Englewood Cliffs, NJ: Prentice-Hall.
Andersen, A.K.
    1995   Vigtige aspekter ved forminskelsen af inkonsistensen mellem forbrugerens positive holdning til at købe økologiske landbrugsprodukter og forbrugerens reelle køb af økologiske landbrugsprodukter (Important Aspects Regarding the Inconsistency Between Consumer Attitudes and Purchase of Organic Products). Unpublished MSc dissertation. Århus, Den.: Handelshøjskolen i Århus.

Andersen, M., and L. Vestergaard
  1998   Markedet for afsætning af økologiske fødevarer. Herunder en afdækning af hvilke faktorer der har betydning for forbrugernes køb/ fravalg af de forskellige økologiske produktkategorier (The Market for Organic Food Products). Unpublished MSc dissertation. Århus, Den.: Handelshøjskolen i Århus.

Andersen, M.S., and R.U. Sprenger, eds.
  2000   *Market-Based Instruments for Environmental Management: Politics and Institutions.* Cheltenham, Eng.: Edward Elgar.

Backman, M., T. Lindkvist, and Å. Thidell
  1995   The Nordic white swan: Issues concerning some key problems in environmental labelling. In *Sustainable Consumption - Report from the International Conference on Sustainable Consumption,* Eivind Stø, ed. Lysaker, Nor.: National Institute for Consumer Research (SIFO).

Bech-Larsen, T.
  1996   Danish consumers' attitudes to the functional and environmental characteristics of food packaging. *Journal of Consumer Policy* 19:339-363.

Beckerus and Rosander HB
  1999   *What Effect Has the Eco-Labelling of Household Detergents Had on Sewage Treatment Plants?* Stockholm, SE: Kemi & Miljö AB.

Bekholm, M., and B. Sejersen
  1997   Mærkeanvendelse og associationer. Personlig omnibus 44. Udarbejdet for Forbrugerstyrelsen (Labeling Use and Associations). Copenhagen, Den.: AIM Nielsen.

Bell, S.J., P.J. Erwin, and C.S. McLeod
  1996   Attitudes, norms and knowledge: Implications for ecologically sound purchasing behavior. In *1996 Winter Educators' Conference: Marketing Theory and Applications,* Vol. 7, E.A. Blair, and W.A. Kamakura, eds. Chicago: American Marketing Association.

Berger, I.E., B.T. Ratchford, and G.H. Haines, Jr.
  1994   Subjective product knowledge as a moderator of the relationship between attitudes and purchase intentions for a durable product. *Journal of Economic Psychology* 15:301-314.

Biel, A., and U. Dahlstrand
  1997   Habits and the establishment of ecological purchase behavior. In *IAREP XXII Conference,* Valencia, Spain: Corpas, C.B.

Buchtele, F., and H.H. Holzmuller
  1990   Die Bedeutung der Umweltverträglichkeit von Produkten für die Kaufpraëferenz (The importance of product's environmental friendliness for buying preferences). *Jahrbuch der Absatz- und Verbrauchsforschung* 36(1):86-103.

Celsi, R.L., and J.C. Olson
  1988   The role of involvement in attention and comprehension processes. *Journal of Consumer Research* 15:210-224.

Cohen, J.B.
  1982   The role of affect in categorization: Towards a reconsideration of the concept of attitude. In *Advances in Consumer Research,* Vol. 9, Andrew A. Mitchell, ed. Ann Arbor, MI: Association for Consumer Research.

Davis, J.J.
  1992   Ethics and environmental marketing. *Journal of Business Ethics* 11:81-87.

Durning, A.T.
  1992   *How Much Is Enough? The Consumer Society and the Future of the Earth.* London: Earthscan Publications Ltd.

Dyer, R.F., and T.J. Maronick

    1988   An evaluation of consumer awareness and use of energy labels in the purchase of major appliances: A longitudinal analysis. *Journal of Public Policy and Marketing* 7:83-97.

Eden, S.

    1994/ Business, trust and environmental information: Perceptions from consumers and retailers.
    1995   *Business Strategy and the Environment* 3(4):1-7.

Enger, A.

    1998   *Miljøargumentasjon i markedsføring (Environmental Arguments in Marketing)*. Oslo: SIFO.

Enger, A., and R. Lavik

    1995   Eco-labeling in Norway: Consumer knowledge and attitudes. In *Sustainable Consumption - Report from the International Conference on Sustainable Consumption*, Eivind Stø, ed. Lysaker, Nor.: National Institute for Consumer Research (SIFO).

Fazio, R.H.

    1986   How do attitudes guide behavior? In *The Handbook of Motivation and Cognition: Foundations of Social Behavior*, R.M. Sorrentino and E.T. Higgins, eds. New York: Guilford Press.

Fiske, S.T., and M.A. Pavelchak

    1986   Category-based versus piecemeal-based affective responses. In *The Handbook of Motivation and Cognition: Foundations of Social Behavior*, R.M. Sorrentino and E.T. Higgins, eds. New York: Guilford Press.

Forbrugerstyrelsen

    1993   *Mærkning rettet til forbrugerne (Labeling Targeting the Consumers)*. Copenhagen: Forbrugerstyrelsen.

Gardner, G.T., and P.C. Stern

    1996   *Environmental Problems and Human Behavior*. Boston, MA: Allyn and Bacon.

Geyer-Allély, E., and J. Eppel

    1997   *Consumption and Production Patterns: Making the Change*. Paris: Organization for Economic Co-operation and Development.

Grunert, S.C., and H. Jørn Juhl

    1995   Values, environmental attitudes, and buying of organic foods. *Journal of Economic Psychology* 16:39-62.

Gutman, J.

    1982   A means-end chain model based on consumer categorization processes. *Journal of Marketing* 46(Spring):60-72.

Hansen, U., and S. Kull

    1994   Öko-Label als umweltbezogenes Informationsinstrument: Begründungszusammenhänge und Interessen (Eco-labels as environmental information tool: Reasoning and interest). *Marketing* 4(4. kvartal):265-273.

Harland, P., H. Staats, and H.A.M. Wilke

    1999   Explaining pro-environmental intention and behavior by personal norms and the theory of planned behavior. *Journal of Applied Social Psychology* 29:2505-2528.

Hastak, M., R.L. Horst, and M.B. Mazis

    1994   Consumer perceptions about and comprehension of environmental terms: Evidence from survey research studies. In *Proceedings of the 1994 Marketing and Public Policy Conference*, D.J. Tingold, ed. Arlington, VA: American Marketing Association.

Heberlein, T.A.

    1972   The land ethic realized: Some social psychological explanations for changing environmental attitudes. *Journal of Social Issues* 28(4):79-87.

Herr, P.M., and R.H. Fazio
1993   The attitude-to-behavior process: Implications for consumer behavior. In *Advertising Exposure, Memory, and Choice: Advertising and Consumer Psychology*, A.A. Mitchell et al., eds. Hillsdale, NJ: Lawrence Erlbaum Associates.

Hoyer, W.D., and D.J. MacInnis
1997   *Consumer Behavior*. Boston: Houghton Mifflin.

Hutton, R.R.
1982   Advertising and the Department of Energy's campaign for energy conservation. *Journal of Advertising* 11(2):27-39.

Jacoby, J.
1984   Perspectives on information overload. *Journal of Consumer Research* 11:569-573.

Juhl, H.J., E. Høg, and C.S. Poulsen
2000   *Forbrugernes Vurdering af Nogle Udvalgte Kvalitetsmærkninger for Hakket Oksekød – Undersøgelsens Design, Gennemførelse og Resultater (The Consumers' Evaluation of Selected Quality Labels for Minced Beef)*. MAPP Working Paper. Århus, Den.: Handelshøjskolen i Århus.

Kampmann, L.
2000   Miljø er vigtigt for danskerne (Environmental consequences are important for the Danes). *Miljømærkenyt* 1(1):3-4.

Kearney, A.R., and R. De Young
1995   A knowledge-based intervention for promoting carpooling. *Environment and Behavior* 27(5):650-678.

Kokkinaki, F.
1997   Involvement as a determinant of the process through which attitudes guide behaviour. In *IAREP XXII Conference*. Valencia, Spain: Corpas, C.B.

Konsumentverket
1993   *Konsumenten og Miljön. Resultat från en undersökning av svenska konsumenters miljömedvetenhet (The Consumer and the Environment. A study of Swedish Consumers' Environmental Awareness)*. Stockholm: Konsumentverket.

1995/ *The consumer and the environment: Results of a survey into awareness of the environ-*
1996   *ment amongst Swedish consumers*. Stockholm: Konsumentverket.

1998   *Allmänhetens kunskaper, attityder och agerande i miljøfrågor (The Public's Knowledge, Attitudes, and Behavior in Environmental Matters)*. Stockholm: Konsumentverket.

Laric, R.J., and D. Sarel
1981   Consumer (mis)perceptions and usage of third party certification marks, 1972 and 1980: Did public policy have an impact? *Journal of Marketing* 45(Summer):135-142.

Lindberg, K.-E.
1998   *Nordisk omnibus svanemerket. Utarbeidet for Nordisk Miljömärkning (Nordic Survey About the Swan Label)*. Oslo: Markeds- og Mediainstituttet.

Mackenzie, J.J.
1997   Driving the road to sustainable ground transportation. In *Frontiers of Sustainability*, R. Dower, D. Ditz, P. Faeth, N. Johnson, K. Kozloff, and J.J Mackenzie, eds. Washington, DC: Island Press.

Matthews, E., C. Amann, S. Bringezu, M. Fischer-Kowalski, W. Hüttler, R. Kleijn, Y. Moriguchi, C. Ottke, E. Rodenburg, D. Rogisch, H. Schandl, H. Schütz, E. Van der Voet, and H. Weisz
2000   *The Weight of Nations—Material Outflows from Industrial Economies*. Washington, DC: World Resource Institute.

Mayer, R.W., and J. Gray-Lee
  1995   Environmental marketing claims in the U.S.A.: Trends since issuance of the FTC guides. In *Sustainable Consumption - Report from the International Conference on Sustainable Consumption,* Eivind Stø, ed. Lysaker, Nor.: National Institute for Consumer Research (SIFO).

McKenzie-Mohr, D., and W. Smith
  1999   *Fostering Sustainable Behavior: An Introduction to Community-Based Social Marketing.* Gabriola Island, British Columbia, Can.: New Society Publishers.

Miljø- og Energiministeriet
  1995   *Natur- og miljøpolitisk redegørelse 1995 (Nature and Environmental Policy Report 1995).* Copenhagen: Miljø- og Energiministeriet.
  1999   *Natur- og miljøpolitisk redegørelse 1999 (Nature and Environmental Policy Report 1999).* Copenhagen: Miljø- og Energiministeriet.

Miljøstyrelsen
  1996   *En styrket produktorienteret miljøindsats. Et debatoplæg (A Strengthened Product Oriented Environmental Effort: Discussion Paper).* Copenhagen: Miljøstyrelsen.

Morris, J.
  1996   *Buying Green: Consumers, Product Labels and the Environment.* Los Angeles: Reason Foundation.

Morris, L.A., M. Hastak, and M.B. Mazis
  1995   Consumer comprehension of environmental advertising and labeling claims. *The Journal of Consumer Affairs* 29:328-350.

Naturvårdsverket
  1997   *En studie hur olika styrmedel påverkat skogsindustrin (A Study of How Various Regulation Means Influence the Forest Industry).* Stockholm: Naturvårdsverket.

Nilsson, O.S., N.P. Nissen, J. Thøgersen, and K. Vilby
  1999   Rapport fra "forbrugergruppen" under Erhvervsministeriets mærkningsudvalg (Report from the "Consumer Committee" of the Ministry of Business' Labeling Committee). In *Mærkning. Mærkningsudvalgets redegørelse (Labeling. Report from the Labeling Committee).* Copenhagen: Forbrugerstyrelsen, Erhvervsministeriet.

Noorman, K.J., and T.S. Uiterkamp, eds.
  1998   *Green Households? Domestic Consumers, Environment and Sustainability.* London: Earthscan.

Nordic Council of Ministers
  2001   *Evaluation of the Environmental Effects of the Swan Eco-Label—Final Analysis.* Copenhagen: Nordic Council of Ministers.

Norwegian Ministry of Environment
  1994   *Symposium: Sustainable Consumption. 19-20 January 1994, Oslo, Norway.* Oslo: Ministry of Environment, Norway.

Organization for Economic Co-operation and Development
  1991   *Environmental Labeling in OECD Countries.* Paris: Organization for Economic Co-operation and Development.
  1997a *Eco-Labeling: Actual Effects of Selected Programmes.* Paris: Organization for Economic Co-operation and Development/GD(97)105.
  1997b *Sustainable Consumption and Production—Clarifying the Concepts.* Paris: Organization for Economic Co-operation and Development.

Pallak, M.S., D.A. Cook, and J.J. Sullivan
  1980   Commitment and energy conservation. In *Applied Social Psychology Annual,* Vol.1, L. Bickman, ed. Beverly Hill, CA: Sage.

Palm, L., and G. Jarlbro
 1999   Nordiska konsumenter om Svanen—livsstil, kännedom, attityd och förtroende (Nordic Consumers About the Swan—Lifestyle, Awareness, Attitude, and Trust), Report 1999:592. Copenhagen: Nordisk Ministerråd.
Palm, L., and S. Windahl
 1998   How Swedish Consumers Interpret and Use Environmental Information: A Study of Quantitative Environmental Product Declarations. Stockholm: Konsumentverket.
Parkinson, T.L.
 1975   The role of seals and certifications of approval in consumer decision-making. Journal of Consumer Affairs 9(1):1-14.
Peattie, K.
 1995   Environmental Marketing Management: Meeting the Green Challenge. London: Pitman Publishing.
 1999   Rethinking marketing: Shifting to a greener paradigm. In Greener Marketing: A Global Perspective on Greening Marketing Practice, M. Charter and M.J. Polonsky, eds. Aizlewood's Mill, UK: Greenleaf.
Peter, J.P., J.C. Olson, and K.G. Grunert
 1999   Consumer Behavior and Marketing Strategy. European ed. London: McGraw Hill.
Pieters, Rik G.M.
 1991   Changing garbage disposal patterns of consumers: Motivation, ability, and performance. Journal of Public Policy and Marketing 10:59-76.
Rogers, E.M.
 1995   Diffusion of Innovations. 4th ed. New York: Free Press.
Roskos-Ewoldsen, D.R., and R.H. Fazio
 1992   On the orienting value of attitudes: Attitude accessibility as a determinant of an object's attraction of visual attention. Journal of Personality and Social Psychology 63:198-211.
Scandia Consult Sverige AB
 1999   Eco-Evaluation of the 'Good Environmental Choice'—Label on Household Chemicals. Stockholm: Scandia Consult Sverige AB.
Shippee, G.E., and W.L. Gregory
 1982   Public commitment and energy conservation. American Journal of Community Psychology 10:81-93.
Sitarz, D., ed.
 1994   Agenda 21: The Earth Summit Strategy to Save Our Planet. Boulder, CO: EarthPress.
Sparks, P., and R. Shepherd
 1992   Self-identity and the theory of planned behavior: Assessing the role of identification with green consumerism. Social Psychology Quarterly 55:388-399.
Stern, P.C.
 1999   Information, incentives, and proenvironmental consumer behavior. Journal for Consumer Policy 22:461-478.
Sujan, M.
 1985   Consumer knowledge: Effects on evaluation strategies mediating consumer judgments. Journal of Consumer Research 12(June):31-46.
Tenbrunsel, A.E., K.A. Wade-Benzoni, D.M. Messick, and M.H. Bazerman
 1997   The dysfunctional aspects of environmental standards. In Environment, Ethics, and Behavior: The Psychology of Environmental Valuation and Degradation, M.H. Bazerman, D.M. Messick, A.E. Tenbrunsel, and K.A Wade-Benzoni, eds. San Francisco: New Lexington Press.
The Swedish Society for Nature Conservation
 1999   Changes in Household Detergents: A Statistical Comparison Between 1988 and 1996. Stockholm: Swedish Society for Nature Conservation.

Thøgersen, J.
    1996a *The Demand for Environmentally Friendly Packaging in Germany.* MAPP Working
           Paper No. 30. Aarhus, Den.: The Aarhus School of Business.
    1996b Recycling and morality. A critical review of the literature. *Environment and Behavior*
           28:536-558.
    1998   *Understanding Behaviours with Mixed Motives: An Application of a Modified Theory of
           Reasoned Action on Consumer Purchase of Organic Food Products.* Working Paper
           No. 98:2. Aarhus, Den.: Department of Marketing, The Aarhus School of Business.
    1999   The ethical consumer. Moral norms and packaging choice. *Journal of Consumer
           Policy* 22:439-460.
    2000a Knowledge barriers to sustainable consumption. In *Marketing and Public Policy Con-
           ference Proceedings 2000,* P.F. Bone, K.R. France, and J. Wiener, eds. Chicago:
           American Marketing Association.
    2000b Psychological determinants of paying attention to eco-labels in purchase decisions:
           Model development and multinational validation. *Journal of Consumer Policy* 23:285-
           313.
Thøgersen, J., and A.K. Andersen
    1996   Environmentally friendly consumer behavior: The interplay of moral attitudes, private
           costs, and facilitating conditions. In *Marketing and Public Policy Conference Proceed-
           ings,* Vol. 6, R.P. Hill and C.R. Taylor, eds. Chicago: American Marketing Associa-
           tion.
Tufte, P.A., and R. Lavik
    1997   *Helse- og miljøinformasjon. Forbrukernes behov for informasjon om skadelige stoffer i
           produkter (Health and Environmental Information: The Consumers' Need for Informa-
           tion About Harmful Substances in Products).* Lysaker, Nor.: Statens Institutt for For-
           bruksforskning, National Institute for Consumer Research (SIFO).
United Nations Conference on Trade and Development
    1995   *Trade, Environment and Development Aspects of Establishing and Operating Eco-La-
           beling Programmes.* Geneva: United Nations Conference on Trade and Development.
United Nations Environment Program
    1994   Elements for policies for sustainable consumption. In *Symposium: Sustainable Con-
           sumption. 19-20 January 1994, Oslo, Norway.* Oslo: Ministry of Environment, Nor-
           way.
U.S. Environmental Protection Agency
    1998   *Environmental Labeling Issues, Policies, and Practices Worldwide.* Washington DC:
           U.S. Environmental Protection Agency.
Van Dam, Y.K., and M. Reuvekamp
    1995   Consumer knowledge and understanding of environmental seals in the Netherlands. In
           European Advances in Consumer Research, F. Hansen, ed. Provo, UT: Association for
           Consumer Research.
Verplanken, B.,G. Hofstee, and H.J.W. Janssen
    1998   Accessibility of affective versus cognitive components of attitude. *European Journal
           of Social Psychology* 28:23-35.
Verplanken, B., and M.W.H. Weenig
    1993   Graphical energy labels and consumers' decisions about home appliances: A process
           tracing approach. *Journal of Economic Psychology* 14:739-752.
Von Weizsacker, E.U., and J. Jesinghaus
    1992   *Ecological Tax Reform.* London, NJ: Zed.
Waller-Hunter, J.
    2000   2020: A clearer view for the environment. *OECD Observer* 221/222:58-60.

6

The Public Health Perspective for Communicating Environmental Issues

Thomas W. Valente and Darleen V. Schuster

The public health field has been engaged in efforts to promote health-related behavior for some time. These efforts include promotion of hygiene behaviors, safety promotion and accident prevention, substance abuse prevention, adoption of healthy lifestyles and eating habits, family planning and contraceptive use, and many other areas. These promotions take many forms and have been accompanied by considerable research on their planning and effectiveness. The purpose of this chapter is to convey some of the experiences and lessons learned from these activities and how they might be applied to environmental issues.

Although experiences from many fields may be relevant, this chapter will focus somewhat on experiences from the field of family planning promotion for three reasons. First, family planning practices have some similarity to environmental behaviors. Second, family planning promotion is one of the larger bodies of research available. Third, the authors have experience working with family planning promotion campaigns. This discussion, however, will be not limited to the family planning literature because our purpose is to provide a broad view from the public health perspective.

The chapter consists of five sections: behavior change theory, mass media campaigns, attitudes and attitude change, interpersonal communication networks, and the importance of evaluation research in behavior change promotion. We provide an overview of the public health experiences for each of these topics, and summarize their implications for the promotion of environmental voluntary measures. A guiding principle for these health communication programs has been that they be theory based.

BEHAVIOR CHANGE THEORY

Theory attempts to explain people's behavior and describe factors that motivate or present barriers to it. It can provide the basis for effective program design and meaningful evaluation by informing the selection of goals, objectives, and techniques to measure them. A theoretical perspective should be stated explicitly to guide the program. Although most behavior change models are individually based, public health research increasingly has recognized the importance of ecological levels of analysis (not to be confused with our more common use of ecology in this volume).

The ecological perspective in this instance refers to the interaction of behavior and environment. Behavior has many determinants and is influenced by multiple levels of social, cultural, and physical environmental factors. Ecological models typically consider the following levels of analysis (McElroy et al., 1988; Green et al., 1996): (1) individual (intrapersonal), (2) interpersonal, (3) institutional, (4) communal, and (5) societal. By including analyses at these levels, researchers can examine and incorporate various sources of influence on behavior in addition to an individual's attributes. The ecological perspective on environmental issues explicitly forces us to look at governmental or organizational policies that present barriers to environmentally sensitive behaviors. For example, many individuals may be positively predisposed to recycling, but fail to comply because their employer lacks a formal policy promoting it.

Table 6-1 describes the ecological levels and the advantages and disadvantages for interventions targeted at each level. Although multipronged, multilevel interventions are considered the most effective, they can be impractical and costly to implement in many settings. The ecological perspective is not a behavior change theory, but rather sensitizes us to the need to consider different influences on behavior. The most common theories used in public health research (Glanz et al., 1997) are the health belief model (Hochbaum, 1958; Rosenstock, 1960; Rosenstock et al., 1988), theory of reasoned action (Fishbein, 1967; Fishbein and Ajzen, 1975; Ajzen and Fishbein, 1980) and theory of planned behavior (Ajzen, 1991; Ajzen and Driver, 1991; Ajzen and Madden, 1986), social cognitive theory (Mischel, 1973; Bandura, 1977, 1986), stages of change or transtheoretical model (DiClemente and Prochaska, 1983), and diffusion of innovations (Rogers, 1995). Of these, diffusion of innovations theory is probably the most commonly used theory in the health promotion and communication arena.

Diffusion of Innovations

Diffusion of innovations theory describes how new ideas, opinions, attitudes, and behaviors spread throughout a community (Katz et al., 1963; Rogers, 1995; Ryan and Gross, 1943; Valente, 1993, 1995; Valente and Rogers, 1995). "Diffusion is the process by which an innovation is communicated through cer-

TABLE 6-1   Ecological Levels of Analysis and Intervention, Their Advantages and Disadvantages

Ecological Level	Advantages	Disadvantages
Individual: Clinics and treatment site	Can be tailored Are direct and immediate Some attempt to use "brief interventions"	Effectiveness dependent on similarity/empathy between patient and provider
Organizational: Worksite, school, etc.	Working in bounded, closed communities More control over the intervention and setting	Effectiveness depends on organizational factors Variability among organizations of the same type
Community: Neighboorhoods, associations	Generally most effective Empowering and sensitive to community dynamics	Take a long time to forge collaboration and work with groups Hard to scale up and replicate
Mass Media: TV, radio, and print	Reach many people Can change societal/normative perceptions Can change some people's behavior	Usually do not change a large percentage Dependent on quality Specific to the situation
Policy: Local and global levels	Can target few people Small changes can have big effects Highly visible	Somewhat unpredictable Replication would be uncertain
Multipronged: Address supply and demand for health-related behavior	Addresses both motivations and barriers of change	Can be expensive Hard to coordinate diverse organizations and activities

tain channels over time among the members of a social system" notes Rogers (1995:5). Diffusion theory has been used to examine the spread of new computer technology, educational curricula, farming practices, family planning methods, medical technology, and many other innovations. Considerable research on the diffusion of family planning practices and fertility preferences has been conducted to date (for a recent review see Casterline and Cleland, 2002). This chapter focuses somewhat on how media campaigns have been used to accelerate the adoption of family planning practices and on the public health communication perspective that has been used to study these programs. Diffusion theory has five major assumptions: (1) adoption takes time; (2) people pass through various stages in the adoption process; (3) they can modify the innovation and sometimes discontinue its use; (4) perceived characteristics of the innovation influence adoption; and (5) individual characteristics influence adoption.

The first two assumptions will be discussed at length. Diffusion of innovations specifies five stages in the behavior change process: knowledge, persuasion, decision, trial, and adoption (Rogers, 1995). The diffusion of innovations stages have been expanded into a hierarchy model (McGuire, 1989) that was adapted specifically to the case of family planning (Rogers, 1973, 1995; Piotrow et al., 1997; Valente et al., 1996). A common outcome variable for health promotion programs can be a score on a composite index indicating the stage or step of behavior change.

Because people become aware of new behaviors at different times, and because they pass through the stages at different rates, there is considerable lag between the first and last adopters of a new behavior. For example, Ryan and Gross (1943) showed that 14 years passed between first and last adopters of hybrid seed corn in two Iowa counties, in spite of this innovation being far superior to the one it replaced. When the spread of new ideas and practices is graphed, it resembles a typical growth or S-shaped curve (see Figure 6-1).

Diffusion theory classifies individuals in terms of their time of adoption relative to a community or population. The first people to try a new practice are called innovators or pioneers. The second group to adopt is called early adopters. These first two groups constitute the first 16 percent of adopters. The next 34 percent of the population to adopt are the early majority, followed by the late majority, then laggards. This classification initially was devised because adoption behavior was thought to follow a normal curve, and thus provided a convenient way to compare research studies (in terms of the characteristics associated with each adopter type). More recent research, however, has acknowledged that diffusion curves often deviate from normality, and adoption behavior more often is classified dichotomous (adopter, nonadopter) or left continuous (time of adoption). The units for measuring time vary considerably because some innovations diffuse in days and others in years or decades. Similarly, some innovations will reach saturation of 100 percent, while others may attain lower levels of penetration. Early in the diffusion of a new behavior, there are few adopters and the growth in new adopters is slow. Research has found that these early adopters often are persuaded more by mass media and other targeted communications that provide information relevant to the behavior. Moreover, these new adopters sometimes are freed from social norms that would otherwise inhibit them from adopting a new behavior. Because new behaviors often are perceived as uncertain and risky, these early adopters often require some form of compensation or rationale for them to adopt.

These two components, stages of adoption and the time it takes for diffusion to occur, are graphed in Figure 6-2, showing projected rates of the spread of awareness, positive attitude, and behavior (Valente, 1993). Expected levels for

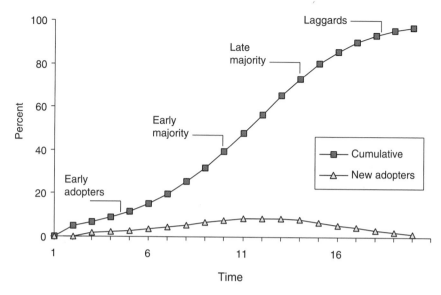

FIGURE 6-1 Typical diffusion curve showing the cumulative percent of adopters and percent of new adopters at each point in time, with adoption categories.

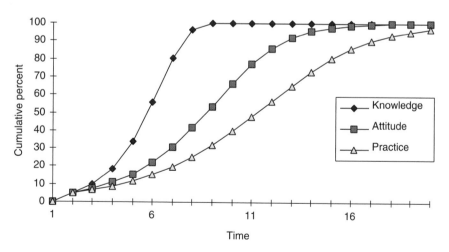

FIGURE 6-2 Typical knowledge, attitude, and practice diffusion curves used to predict the rate of diffusion and the average time between knowledge and practice.

each can be determined by looking at any point in time and expected time between awareness and use by taking the time difference between the two curves. The general model of awareness leading to positive attitudes and eventually to use has been termed the learning hierarchy. Others (Chaffee and Roser, 1986; Valente et al., 1998) have argued that alternative sequences are possible. For example, some behaviors may be adopted first, followed by positive attitudes or knowledge of the benefits. Teeth brushing, for example, usually is adopted because we are forced to do so as young children; only later do we learn the benefits, and still later develop a positive attitude toward it. The same can be said with energy conservation: The benefit of lower electric bills creates positive attitudes toward this practice. Thus, although the learning hierarchy, as depicted in Figure 6-2, may hold for some innovations, other behavior change sequences will characterize most behaviors, particularly those for which evidence on their relative advantage is not readily available.

Regardless of the behavior change sequence, health promoters have tried to accelerate behavior change by creating awareness (increasing knowledge), improving attitudes, and directly encouraging individuals to adopt healthy behaviors. Two specific functions of media campaigns have been to spread knowledge rapidly so that its curve grows quickly and to shorten the Knowledge, Attitude, and Practice (KAP)-gap, the time between awareness and use. Some argue that media campaigns are most effective early in the diffusion process since there are few other users whom potential adopters can turn to for advice. Later in the diffusion process, media campaigns serve to place the behavior back on the public agenda (McCombs and Shaw, 1972) and stimulate further interpersonal communication. The other three assumptions of the diffusion model can be used to guide message development. For example, because the perceived risk to adoption slows diffusion, promotion programs can attempt to reduce risk. Programs can also emphasize the perceived characteristics of environmental behaviors, for example, by emphasizing their compatibility.

Empirical support for the diffusion theory is spread throughout the behavioral sciences, although the most definitive results came from early studies of farmer adoption of new agricultural practices. It has been used extensively to study the diffusion of contraception and family planning in developing countries and in the United States to study adoption of many different health-related behaviors. In a review of behavior theories used in the mid-1990s, Glanz and colleagues (1997) note that it was the 10th most used theory. Like other behavioral theories, diffusion variables alone often explain less variation in behavior than desired. The theory works best when potential adopters can afford and have easy access to the innovations being promoted, and works less well when purely economic considerations influence adoption decisions.

MASS MEDIA CAMPAIGNS

Early studies on communication campaign effects highlight both successes (Cartwright, 1949; Mendolsohn, 1973; Rogers and Storey, 1987) and failures (Hyman and Sheatsley, 1947; National Public Radio, 1996; Udry et al., 1972). Communication campaigns fail to achieve their hoped-for results for many reasons. Campaigns deemed relatively unsuccessful have, in part, been attributed to unrealistic beliefs in the ability of the media to engender significant and sustained behavioral change. The media can be a very powerful influence on people's attitudes and beliefs, but typically not on behaviors because changing behavior often requires relinquishing a rewarding behavior and replacing it with another that provides significantly less pleasure or rewards. However, in the event that changes in action are achieved, such as earthquake preparedness or ultraviolet radiation protection, the effects are often of relatively short duration. This is highly characteristic of campaigns targeting habitual behaviors such as smoking, drinking, and sexual behaviors. It is unlikely that someone will stop smoking as a result of viewing a single or series of public service announcements. More commonly, the media are effective in priming audiences for change, while other cues to action (such as personal contact) are needed for individuals to implement the desired behaviors. The media have a great capacity to expose large numbers of people to prevention messages. Therefore, media campaigns often are used initially to increase public awareness of a problem, its determinants, and strategies for change, while other intervention activities are used to change behaviors.

Studies conducted by Lazarsfeld and colleagues (Berelson et al., 1954; Katz and Lazarsfeld, 1955; Lazarsfeld et al., 1948; Merton, 1968; and see Eulau, 1980, for a review) propose a classic two-step flow model of communications (Katz, 1957, 1987; see Gitlin, 1978, for critique). This model of interpersonal communications posits that opinion leaders use the mass media for information more than opinion followers, then these leaders share their opinions with these followers. Consequently, many scholars have argued that the mass media are effective at disseminating information and achieving awareness of campaign messages, but that interpersonal communication is necessary for motivating behavior change (Hornik, 1989; Valente et al., 1996; Valente and Saba, 1998). This adage has directed many projects to use the mass media to promote new ideas, and products, then to rely on outreach and peer education programs for adoption.

The use of a combinatory approach is illustrated in one of the most successful health behavior change campaigns incorporating a significant mass media component: the Stanford Heart Disease Prevention Program (Farquhar et al., 1977). Inspired by the vision of a cardiologist and a communication scholar, a health communication and education program was developed to reduce individual risk factors associated with heart disease (e.g., smoking, obesity, stress, lack of exercise). This large-scale intervention combined mass media channels (e.g., television, radio, newspapers, mass-distributed print media) with interpersonal communication methods (e.g., group training classes for smoking cessation and aerobic exercise) to influence knowledge, attitudes, and risk-related behaviors. Significant increases in risk reduction knowledge were attributed to the campaign's integrated design and novelty of information dissemination (such as mass-distributed tip sheets and self-help kits). The campaign's widespread exposure sparked further information seeking (e.g., tip sheets encouraged the use of booklets) and interpersonal discussions of cardiovascular disease and related issues (Schooler et al., 1993). In this example, mass media channels were employed in the beginning of the campaign to increase the public's awareness of the need to change, while interpersonal channels were used to present reinforcing materials and persuade people to engage in recommended risk reduction behaviors. In spite of its fame, the Stanford program is credited only with modest increases in behavior, as the comparison communities quickly matched behavioral levels initially seen in the intervention ones.

Environmental awareness interventions incorporating mass media and interpersonal approaches have been shown to be effective in enhancing knowledge and improving short-term health-protective behaviors (Campbell et al., 2000). As an example, Dietrich and colleagues (1998) examined the effects of a multicomponent intervention designed to change children's sun protection behaviors. Messages encouraging solar protection were delivered to children, families, and caregivers through counseling, educational sessions, displays, educational materials, posters, and sunscreen samples. Based on observations at beach recreation areas, significantly more children in the intervention towns used sunscreen than in the control towns. Other studies of solar protection behaviors combined mass media messages (newspaper, radio, and television) with the dissemination of educational materials to increase melanoma awareness and detection. For instance, Graham-Brown (1990) reported significant increases in new patient visits at community clinics and the detection of melanomas following a public education campaign promoting the medical assessment of potentially dangerous skin lesions. Similar results were obtained in an Australian campaign, where annual melanoma detection rates increased significantly, from 130 diagnosed cases before a multimedia campaign to 189 during the campaign (Pehamberger et al., 1993). Such studies provide evidence for the effectiveness of mass media public education campaigns in increasing melanoma awareness and related solar protection behaviors.

ATTITUDES AND ATTITUDE CHANGE

Although the ultimate goal of an intervention is to change behavior (e.g., to wear sunscreen, to recycle, to rideshare), this is often a difficult task. Although it is relatively easy to raise awareness of a health or environmentally relevant behavior, this is not the case for attitudes that are often fairly well entrenched. Attitudes are important not only because of their presumed ability to direct behaviors in some instances, but because they also serve many important functions for individuals (Katz, 1960). Besides summarizing a person's beliefs about a topic (knowledge function), attitudes can serve a value-expressive function, which occurs when holding a particular attitude permits us to convey an important value or principle to others. For example, the person who has a preference for electric cars because their use demonstrates an important concern about minimizing pollution has an attitude that serves a value-expressive function. Likewise, attitudes also may serve a utilitarian function, where the adoption of certain attitudes helps people gain rewards and avoid punishments (Schultz, this volume, Chapter 4). Individuals may favor the use of nontoxic chemicals in clothing and environmentally safe trash bags, for example, in an attempt to gain approval from important others such as family, friends, and neighbors. Considering the important functions served by attitudes, it follows that a central goal of media campaigns is to promote positive attitudes toward recommended behaviors.

Attitudes toward a behavior and attitudes toward the process of adopting a behavior can be important predictors of adoption (Petty and Cacioppo, 1981; Fishbein and Ajzen, 1975). Attitudes can be complex, comprising one's attitude toward the behavior, toward products and actions associated with the behavior, and toward perceptions of normative behavior. For example, many people may believe that energy conservation is beneficial and hold a positive attitude about fuel-efficient cars, but still buy a larger vehicle because of normative expectations in their neighborhood. Additionally, there may be a perception that engaging in a particular behavior is beyond one's control, most likely because of the presence of insurmountable external factors (e.g., lack of financial resources) (Ajzen, 1991). Despite holding positive attitudes and normative beliefs surrounding the purchase of a fuel-efficient car, an individual still may fail to purchase one due to financial constraints. Although favorable attitudes toward a behavior are important predictors of adoption, the perception of behavioral control is of equal importance.

INTERPERSONAL COMMUNICATION NETWORKS

Although media campaigns are conceptualized as broadcasts to a population of disconnected individuals, the audience is a web of human relations connected to one another in complex and nonrandom ways. Consequently, campaign messages are not received in a vacuum, but rather are filtered through these social networks. People often consume messages with others, directly influencing the manner in which messages are interpreted. Furthermore, talking to others about health promotion messages may cause them to reinterpret them. Consistently, one goal of a campaign is to generate interpersonal discussion on the topic in an attempt to set the public's agenda.

One of the most significant correlates to behavior and behavior change is the perception of peer approval (Valente et al., 1997; Valente and Saba, 1998; Alexander et al., 2001), otherwise referred to as social norms. Individuals deciding on the appropriateness of certain behaviors make social comparisons and use peers as reference points when making decisions, particularly when an ambiguous situation arises. Thus, the fact that peers influence behavior is not surprising. Measuring this peer influence, however, presents challenges, and findings on its influence are not uniform.

In most studies, peer influence and perceptions of peer behavior often were measured by asking people, "To what degree do your friends approve of X?" Response categories were often likert scales that had a positive correlation with behavior. Unfortunately, the nature of this correlation is unclear because respondents may be projecting their beliefs on others, or because they practice it, they think their friends do as well. Social network techniques have been developed to better measure peer influence.

Social networks consist of the friends, colleagues, and family members in a person's immediate social circle, and are measured by asking respondents to provide the names or initials of their friends or those people with whom they discuss personal matters (Burt, 1980; Marsden, 1990; Valente and Saba, 1998; Valente and Vlahov, 2001). Respondents then are asked questions about the persons they named: (1) whether they approve of the behavior, (2) whether they practice it, and (3) whether they talked about it. This measure provides a more refined indication of which friends support and/or practice the behavior. Specific characteristics of these friends that also can be linked to the behavior include socioeconomic level, attitude toward the behavior, and practice of it (Valente and Saba, 2001). However, personal network data still may be prone to projection bias. Figure 6-3 presents a general evaluation framework.

Sociometric network methods overcome this bias by collecting data from all members of a community, such as an organization, a school, a rural village, or a neighborhood. Links between individuals in the network are measured so that a map of the community can be drawn and individual positions within the network determined (Burt, 1980; Marsden, 1990; Rogers and Kincaid, 1981; Scott, 2000; Valente, 1995; Wasserman and Faust, 1994). An individual's position in the network may influence behavior. Furthermore, because there are reports on every person's behavior, and links between him and her, it is possible to measure how many people in each person's network practice the behavior. Some people, by virtue of their connections, will be surrounded by others that engage in the behavior, while others will be surrounded by few who do. Network exposure is highly correlated with behavior, and its measurement does not suffer from projection bias. Using this technique, for example, Valente and others (1997) showed that women were more likely to practice contraception if they thought

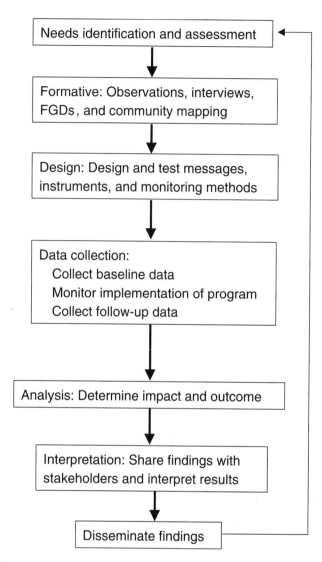

FIGURE 6-3 Health promotion evaluation framework.
Source: Valente (2002).

their friends used it, regardless of whether the friends' self-reports showed they practiced contraception.

Although the correlation between network exposure and behavior is high, there are many individuals who adopt a new behavior before a majority of their network does so. Indeed, in order for diffusion to occur, some people must be willing to initiate change when no or few others in the community or in their network have done so. These early adopters can be labeled low-threshold adopters because their resistance is low. Low-threshold adopters have reported greater use of mass media for information and greater use of nonpersonal sources of information for decision making (Valente, 1995). Promotional programs can increase their effectiveness by appealing to these low-threshold adopters.

Promotion of environmental behaviors should consider the social network context of the audience. People are likely to consult with friends and neighbors to gauge what is appropriate behavior. Perceptions of what is normative may or may not be accurate, but they still drive behavior. Changing a norm is likely to require both mass and interpersonal media. For example, Burns (1991) used neighborhood block leaders to promote recycling, recognizing that interpersonal persuasion is likely to be the best strategy for promoting behavior change (Darley and Beniger, 1981).

EVALUATION

Theoretical models, past experience, and logic provide guidance on how to launch campaigns to promote knowledge, attitudes, and practices. But every behavior, every culture, and every campaign presents unique challenges (Stern, this volume, Chapter 12; Schultz, this volume, Chapter 4). Coping with these challenges requires research—formative, process, and summative. Although one can launch campaigns without conducting research, the odds of success are increased greatly when research is used to set objectives, segment the audience, and understand the behavior from the audience's perspective (Valente, 2002). Furthermore, without research, once the campaign is completed, no one will know whether it worked or not, the reasons why, and whether it really had a significant impact.

Formative research is usually qualitative and is conducted to determine current perceptions, motivations, barriers, and language used to describe the behavior from the audience's point of view. Focus group discussions, in-depth interviews, and observations can be used to learn how to position the behavior in the audience's mind and what types of appeals are likely to be successful. Process research monitors program implementation to track audience exposure to campaign messages. This type of research usually is conducted with viewer logs, counts of the distribution of materials, and ratings. Summative research is quantitative and conducted to determine whether the intervention was effective, and if so, for whom and to what degree. Summative research usually consists of population-based surveys designed to quantify overall impact.

Rules and procedures for evaluating health promotion interventions are well specified (Rossi et al., 1999; Valente, 2001, 2002). The difficulty lies in the fact that every evaluation presents its own demands in terms of the tradeoff between rigor and cost. Although randomized control trials are the "gold standard" for evaluating the impact of an intervention, they are rarely feasible for community- or population-based programs. Tradeoffs between rigor and feasibility are inevitable, and are best addressed by informed researchers who can control relevant threats to validity.

In health communication campaign evaluation, the diffusion/hierarchy steps to behavior change have been used to formulate campaign objectives. These objectives generally stipulate larger changes in knowledge and attitudes and modest yet significant changes in behavior. Although deviations from this pattern have been and can be expected (Valente et al., 1998), it provides a reasonable guide for setting goals. Meta-analysis studies by Snyder (2001) have shown five to nine percentage-point changes in behavior attributable to mass media campaigns.

LESSONS LEARNED

The history of campaigns and public health interventions has yielded several lessons learned about what works, how to improve interventions, and most important, why we often think that interventions do not work. Some researchers and many policymakers argue that interventions to promote behavior change have, by and large, failed. The reasons for concluding lack of success originate from a variety of factors. First, we often fail to recognize that behavior change is a process that takes time—a long time—and we rarely have the patience to wait. Most communication campaign studies collect postcampaign data immediately following a broadcast to capitalize on higher recall levels, but fail to wait for effects on behavior to emerge. Second, most designers have unrealistic expectations regarding the effect sizes to be expected from promotional campaigns. Most mass media campaigns realistically can be expected to increase behavior by one to five percentage points over baseline levels. Although these effect sizes may seem small, they represent a large absolute impact when translated to the number of people reached. Third, we often conclude campaigns have failed because studies designed to test their effects have not collected data from sufficiently large samples to detect these small effects (Borenstein et al., 1997; Kraemer and Thieman, 1987; Valente, 2002). Fourth, the variety of campaigns and campaign objectives has given rise to a diversity of measures making comparisons across studies difficult (but see Snyder et al., 2001; Freimuth and Taylor, 1998). Finally, many interventions lack a theoretical foundation. Designers and researchers have not worked in concert to follow accepted behavior change models, but rather have expended more effort on addressing tension over what is "creative" versus "effective."

Nonetheless, many campaigns have achieved some success, and some prescriptions for creating successful campaigns can be described. These suggestions apply to the message content, campaign strategy developed, the choice of media, timing, dose, and so on. In terms of strategy, first, use active strategies rather than passive ones, by engaging the audience in street fairs, group activities, or neighborhood events. Second, use multifaceted interventions rather than unifaceted ones. Different media have different strengths, and a good campaign uses a variety of media to disseminate messages. Third, create continuous rather than static (one-shot) programs. Rarely does a single campaign provide the needed persuasion to change audiences; a more effective strategy is to use a series of strategically planned, integrated communications. Fourth, consider intervening on multiple ecological levels by targeting individual behavior, forming appropriate policies, and reducing barriers to behavior change. Finally, use interpersonal media rather impersonal media whenever possible. Humans respond to humans, and adding a personal dimension to a campaign can be beneficial.

In terms of messages, first, provide positive reinforcements to the behavior. People respond to being rewarded, even if the reward is nominal in nature. Acknowledge positive behavior: There is nothing like a pat on the back. Second, provide those rewards immediately, rather than delaying them. Link the reward more closely with the behavior. Third, provide role models with which the audience identifies to stimulate observational learning. Humans learn by imitation, and role models enable people to vicariously enact new behaviors, breaking down barriers and providing solutions to overcome those barriers. Fourth, use campaigns to change mediating variables such as outcome expectations, self- and collective efficacy (confidence in the ability to perform the behavior), motivations, and beliefs. Often campaigns are effective at priming the audience for behavior change by modifying these mediators, rather than creating overt behavior. Such modifications are important to accelerating change.

Many implications concerning campaign design, implementation, and evaluation can be gleaned from past research. Although these prescriptions may seem daunting for most, three simple rules will help keep both designers and researchers focused. Nothing is for certain, but observations presented in this chapter may help avoid some past mistakes. First, pretest! Pretest! Pretest! Every message, piece of material, and survey instrument should be tested with the audience prior to implementation. Second, formative research will provide clues as to what is needed and how to frame the intervention. Finally, keep the audience involved through formative and process research, and use researchers trained to translate behavioral research into message design.

CONCLUSIONS

The field of public health provides us with many lessons learned and caveats to keep in mind. From an economic point of view, it would seem rational to simply lower the costs to adoption and provide incentives to behavior change in order to bring about a public good. If we want more people to exercise, we lower the cost of exercising, and increase the incentive to do so. In the context of family planning for population control, for example, one might argue that we should provide economic incentives for women to use contraception and find an equilibrium point at which the incentives are sufficient to create the right population growth.

From a public health as well as an ethical vantage point, however, these arguments confront the reality that we are then imposing burdens on decision making among those who are less able financially to resist such incentives. Consequently, the behavioral burden will fall on those who need the economic inducements. These incentives thus do not appeal to an individual's altruism or enlightened self-interest; instead, the incentives appeal to the pocketbook. Although economic incentives are motivational, they are not always the best avenues to sustained behavior change.

Instead, we prefer to educate and persuade our audiences. Empowerment is the final outcome of successful communication when individuals, armed with information, take charge of their own lives. The goal is to create an informed public capable of making rational choices in their self-interest that still benefit the public good. In the case of contraception, most women (and men) in high-fertility countries report wanting fewer children than they have. The reported ideal family size throughout the world has dropped dramatically in a relatively short period of time. As people become educated about the consequences of unchecked fertility, and learn about options to control it, they generally make choices in the aggregate that are beneficial to society.

Environmental behaviors are likely to discover similar patterns. Economic self-interest is an important motivator, and clearly individuals need to provide food and shelter for themselves. Economic incentives may drive many behaviors that are environmentally damaging and for which enforcement and coercion are necessary. Use of pesticides and herbicides may be environmentally damaging, but to someone who needs to provide a livelihood, such considerations may seem tangential. Some people, however, are willing to change behaviors for altruistic reasons if the benefits are communicated clearly to them and the barriers to practice are not excessive.

Our admonition, however, is to expect gradual changes in public perceptions and behaviors. Any given intervention designed to promote environmental behavior is likely to have modest effects on action. Interventions can, however, inform publics and seed changes in attitudes that will continue to pay behavior change dividends later. These interventions, if accompanied by evaluation research, will inform policymakers of successful elements and enable continued planning for future efforts at behavior change.

Importantly, the research will provide fodder to further engage the audience by elevating the environment on the public's agenda. For example, research findings for promotional study can be released to the media and disseminated to a wider audience. This activity can further stimulate changes in other communities not directly affected by the initial campaign. Moreover, once a campaign is completed, diffusion through interpersonal contacts is likely to continue, and promotional designers need to provide materials that can be used by laypersons to stimulate action by others. The social network needs to be charged.

Evaluation of programs will help detect unanticipated roadblocks. It also creates the need to set goals and objectives. Once set, strategic plans can be developed to reach these goals. The plans are likely to consist of media and message development activities that build on the lessons learned as described earlier. In the end, what will matter is whether the public is adequately informed of their options and the consequences of their behavior. Such empowerment will bring about sustained behavior change in the form of an enlightened public willing, in small ways, to promote the greater good.

REFERENCES

Ajzen, I.
  1991   The theory of planned behavior. *Organizational Behavior and Human Decision Processes* 50:179-211.
Ajzen, I., and B.L. Driver
  1991   Prediction of leisure participation from behavioral, normative and control beliefs: An application of the theory of planned behavior. *Leisure Sciences* 13:185-204.
Ajzen, I., and M. Fishbein
  1980   *Understanding Attitudes and Predicting Social Behavior.* Englewood Cliffs, NJ: Prentice-Hall.
Ajzen, I., and T.J. Madden
  1986   Prediction of goal directed behavior: Attitudes, intentions, and perceived behavioral control. *Journal of Experimental Social Psychology* 22:453-474.
Alexander, C., M. Piazza, D. Mekos, and T.W. Valente
  2001   Peers, schools, and adolescent cigarette smoking: An analysis of the national longitudinal study of adolescent health. *Journal of Adolescent Health* 29:22-30.
Bandura, A.
  1977   *Social Learning Theory.* Englewood Cliffs, NJ: Prentice-Hall.
  1986   *Social Foundations of Thought and Action: A Social Cognitive Theory.* Englewood Cliffs, NJ: Prentice-Hall.
Berelson, B., P.F. Lazarsfeld, and W. McPhee
  1954   *Voting: A Study of Opinion Formation in a Presidential Campaign.* Chicago, IL: University of Chicago Press.
Borenstein, M., H. Rothstein, and J.Cohen
  1997   *Power and Precision: A Computer Program for Statistical Power Analysis and Confidence Intervals.* Teaneck, NJ: Biostat.
Burns, S.
  1991   Social psychology and the stimulation of recycling behaviors: The block leader approach. *Journal of Applied Social Psychology* 21:611-629.
Burt, R.
  1980   Models of network structure. *Annual Review of Sociology* 6:79-141.
Campbell, M., D. Buckeridge, J. Dwyer, S. Fong, V. Mann, O. Sanchez-Sweatman, A. Stevens, and L. Fung
  2000   A systematic review of the effectiveness of environmental awareness interventions. *Canadian Journal of Public Health* 91(2):137-143.

Cartwright, D.
   1949   Some principles of mass persuasion: Selected findings of research on the sale of United States War Bonds. *Human Relations* 2:253-267.
Casterline, J.B., and J. Cleland
   2002   *Diffusion Processes and Fertility Transition: Selected Perspectives.* Committee on Population. J.B. Casterline, ed. Washington, DC: National Academy Press.
Chaffee, S. H., and C. Roser
   1986   Involvement and the consistency of knowledge, attitudes, and behaviors. *Communication Research* 13:373-399.
Darley, J.M., and J.R. Beniger
   1981   Diffusion of energy conserving innovations. *Journal of Social Issues* 37(2):150-171.
DiClemente, C.C., and J.O. Prochaska
   1983   Self change and therapy change of smoking behavior. A comparison of processes of change in cessation and maintenance. *Addictive Behavior* 7:133-142.
Dietrich, A.J., A.L. Olson, C.H. Sox, M. Stevens, T.D. Tosteson, T. Ahles, C.W. Winchell, J. Grant-Peterson, D.W. Collison, and R. Sanson-Fisher
   1998   A community-based randomized trial encouraging sun protection for children. *Pediatrics* 102(6):e64.
Eulau, H.
   1980   The Columbia studies of personal influence. *Social Science History* 4:207-228.
Farquhar, J., N. Maccoby, P. Wood, and J. Alexander
   1977   Reducing the risk of cardiovascular disease. *Journal of Community Health* 3:100-114.
Festinger, L.
   1954   A theory of social comparison processes. *Human Relations* 7:117-140.
Fishbein, M., ed.
   1967   *Readings in Attitude Theory and Measurement.* New York: Wiley.
Fishbein, M., and I. Ajzen
   1975   *Belief, Attitude, Intention and Behavior: An Introduction to Theory and Research.* Boston: Addison-Wesley.
Flay, B.R.
   1977   Mass media and smoking cessation: A critical review. *American Journal of Public Health* 77(2):153-159.
Freimuth, V.S., and M. Taylor
   1998   Are Mass Mediated Health Campaigns Effective? A Review of the Empirical Evidence. Paper prepared for the National Heart, Lung, and Blood Institute, National Institutes of Health, Bethesda, MD.
Gitlin, T.
   1978   Media sociology: The dominant paradigm. *Theory and Society* 6:205-253.
Glanz, K., F.M. Lewis, and B.K. Rimer, eds.
   1997   *Health Behavior and Health Education.* San Francisco: Jossey-Bass.
Graham-Brown, R.A.C., J.E. Osborne, S.P. London, A. Fletcher, D. Shaw, B. Williams, and V. Bowry
   1990   The initial effects on workload and outcome of a public education campaign on early diagnosis and treatment of malignant melanoma in Leicestershire. *British Journal of Dermatology* 122:53-59.
Green, L.W., L. Richard, and L. Potvin
   1996   Ecological foundations of health promotion. *American Journal of Health Promotion* 10:270-281.
Hochbaum, G.M.
   1958   *Public Participation in Medical Screening Programs: A Sociopsychological Study.* Publication No. 572. Washington, DC: U.S. Public Health Services.

Hornik, R.
  1989 Channel effectiveness in development communication programs. Pp. 309-330 in *Public Communication Campaigns,* 2nd ed. R. Rice and C. Atkin, eds. Newbury Park, CA: Sage.
Hyman, H.H., and P.B. Sheatsley
  1947 Some reasons why information campaigns fail. *Public Opinion Quarterly* 11:412-423.
Katz, D.
  1960 The functional approach to the study of attitudes. *Public Opinion Quarterly* 24:163-204.
Katz, E.
  1957 The two-step flow of communication: An up-to-date report on a hypothesis. *Public Opinion Quarterly* 21:61-78.
  1987 Communication research since Lazarsfeld. *Public Opinion Quarterly* S57.
Katz, E., and P.F. Lazarsfeld
  1955 *Personal Influence: The Part Played by People in the Flow of Mass Communications.* New York: Free Press.
Katz, E., M.L. Levine, and H. Hamilton
  1963 Traditions of research on the diffusion of innovations. *American Sociological Review* 28:237-253.
Kraemer, H.C., and S. Thieman
  1987 *How Many Subjects?: Statistical Power Analysis in Research.* Newbury Park, CA: Sage.
Lazarsfeld, P.F., B. Berelson, and H. Gaudet
  1948 *The People's Choice.* 2nd ed. New York: Columbia University Press.
McCombs, M.E., and D.L. Shaw
  1972 The agenda-setting function of mass media. *Public Opinion Quarterly* 36:176-187.
Marsden, P.V.
  1990 Network data and measurement. *Annual Review of Sociology* 16:435-463.
McElroy, L., D. Bibeau, A. Steckler, and K. Glanz
  1988 An ecological perspective on health promotion programs. *Health Education Quarterly* 15:351-377.
McGuire, W.J.
  1989 Theoretical foundations of campaigns. Pp. 43-65 in *Public Communication Campaigns.* 2nd ed. R. Rice and C. Atkin, eds. Newbury Park, CA: Sage.
Mendolsohn, H.
  1973 Some reasons why information campaigns can succeed. *Public Opinion Quarterly* 37:50-61.
Merton, R.
  1968 *Social Theory and Social Structure.* New York: Free Press.
Mischel, W.
  1973 Toward a cognitive social learning reconceptualization of personality. *Psychological Review* 80:252-283.
National Public Radio
  1996 All Things Considered. Public service announcements, radio broadcast, National Public Radio, Washington, DC, April 16.
Pehamberger, H., M. Binder, S. Knollmayer, and K. Wolff
  1993 Immediate effects of a public education campaign on prognostic features of melanoma. *Journal of the American Academy of Dermatology* 29:106-109.
Petty, R.E., and J.T. Cacioppo
  1981 *Attitudes and Persuasion: Classic and Contemporary Approaches.* Dubuque, IA: Brown.

Piotrow, P.T., D.L. Kincaid, M.J. Hinden, C.L. Lettenmaier, I. Kuseka, T. Silberman, A. Zinanga, F. Chikara, D.J. Adamchak, and M.T. Mbizvo
   1992   Changing men's attitudes and behavior: The Zimbabwe male motivation project. *Studies in Family Planning* 23(6):365-375.
Piotrow, P.T., J. Rimon, D.L. Kincaid, and W. Rinehart
   1997   *Family Planning Communication: Lessons for Public Health.* New York: Praeger.
Rogers, E.M.
   1973   *Communication Strategies for Family Planning.* New York: Free Press.
   1995   *Diffusion of Innovations.* 4th ed. New York: Free Press.
Rogers, E.M. and D.L. Kincaid
   1981   *Communication Networks: A New Paradigm for Research.* New York: Free Press.
Rogers, E.M., and J.D. Storey
   1987   Communication campaigns. Pp. 817-846 in *Handbook of Communication Science,* C.R. Beniger and S.H. Chaffe, eds. Newbury Park, CA: Sage.
Rosenstock, I.M.
   1960   What research in motivation suggests for public health. *American Journal of Public Health* 50:295-301.
Rosenstock, I.M., V.J. Strecher, and M.H. Becker
   1988   Social learning theory and the health belief model. *Health Education Quarterly* 15(2):175-183.
Rossi, P.H., H.E. Freeman, and M. Lipsey
   1999   *Evaluation: A Systematic Approach.* 6th ed. Newbury Park, CA: Sage.
Ryan, B., and N.C. Gross
   1943   The diffusion of hybrid seed corn in two Iowa communities. *Rural Sociology* 8(1):15-24.
Schooler, C., J.A. Flora, and J.W. Farquhar
   1993   Moving toward synergy: Media supplementation in the Stanford Five-City Project. *Communication Research* 20(4):587-610.
Scott, J.
   2000   *Network Analysis: A Handbook.* 2nd ed. Newbury Park, CA: Sage.
Snyder, L.B.
   2001   How effective are mediated health campaigns? Pp. 181-190 in *Public Communication Campaigns,* (3rd ed.) R.E. Rice and C.K. Atkin, eds. Newbury Park, CA: Sage.
Snyder, L.B., M.A. Hamilton, E. Mitchell, J. Kiwanuka-Tondo, F. Fleming-Milici, and D. Proctor
   2001   The effectiveness of mediated health communication campaigns: meta-analysis of commencement, prevention, and cessation behavior campaigns. In *Meta-Analysis of Media Effects,* R. Carveth and J. Bryant, eds. Mahwah, NJ: Lawrence Erlbaum Associates.
Udry, J.R., L.T. Clark, C.L. Chase, and M. Levy
   1972   Can mass media advertising increase contraceptive use? *Family Planning Perspectives* 4(3):37-44.
Valente, T.W.
   1993   Diffusion of innovations and policy decision-making. *Journal of Communication* 43:30-45.
   1995   *Network Models of the Diffusion of Innovations.* Cresskill, NJ: Hampton Press.
   2001   Evaluating communication campaigns. Pp. 105-124 in R.E. Rice and C.K. Atkin (eds.). *Public Communication Campaigns* (3rd ed.). Newbury Park, CA: Sage.
   2002   *Evaluating Health Promotion Programs.* New York: Oxford University Press.
Valente, T.W., P. Paredes, and P.R. Poppe
   1998   Matching the message to the process: Behavior change models and the KAP gap. *Human Communication Research* 24:366-385.

Valente, T.W., P.R. Poppe, and A.P. Merritt
  1996   Mass media generated interpersonal communication as sources of information about family planning. *Journal of Health Communication* 1:259-273.
Valente, T.W., and E.M. Rogers
  1995   The origins and development of the diffusion of innovations paradigm as an example of scientific growth. *Science Communication: An Interdisciplinary Social Science Journal* 16(3):238-269.
Valente, T.W., and W.P. Saba
  1998   Mass media and interpersonal influence in the Bolivia National Reproductive Health Campaign. *Communication Research* 25:96-124.
  2001   Campaign recognition and interpersonal communication as factors in contraceptive use in Bolivia. *Journal of Health Communication* 6(4):1-20.
Valente, T.W., and D. Vlahov
  2001   Selective risk taking among needle exchange participants in Baltimore: Implications for supplemental interventions. *American Journal of Public Health* 91:406-411.
Valente, T.W., S. Watkins, M.N. Jato, A. Van der Straten, and L.M. Tsitsol
  1997   Social network associations with contraceptive use among Cameroonian women in voluntary associations. *Social Science and Medicine* 45:677-687.
Wasserman, S., and K. Faust
  1994   *Social Networks Analysis: Methods and Applications.* Cambridge, Eng.: Cambridge University Press.

7

Understanding Individual and Social Characteristics in the Promotion of Household Disaster Preparedness

Dennis S. Mileti and Lori A. Peek

The object of social marketing is to increase the prevalence of a target behavior in a specific population. Hazards education is one form of social marketing; it attempts to increase protective actions by people, households, and groups through the presentation of information about a hazard and the risk it poses. This type of education often fosters a sense of doubt and insecurity, causing people to wonder about their environment and to question their safety in it. A good hazards education project gives people something to think about and to discuss with friends, family, and colleagues. It causes them to seek more information to answer their questions, and specialists need to be ready with clear information and answers when the questions are asked.

Most successful social marketing campaigns follow a similar model: They begin by showing the risks or problems associated with particular behaviors, then present the benefits associated with altering those same behaviors. For example, some of the most widely used social marketing campaigns have encouraged people to stop smoking for their health, fasten seatbelts to save lives, and recycle to reduce waste and improve environmental quality. The major themes these campaigns share is that they (1) raise questions in the minds of their audiences, (2) offer fairly simple answers, and (3) have authorities available over time to reinforce the message. Social marketing campaigns often posit problems or suggest areas for positive change in social life, repeatedly informing the audience of ways to improve. Although marketing may involve colorful pamphlets, eye-catching posters, and provocative public interest announcements on TV and radio, even more valuable is an understanding of the dynamics of human behavior, effective ways to change it, and a systematic approach to carrying it out over time.

Social scientists in the United States have systematically studied human response to natural and technological disasters since the early 1950s (Quarantelli, 1991). Indeed, the past five decades of research resulted in an extensive body of applied and scholarly literature, which documents human preparation for, response to, and recovery from hazards and disasters (Drabek, 1986; Cutter, 1994; Mileti, 1999). Researchers have employed a variety of quantitative and qualitative methods in an effort to accurately record and analyze individual and group actions. In this chapter we focus on what has been learned about how, why, and when people prepare for natural hazards and disasters, with specific attention given to empirical findings as related to natural hazards social marketing and/or public education campaigns. The natural hazards literature reviewed in this chapter references a wide range of environmental extremes, including floods, earthquakes, tornadoes, hurricanes, and tsunamis.

WHO PREPARES AND WHO DOES NOT

Certain personal and social characteristics of individuals and households make them more or less likely to heed information about hazards and do something to increase their safety (Oliver-Smith, 1996; Lindell and Perry, 2000). Previous experience with a natural disaster, higher levels of formal education, middle age, and having family members who live in the same area may make people more apt to take protective actions (Mileti and Darlington, 1997). For example, a middle-aged person whose house was seriously damaged in the Northridge earthquake is likely to live in a bolted and braced home today. On the other hand, a young unmarried male is less likely to take precautionary measures. A 1989 survey that asked people what they did during the Loma Prieta earthquake revealed that most 20-something males did not try to protect themselves from injury while the shaking was going on (O'Brien and Mileti, 1992).

Social marketing certainly does not change one's ascribed characteristics (such as race, gender, age), but rather utilizes knowledge of these characteristics to deliver information to various groups to generate questions about risk, options, and actions. Good information can encourage people to ask questions about their environment and search for more information; this is the first step in the long journey to changed behavior and increased protection.

Research into the social psychology of perceptions and belief indicates that—as counterintuitive as it may seem—perceived risk does not contribute directly to taking protective action (Slovic, 1989, 2000). Because human risk perception does not always follow from objective estimates and definitions of risk, and human and societal action to mitigate risk often can be inconsistent with estimated scientific probabilities (Tweedale, 1996), professional risk estimators often are frustrated in their attempts to motivate people and societies into what would constitute appropriate action from their point of view (Mileti et al., 1992). Furthermore, just because individuals report high levels of risk awareness does not necessarily mean that they internalize that risk. For example, Mileti and Fitzpatrick (1993) found that 80 percent of survey respondents believed that they would experience a Parkfield earthquake, but only about one-third thought it would harm them, their families, or their property.

Moreover, people do not think in probabilities. Typically, the human thought process about future events is binary: it will happen/it won't happen; it will affect me/it won't affect me. Elaborate probability estimates for a hazard most often do not change this binary type of thinking (Mileti et al., 1992). The official probability will be added to other pieces of information, beliefs, and experiences, and may—if accompanied by continuous, credible information over time—inspire some questioning and fact seeking in the future (Mileti and Sorensen, 1990).

Marketing experts and educators have learned through personal experience and the research literature that people generally are not motivated by lectures on why they should do something (Mileti and Sorensen, 1990). Neither moral exhortations nor discourses on ethical or legal imperatives tend to produce major behavioral changes in the average citizen or household. People are more apt to follow an agenda if they work out a solution themselves, with helpful information from specialists (Mileti et al., 1990). Not surprisingly, most people are motivated to change their behavior when they think a behavior change is their own idea.

WHAT HAS WORKED IN HAZARDS MARKETING

Much research has been done in a variety of disciplines on how human behavior can be changed. However, relatively few empirical studies have been made to measure the impact of nonemergency hazards education on public risk perception and subsequent risk reduction behavior (for exceptions, see Haas and Trainer, 1974; Ruch and Christenson, 1980; Palm, 1981).

One study in the early 1980s assessed the public response of Los Angeles residents to news coverage of the Palmdale uplift, a rare geological phenomenon in an area along the San Andreas fault that was believed, between 1976 and 1979, to be a precursor to an earthquake (Turner et al., 1986). Social scientists surveyed hundreds of people to determine where they received their information on earthquakes, how they interpreted what they received, and ultimately, what they did about this new information. The researchers did not look specifically at social marketing in this study per se, but instead focused on how the mainstream media conveyed the news of the threat. They concluded that scientists and the media should make available credible information regarding an event that provokes widespread curiosity. Otherwise, when reliable information is not available, rumor fills the gap.

Another major finding that resulted from the aforementioned study (Turner et al., 1986), with respect to household disaster preparedness, was as follows: Although increases in mass media attention to the earthquake threat does raise public awareness of various earthquake issues, it is the active involvement of individuals in the discussion of these topics, through social ties in their neighborhoods and communities, that overcomes the passivity that often characterizes the receipt of such information. The researchers concluded that the presence of individual and group interest and involvement made it more likely that people actually would take action to lessen their household vulnerability.

In the late 1980s, another research effort analyzed the effectiveness of a pamphlet is raising awareness of earthquake risk among residents in communities near Parkfield, California (Mileti et al., 1990). The U.S. Geological Survey had announced that the Parkfield segment of the San Andreas fault in central California was likely to experience a moderate earthquake between 1986 and 1993. The California Office of Emergency Services mailed a comprehensive pamphlet to residents in the affected area that described the probabilities and the possible impacts of the quake and recommended certain actions to reduce damages. The study evaluated which pieces of information moved residents to take protective action.

Some of the study findings have been used as the basis for hazards marketing and education programs: (1) complicated phenomena must be explained in nontechnical terms; (2) information must come from various credible sources; (3) consistent information should be repeated in many different media; (4) messages on TV and radio are somewhat effective, but people like to have a written document to which they can refer as they think about their risk; (5) information should tell people what they can do before, during, and after a disaster; and (6) discussion with peers helps people to believe the information and act on it.

In the early 1990s, a similar study concerned a publication in the Bay Area that explained in lay language the findings of a scientific report on earthquake probabilities (Mileti et al., 1993). Following their release of a very technical report, the U.S. Geological Survey thought it wise to explain to the public what it meant and what they ought to do about it. In concert with a number of other agencies, a booklet was developed and distributed to millions of residents as a Sunday newspaper insert. Shortly after, researchers queried a large number of readers about their responses to the booklet and its information.

The findings of this research added to the collection of rules of hazards marketing and education in several ways. When clearly informed about risk, people can comprehend the basics and remember what they read. Following from this, we know that people who understand that there is something they can do to reduce vulnerability (i.e., bolt and brace their homes to protect their property from earthquake damage) are more apt to act than those who are unaware of safety measures that can be taken. Another finding was that people consistently search out more information to validate what they've already heard. Many people, households, and organizations reported that they took actions after reading the insert, not only because it made them aware of specific actions to take, but also reinforced things they had already heard elsewhere.

At almost the same time, a different but complementary investigation was underway, also in the Bay Area (Bolton and Orians, 1992). This one asked people about their preferred sources of information on earthquake risk and mitigation. Though this study did not set out to determine whether the information actually changed behavior, its findings are instructive and corroborate the observations of earlier research. In general, people prefer public education programs that convey scientific and technical information from credible authorities; communicate the information clearly; present it attractively; and disseminate it through various community or professional networks.

Educational organizations with a high-profile presence in the area over time were more trusted than those without a credible track record. Deemed unsuccessful were educational programs that did not feature specialists, did not adapt the material to their constituents, and took only an impersonal mass mailing approach. The Bolton and Orians (1992) study highlighted the error of assuming a very homogeneous "public" and advocated tailoring information materials to the many special groups in an area. For example, the approach to, and materials for, middle-class homeowners should be different from those for renters, and those for school districts should not be like those for large corporations, according to study recommendations.

A study of public education outside California was undertaken by a professional staff member of the American Red Cross in affiliation with the University of Maryland (Lopes, 1992). This study included 60 slides illustrating disaster damage and 60 additional slides that did not include any images of disaster damage. The study of correct action message content and images (i.e., "the right thing to do") supported the widely held notion that too much gloom and doom is just as bad as no information at all. A few well-chosen images of destruction have a useful impact on most people early in a presentation. However, when verbal messages on how to prepare are juxtaposed with photos of impacted structures, people have trouble dealing with the verbal/visual mismatch. People tend to remember the visual message more clearly than the verbal, and repeated images of damage sometimes convince people there is nothing they can do about the hazard. Far more effective are coordinated verbal and visual representations of what to do and how. Finding the right mix of information on potential losses and on effective actions is critical to the success of social marketing.

One last study bears mentioning; it concerned public response to a spurious earthquake prediction on the New Madrid fault in the central United States (Farley, 1998). The findings confirmed the need for governments and scientists to place accurate information before the public to counter inaccuracies that may be receiving media attention. When Iben Browning—a scientist, albeit not an earth scientist—predicted a large quake on the New Madrid fault on December 3, 1990, countless people believed him and reacted accordingly. The populace in the heartland, which had never been taught much about earthquakes, did not have the analytical tools to question Browning's prediction. Credible scientists and government spokespersons were slow to disagree with Browning, perhaps because they hadn't learned the lesson of the Palmdale uplift study mentioned earlier. Once they responded and released accurate information, however, the "prediction" provided an opportunity for solid public education.

THE WINDOW OF OPPORTUNITY

Both empirical research and seasoned observation support the golden rule of social marketing for hazards: All of the sophisticated materials and behavior modification techniques do not have the force of one major disaster to change both behavior and public policy, at least in the short term. Losing something in a disaster, or knowing someone who did, has inspired many people and households to take protective actions. During the well-known "window of opportunity" that opens following a disaster, abundant information from various credible sources in the affected locale will increase the chances for behavior change (Mileti et al., 1993).

However, although people and households are more apt to alter behavior after disaster strikes, change is most likely when educators have already worked to make sure the problem is recognized, the solution is known, and some advocates are already in place. Educators are aware that they must not wait for the window to open, but rather must build a sustained advocacy program beforehand. Not working consistently and constantly may result in waiting forever.

Advocates can also take advantage of a window opening someplace else. After the 1995 earthquake in Kobe, Japan, for example, there was a fleeting but pronounced interest in earthquake risk in both the Bay Area and Seattle—each with a built environment and setting similar to Kobe. A number of earthquake organizations on the west coast seized this golden opportunity to draw comparisons between the Kobe quake and expected impacts due to local tremblers.

Experts must use the opportunity while they can, for the window is not open long. The fleeting interest wanes. A population that jams the phone lines requesting hazards loss reduction information in January of one year will not be doing so the next. A public policymaker's memory and attention are even shorter than the public's. Typically, he or she will not keep hazards mitigation high on the list of big issues for more than 2 to 3 months.

Following are suggestions for public education based on what has been learned about hazards education, derived from the systematic research mentioned earlier, and from experience with social marketing campaigns and education programs. First, the ideal message is explained, then ways for delivering it are recommended.

THE IDEAL MESSAGE

If hazards educators could develop the ideal message to educate the public about hazards, that message would include several important elements, including accessible information, consistent information, media-ready packaging, clear explanation of critical issues, specification of whom is most at risk, and clarity surrounding the level of certainty of the message. Moreover, educators must account for individual characteristics and social elements in designing hazards social marketing campaigns. The following paragraphs elucidate these components of an ideal educational message.

Make Hazards Information Accessible

Reading in the newspaper the technically sophisticated and generally incomprehensible statements of scientists, engineers, or actuaries will not give most people an elementary understanding of hazards and likely impacts on their lives. Simple language in manageable amounts is absolutely necessary. Though credentialed spokespersons are one of the most important sources of information, specialists who speak only in the jargon of their discipline will not be effective. Authoritative interpreters of technical information should be cultivated, encouraged, and paid well. Fit the specialist to the topic: Geologists and seismologists should talk about earth sciences; engineers and architects should talk about structures; and firefighters and emergency responders should talk about home safety and neighborhood organization.

Keep the Information Consistent

Because most people are exposed to information through a number of media and from various sources, the information must be consistent in order to be credible. Inconsistent information confuses people and allows them to discount some or all of it. Experts should work together, across jurisdictions and organizations, to see that messages are similar. For example, numerous organizations—state agencies, the Red Cross, school authorities, and media outlets—in California met in the immediate aftermath of the Loma Prieta quake to discuss and agree on the wording all of them would use for the "Drop, Cover, and Hold" message. The essence of the message was that when the first signs of an earthquake are felt, people should get down (Drop), move under a heavy table or desk (Cover), and stay there until the shaking stops (Hold).

Package Information for the Media

One of the hallmarks of an effective social marketing program is to have plenty of material on hand when the TV and radio stations start calling and the feature writer from the paper shows up looking for the local angle. For example, if the issue is a vulnerable housing type, provide clear guidance about what homeowners should do so the newspaper can run the information next to its article. Get photos, maps, and checklists ready so the hazards education article makes it in under the deadline and gains its rightful place on the front page of the paper.

Three Critical Issues

The message presented to the public should clearly explain the following three critical issues on which good hazards education rests. If any of these three components are lacking or missing completely, the education initiative may not be as effective and ultimately could fail.

1. Describe potential losses. Generally, people can't imagine the impact a hazard could have on their community, their house, or their place of work, so they must be assisted by descriptions of other disasters, pictures, scenarios, or computer-based loss estimation maps. The essence of this task is working to overcome the human tendency to conclude that it can't happen here or it won't happen to me. The more relevant the description can be to the situation of the audience, the more likely it is that they will attend to the hazards risk. A good marketer can find the local angle in a disaster—even in a far-off land—and work it.

2. Discuss the potential timeline. Once people understand that it could, indeed, happen here, they must be further convinced that it may happen to them at some point in time. Although tools are becoming increasingly sophisticated for the physical and statistical estimation of seismic risk (Wisner, 1999), few people, other than statisticians, understand odds ratios. Thus, most people want to know the likelihood of a disaster in an uncomplicated sort of way in a finite amount of time. This is where understanding the social elements of risk perception and action-oriented behavior becomes ever more important. Probability estimates will not, in themselves, motivate people to take action (Tweedale, 1996), but the information will assist in creating the uncertainty that is so important to behavior change. Disaster prediction is a very inexact science, but where scientists have some understanding of the behavior of physical systems, they should offer these rough forecasts.

3. Explain how to diminish losses. A person with a clear picture of his or her possible losses must be quickly offered suggestions and directions for how to reduce them. Without these blueprints, people can fall prey to a fatalistic inertia. Appropriate assistance may take many forms: a how-to video for homeowners on strengthening the disaster resistance of their homes; evacuation guidelines for schools; a business resumption planning process for a corporation or a city government; encouragement and help from a neighborhood emergency response team; or recommended policy changes for a water system. People can be guided to mitigation in endless ways.

Specify Who Is at Risk

An ideal message for marketing, planning, and educational purposes clearly specifies who is most at risk in a disaster (Key, 1986). For example, explaining the relative weaknesses of various building types will help people understand that they might be injured if they live or work in them. Such information also will help emergency planners anticipate response needs. Beyond physical effects, people should be helped to recognize that they will be economically damaged, socially isolated, psychologically troubled, and just plain inconvenienced. Detail the exact impacts of the disaster on all groups in the community, on utilities, on transportation systems, and on governmental and nonprofit organizations responsible for public health and well-being.

Clarify the Level of Certainty

In preparing an educational message, one must be honest and clear about the level of certainty in predicting the incidence and effects of a hazard. Any scenario of a future event is a best guess. Overstating the risk or inflating the probability of a disaster inoculates people against belief just as surely as inconsistency. Predictions of catastrophe strike some people as too extreme to be credible; they terrify others. Neither group will be likely to accept the information as deserving of further questioning or attention. More than one social marketing project has painted too dire a picture and compromised its credibility.

Consider Personal and Social Characteristics of the Audience

Finally, in developing the ideal message, it is imperative to keep in mind that the message should be designed from the perspective of the target audience. Research has clearly shown the importance of personal characteristics (i.e., knowledge, attitudes, beliefs) as well as situational predictors and contextual factors in increasing awareness of an issue and ultimately influencing behavioral change (Schultz, this volume, Chapter 4; Stern, this volume, Chapter 12). Valente and Schuster (this volume, Chapter 6) remind us that messages are not received in a space devoid of social interaction. Rather, members of the audience are connected through a web of social networks, which impact interpersonal communication and social norms. The resultant beliefs and attitudes toward a behavior and toward adoption of that behavior are important predictors of whether change will occur. Because the ultimate goal of intervention is to positively alter behavior, it is imperative to consider personal as well as social characteristics, networks, communication patterns, and shared perceptions.

WAYS TO DELIVER THE MESSAGE THAT WORKS

Marketing is a complicated process—on both the delivery end and the receiving end. Campaigns must be coherent and collaborative; their information must be credible and understandable; and the information must reach its intended audience.

In that statement is a prescription for close cooperation among technical specialists and educators; constant communication among educational organizations; and sophistication and creativity in the message translation and communication.

Include Multiple Credible Sources of Information

People are more likely to attend to information if it comes from a group or a person they trust (Key, 1986). Depending on age, education, class, and ethnicity, different people trust different sources. Some people want to hear about hazards from scientists; others believe only what the Red Cross tells them; still others search for data sources online. It is important to use various sources to reach the maximum number of groups in the community.

Assume the Public Is Diverse

It is important to recognize that the public is diverse, and thus information must be tailored to the needs of each group (Turner et al., 1979). For example, the elderly have special needs, so create materials for them that speak to those needs. Don't ignore non-English speakers; write information in multiple languages or get materials translated by knowledgeable local speakers of those languages. Some cultural groups choose not to read information for reasons unrelated to literacy; to reach them, use radio and TV, word of mouth, or pictographic images. Use the media that serve multilingual populations.

Use Multiple Media Sources

Now that we have experienced the technology revolution, there are a multitude of media sources available for information dissemination. You can bounce a fact about hazards risk off satellites, include it in electronic data networks, feature it on interactive computer games, add it to distance learning curricula, and project it onto the screen of the nearby theater. Indeed, recent findings that have come out of the Disaster Research Center's work in evaluating Project Impact for the Federal Emergency Management Agency (FEMA) have shown that mainstream media sources are not the only channel that can be used to reach the public or various segments of the public. In fact, researchers have found that one newly created and useful venue for educational campaigns is "disaster fairs" where local merchants and service businesses advertise what they can do for or provide to homeowners to assist them with preparedness and mitigation (Disaster Research Center, 2001). Officials and disaster experts in attendance at these disaster fairs provide information on hazards and the potential consequences of a catastrophic event in the local area.

Moreover, a variety of spokespersons who can be trusted by the public should be used as well (Key, 1986): today, the Red Cross spokesperson on radio;

tomorrow, cartoon characters on TV; next week, a scientist on the Internet. Effective social marketing programs should have the staff to constantly work the media angles and maintain contact with media personalities.

Use Appropriate Media

Always use media appropriate to the target audience (Sorensen, 1983). The Internet is a marvelous tool, but it is not available or utilized by everyone. For example, text that can be downloaded from a web site is not the way to reach a non-English-speaking or low-income audience. Information for those groups can be disseminated through the community organizations and social service agencies that regularly work with that audience. Conversely, technologically sophisticated packaging gets middle- and upper class, computer-using audiences where they live.

Make the Information Easily Accessible

On an ongoing basis, successful social marketing works to motivate a few people to do something to reduce risk (Mileti et al., 1990). Their activities contribute to the slow, incremental process of reaching others as well. Experts must not frustrate their public. Information should be ready and accessible at the time someone is motivated to ask for it. There is not space in any single marketing or educational document to list all the safety hints, guidelines, model ordinances, neighborhood response plans, exemplary policies, and case studies that have been developed by innumerable agencies and organizations. In many cases, the wheels already have been invented. Adapt them and translate them for use.

Ensure Incremental Information Dissemination

Because learning is incremental, information dissemination should be, too (Sorensen, 1983). Organize the information presented to highlight related themes successively. Some education organizations or emergency services agencies distribute to participating communities monthly newsletters with reproducible masters on different aspects of hazards safety and preparedness. For example, in January, the spotlight is on fastening bookcases and file cabinets for earthquake safety; in February, it moves on to another topic.

Make the Approach Interactive and Experiential

We know that adults learn by comparing new information to what they already know, by thinking through and discussing the new concept or practice, and by doing. They do not sit passively, digest everything they hear or read, and then act. Thus, it is important to use models, visual aids, fancy media, and/or peer group discussions. The audience should be engaged, rather than receive a lecture.

Use Disasters as Learning Opportunities

Disasters can be used as important learning opportunities (Lopes, 1992; Russel et al., 1995). Send elected officials, government functionaries, corporate officials, school superintendents, various professionals, and community organizers to view disaster damage and organizational response. Have them report the lessons they derive for their community, business, school district, or practice. Such people typically return from their reconnaissance with better vision and more active imagination than they had before they left. They have seen the truth and can communicate it to many others. They are motivated to do something, and frequently can encourage others with their commitment.

Emphasize the Role of the Individual

The role of the individual in sparking behavior change never should be minimized or overlooked. There are many examples of disaster champions who singlehandedly prod and cajole their organizations, schools, neighborhoods, or governments into taking action. These individuals are both tenacious in their efforts to stimulate change and passionate in their belief that change is necessary. Finding and motivating such an individual sometimes can be the key to a successful social marketing campaign.

Include an Evaluation Component

Some sort of evaluation component should be built into any social marketing or public education campaign. When you assess the efficacy of your materials and approaches, you can revise what doesn't work. Share that knowledge with other experts, advocates, and educators, so campaigns across the country can benefit from your experiences. Last, but not least, use your data to justify continued or increased financial support.

Provide Long-Term Support

If your organization funds a social marketing program, continue that support over many years. If you run a marketing program, keep it highly visible and recognizable in the community. Programs that deliver helpful information over the years see their credibility and effectiveness grow (Kunreuther, 1978; Turner et al., 1981). Don't decrease the program's effectiveness by altering missions, or by changing logos or names. Be patient, and understand that good social marketing is a long haul.

HOW MUCH CHANGE CAN SOCIAL MARKETING ALONE ELICIT?

The research literature on the effectiveness of public hazards marketing campaigns reports the full gambit of impacts; they range from no behavior change to a relatively great deal of public and household behavior change to reduce losses from future disasters. This variation likely exists due to variation in the types of campaigns conducted. For example, some campaigns have lasted only a short time, used singular media approaches, and delivered messages weak in content. Other marketing campaigns have lasted for protracted periods of time, several years, for example, employed multiple media to communicate with people, and delivered messages that informed on the full range of topics important to include in education. The former have not been very effective, if they were effective at all, while the latter have produced diverse protective and mitigative behaviors by the targeted public.

REFERENCES

Bolton, P.A., and C.E. Orians
  1992   *Earthquake Mitigation in the Bay Area: Lessons from the Loma Prieta Earthquake.* Seattle, WA: Battelle Human Affairs Research Centers.
Cutter, S., ed.
  1994   *Environmental Risks and Hazards.* Englewood Cliffs, NJ: Prentice-Hall.
Disaster Research Center
  2001   *Disaster Resistant Communities Initiative: Evaluation of the Pilot Phase Year 2.* Newark, DE: University of Delaware.
Drabek, T.E.
  1986   *Human System Responses to Disaster: An Inventory of Sociological Findings.* New York: Springer-Verlag.
Farley, J.E.
  1998   *Earthquake Fears, Predictions, and Preparations in Mid-America.* Carbondale and Edwardsville: Southern Illinois University Press.
Haas, J.E., and P. Trainer
  1974   Effectiveness of the tsunami warning system in selected coastal towns in Alaska. In *Proceedings of the Fifth World Conference on Earthquake Engineering,* Rome, Italy. Pasadena: California Institute of Technology.
Key, N.
  1986   Abating risk and accident through communication. *Professional Safety* (November):25-28.
Kunreuther, H.
  1978   *Disaster Insurance Protection: Public Policy Lessons.* New York: John Wiley and Sons.
Lindell, M.K., and R.W. Perry
  2000   Household adjustment to earthquake hazard: A review of the research. *Environment and Behavior* 32:461-501.
Lopes, R.
  1992   *Public Perception of Disaster Preparedness Presentations Using Disaster Damage Images.* Working Paper No. 79. Boulder: Natural Hazards Research and Applications Information Center, University of Colorado.

Mileti, D.S.
   1999   *Disasters by Design: A Reassessment of Natural Hazards in the United States.* Washington, DC: Joseph Henry Press.
Mileti, D.S., and J.D. Darlington
   1997   The role of searching in shaping reactions to earthquake risk information. *Social Problems* (February):89-103.
Mileti, D.S., J.D. Darlington, C. Fitzpatrick, and P.W. O'Brien
   1993   *Communicating Earthquake Risk: Societal Response to Revised Probabilities in the Bay Area.* Fort Collins: Hazards Assessment Laboratory, Colorado State University.
Mileti, D.S., and C. Fitzpatrick
   1993   *The Great Earthquake Experiment: Risk Communication and Public Action.* Boulder, CO: Westview Press.
Mileti, D.S., C. Fitzpatrick, and B. Farhar
   1990   *Risk Communication and Public Response to the Parkfield Earthquake Prediction Experiment.* Fort Collins: Hazards Assessment Laboratory, Colorado State University.
   1992   Fostering public preparations for natural hazards: Lessons from the Parkfield Earthquake prediction. *Environment* (April):16-39.
Mileti, D.S., and J.H. Sorensen
   1990   *Communication of Emergency Public Warnings: A Social Science Perspective and State-of-the-Art Assessment.* Oak Ridge, TN: Oak Ridge National Laboratory.
O'Brien, P., and D.S. Mileti
   1992   Citizen participation in emergency response following the Loma Prieta Earthquake. *International Journal of Mass Emergencies and Disasters* (March):71-89.
Oliver-Smith, A.
   1996   Anthropological research on hazards and disasters. *Annual Review of Anthropology* 25:303-328.
Palm, R.
   1981   *Real Estate Agents and Special Study Zone Disclosure.* Boulder: Institute of Behavioral Science, University of Colorado.
Quarantelli, E.L.
   1991   Disaster research. Pp. 681-688 in *Encyclopedia of Sociology,* 2nd ed., E.F. Borgatta, ed. New York: MacMillan Reference USA.
Ruch, C., and L. Christensen
   1980   *Hurricane Message Enhancement.* College Station: Texas Sea Grant College Program, Texas A&M University.
Russell, L.A., J.D. Goltz, and L.B. Bourque
   1995   Preparedness and hazard mitigation actions before and after two earthquakes. *Environment and Behavior* 27(6):744-770.
Slovic, P.
   1989   Perception of risk. *Science* 236:280-285.
   2000   *The Perception of Risk.* London: Earthscan Publications.
Sorensen, J.H.
   1983   Knowing how to behave under the threat of disaster: Can it be explained? *Environment and Behavior* 15:438-457.
Turner, R., J. Nigg, and D.H. Paz
   1986   *Waiting for Disaster: Earthquake Watch in Southern California.* Berkeley: University of California Press.
Turner, R., J. Nigg, D.H. Paz, and B. Young
   1979   *Earthquake Threat: The Human Response in Southern California.* Los Angeles: Institute for Social Science Research, University of California.

1981 *Community Response to Earthquake Threat in Southern California.* Los Angeles: Institute for Social Science Research, University of California.

Tweedale, M.
1996 The nature and handling of risk. *Australian Journal of Emergency Management* 11(3):2-4.

Wisner, B.
1999 Social Aspects of Earthquake Management. Unpublished paper. United Nations/International Decade for Natural Disaster Reduction (UN/IDNDR) Risk Assessment Tools for Diagnosis of Urban Areas Against Seismic Disasters (RADIUS) Workshop, Geneva, Switzerland, July 5-9.

8

Lessons from Analogous Public Education Campaigns

Mark R. Rosenzweig

A fundamental challenge in environmental policy is to alter the private actions of individuals and institutions so that the social costs and benefits of the consequences of those actions are optimally balanced. In many cases, the net private benefits from an action exceed the net social benefits. When this divergence is confined to the decisions of a limited number of actors—large firms—it is possible for a public agency to effectively regulate the firms' behavior so as to align social and private benefits.

When millions of individuals are the agents whose cumulative behavior has an important environmental impact, it is sometimes impractical to attempt to directly enforce behavioral restrictions. One example is the proper disposal of batteries. Monitoring this behavior is not feasible. It is administratively possible to place a tax on batteries to align social and private costs, but such a tax may be politically unpopular. This is not to say that "ecological" taxes are always politically unacceptable, as such taxes have been put in place in European countries. But it is clear there are limits to individual regulation and to taxation as mechanisms for achieving public policy goals, perhaps particularly in the United States, so that alternative approaches to altering behavior may be warranted.[1] One alternative approach is a program of public education.

Chapters 6 and 7 provide examples of public education programs, most of which are outside the environmental arena. The issue is whether we can draw inferences from the experiences described in those chapters to formulate public education campaigns in the environmental realm. Mileti and Peek's chapter provides the lessons learned from efforts to improve "disaster preparedness" among populations at high risk. Valente and Schuster's chapter describes a number of public education campaigns, mostly focusing on improving health or the efficiency by which households control fertility. Although both the target behaviors and the methods of information delivery that are the foci of each chapter appear quite different, they share important features. First, the campaigns are strictly informational, either about the benefits of changing behavior or about the behaviors of "peer" groups. There is no attempt to change people's values. The presumption is that experts have information that the population does not, and that the transmission of this information therefore will improve welfare. Second, and relatedly, the campaigns emphasize private benefits. Individuals are provided information without reference to externalities or to the collective benefits that exceed the sum of private benefits.[2] Individuals and families presumably want to avert the consequences of disasters, reduce the risk of heart disease, and control family size. As a consequence, they have incentives to be better informed—they will be interested in what is being delivered.

I will briefly discuss and evaluate each of the cases discussed in Chapters 6 and 7 by considering a set of questions: First, is there evidence that the campaigns actually changed people's behavior? Second, is there evidence that the campaigns were cost-effective, in the sense that the total costs of the campaign did not exceed the total benefits? Third, were the campaigns described the most cost-effective means of achieving the goals, thereby deomonstrating global cost-effectiveness? Finally, I will assess to what extent these campaigns are helpful in providing solutions for protecting the environment.

INFORMATION DISSEMINATION IN DISASTER-PRONE AREAS

Chapter 7 provides a clear example demonstrating that how a public education campaign is carried out matters, and that such a campaign can be effective in altering people's behavior. Anyone interested in improving disaster preparedness through a public education campaign should read Chapter 7; it provides clear information on what to do and what not to do. However, the chapter does not attempt to describe the costs or benefits of the campaign. Furthermore, there is little discussion of its rationale. In particular, it is not clear how markets have failed such that people's decisions in a risky environment are suboptimal. There are two types of actions related to risk. First, there are actions taken at the time of an adverse natural event. Second, there are actions taken prior to disastrous events that reduce an individual's vulnerability to disaster, such as bracing a house or moving out of a risky area. Because in this case individuals face all of the costs of not being prepared, they presumably have the appropriate incentives to make whatever risk-reducing costly preparations are in their interest. Or do they?

One reason individuals may not be optimally preparing for disasters is that they are effectively protected against the cost of their risk exposure—they expect that if their house is destroyed by a flood, they will be financially "bailed out." Government programs that provide emergency assistance, for example, reduce incentives for individuals to reduce risk ex ante. Such public bailouts also drive out private insurance. Yet private insurance companies have incentives to set premiums to reflect risk, inclusive of risk-mitigating actions taken ex ante by policyholders. For example, premiums presumably would be less for braced houses, therefore providing an incentive for people to undertake bracing. Many of these risk-reducing remedies are clearly visible, and thus easily monitored. The point is that removing barriers to insurance markets may be much more cost-effective than education campaigns in improving disaster preparedness. We cannot evaluate a campaign solely by whether it alters behavior.

INFORMATION DISSEMINATION AND FERTILITY CHANGE

Chapter 6 provides an excellent overview of the issues involved in evaluating public campaigns in the first sense—do they alter behavior? The chapter demonstrates why this is not easy to do, and is sensitive to the pitfalls of inferring causation from data and to the possibility of alternative interpretations of statistical findings. The chapter would have been more interesting if it had focused on the details of one of the campaigns. Chapter 6 gives special attention to family planning campaigns. Family planning campaigns are perhaps more relevant to environmental issues than campaigns designed to change people's diets or exercise habits. Population growth is viewed by many as relevant to environmental degradation. Fertility decisions taken by families may not fully reflect the social benefits and costs to the extent that the size and growth rate of the population has a direct impact on the environment. This is indeed one of the rationales for the subsidization of family planning efforts, inclusive of both the subvention of the tools of private fertility regulation (contraceptives) and public education campaigns.

Public education campaigns directed to altering contraceptive behavior often have had no impact on behavior. What does this lack of behavioral change tell us that is useful? I believe it suggests that campaigns will fail if there is an incorrect diagnosis of the fundamental problem the campaign is attempting to solve—in this case, high fertility. Many family planning education campaigns are purely informational, providing information on the tools of fertility control. Researchers have found that many households are essentially ignorant of modern family planning methods and practices, and conclude that lack of information is the barrier to reducing fertility. This may be a false inference, however, because it ignores the fact that the information people have is the result of choice, reflecting the costs and benefits of acquiring the information. Residents of Manhattan do not know much about car repair, or in some cases even how to drive. However, that is not why they do not generally own cars; they do not own cars because cars are expensive to maintain and cheap alternatives are available. Similarly, if households in Bolivia, for example, find it optimal to have large families, based on their preferences or on an evaluation of the costs of children and their economic benefits, then they have little incentive to inform themselves about efficient ways of reducing fertility. Ignorance thus may be a symptom of more fundamental features of the Bolivian economy or society, not a cause. Providing family planning information to rural Bolivians therefore would have little effect on behavior if the private value of the information to them is low.

ALTERNATIVE ANALOGUES

The success of the disaster preparedness campaign in terms of altering behavior and the lack of effect of the many family planning campaigns apparently tell us that if information is valuable to individuals they will use it, although it is not clear from the fact that behavior is altered that the campaign is cost-effective relative to alternatives. Conversely, if information provides little private benefit, then an information-based campaign will be ineffective in any sense, even if there is clear evidence that people do not have the information. It is not at all clear that providing accurate information about the consequences of behaviors is the key ingredient that will reduce the environmental damage caused by particular private actions. For example, recycling provides few private benefits and clearly has private costs. Moreover, the specific social consequences of whether an individual recycles or not are minuscule, so making people aware of the specific damage they cause by not recycling would hardly alter behavior, as is suggested by the findings in Chapter 4. The recycling example also shows, however, that information on the behavior of peers may alter individual behavior.

Additional analogous situations may offer more relevant lessons for altering environmentally related behavior, including those in which private actions have little private return, but involve large collective effects. Two examples come to mind. The first example is voting. One person's vote does not count for much. Yet many people vote, so there is hope that people will act with the collective good in mind. Many nonpartisan campaigns have been undertaken to increase voter turnout; perhaps these provide some valuable lessons for changing behavior when it is not in an individual's pure self-interest. Again, however, if we are interested in global cost-effectiveness, it is not clear why market-based incentives are not more effective than "campaigns"—why not give tax benefits to those who vote, or pay voters directly? The second example is in survey research. Again, there is little private gain to anyone participating in a survey, but if no one volunteers to participate, there is an important societal loss. Research exists on augmenting survey participation rates that may be relevant to environmental education campaigns designed to alter behavior when the private benefits of doing so are negative. Among the findings of this research is that influence techniques such as frequent prompts and reference to the behavior of others (which may provide information on social norms) do alter behavior, along with direct payments.

Finally, if purely information "education" campaigns emphasizing private gains or peer behavior are not effective, then it may be necessary to change people's values to render behavior more socially beneficial. However, setting in place governmental efforts that go beyond the dissemination of scientifically valid information to purposively changing the values of citizens raises ethical questions that are beyond my expertise. However, I believe they should be of concern in considering nonregulatory and nonfinancial alternatives to improving the environmental impact of the choices people make.

NOTES

1 For a discussion of the political feasibility of ecological tax reform, see Von Wiezsacker and Jesinghaus (1992).

2 Campaigns in the environment arena also emphasize private benefits. For example, campaigns to improve U.S. household energy efficiency involved marketing the idea that households would benefit by reducing their monthly energy costs.

REFERENCE

Von Wiezsacker, E.U., and J. Jesinghaus
  1992 *Ecological Tax Reform.* London, Eng.: Zed Books.

9

Perspectives on Environmental Education in the United States

John Ramsey and Harold R. Hungerford

This chapter addresses the question of what environmental education (EE) is, explores some of its critical challenges, and describes an effective, long-standing curricular approach to environmental education and its research implications.

OVERVIEW OF ENVIRONMENTAL EDUCATION

Let's begin with the concept of environmental education. According to Stapp (1969), environmental education is aimed at producing a citizenry that is knowledgeable about the biophysical environment and its associated problems, aware of how to solve these problems, and motivated to work toward their solution. An important element implied by this definition is a problem-solving approach, perhaps characterized as informed decision making in a democratic society at both personal and societal levels. Disinger (1983), Harvey (1977), Simmons (2000), and others state a similar conceptualization. This concept is congruent with the progressive philosophy of American education, a tenet of which is the fostering of citizenship participation in a democracy.

The progressive, "responsible citizen" approach to environmental education is taken by the North American Association for Environmental Education (NAAEE), the largest environmental education organization in the United States. This organization incorporated the problem-solving approach into a national policy document, *Excellence in Environmental Education: Guidelines for Learning* (NAAEE, 1999a). This document, which operationalized critical knowledge, skills, and dispositions, can be viewed as the field's standards. Modeled on other recent national education policy guidelines, such as the mathematics and science education standards, it draws heavily not only on the work of the American authors cited earlier but also on two major international policy documents, the Belgrade Charter (1975) and the Tbilisi Declaration (1977). Both of these international policy documents were developed as a result of the United Nations' interest in human activity and the environment.

A significant amount of controversy remains about the definition of environmental education. Some writers express the need for an ecology-based approach rather than the problem-solving one implicit in the technology-capitalism dimension of Western society. They claim that ecology must be the basis for human activity and that ecological parameters cannot simply be factored into an economic equation of costs and benefits. European writers and others take a postmodern approach, emphasizing individual development and opposing a systemic, outcomes-based approach. These writers decry top-down, prescriptive policies and behavior-based curricula. Disinger (1983) provides a more complete treatment of the definitional aspects of environmental education. Regardless of definition, the following characteristics appear to be essential elements in most environmental education perspectives. Environmental education

- is based on knowledge of ecology and social systems, drawing on disciplines in the natural sciences, social sciences, and humanities;
- reaches beyond biological and physical phenomena to consider social, economic, political, technological, cultural, historic, moral, and aesthetic aspects of environmental issues;
- recognizes that the understanding of feelings, values, attitude, and perceptions at the center of environmental issues is essential to analyzing and resolving these issues; and
- emphasizes critical thinking and problem-solving skills needed for informed, reasoned personal decisions and public action (Disinger and Monroe, 1994).

The major challenges to effective environmental education in the United States are interrelated, and so there is no significance implied by the discussion order that follows. One major challenge for the field is that it lacks a formal niche in the K-12 curriculum, suggesting that it is not in the mainstream of American education. This situation arises in part from the decentralized nature of the U.S. school system, with each state and school district declaring its own independent curriculum. It is also related to the multidisciplinary nature of the field, a characteristic that makes it difficult for environmental education to fit into a disciplinary curricular system that is responding more and more to "basics-only" demands for accountability rather than to the broad dimensions of a liberal, general education. Environmental education is either ignored or viewed by mainstream educators as a supplement to the curriculum that must justify its inclusion by enriching other subjects, such as history and science. In our view, the role of environmental education in American education will remain marginal unless a K-16 curricular niche is established for it.

It is not surprising that today's teachers are not prepared to teach environmental education. Neither the formal education curricula nor teachers' professional training experiences have prepared them for this instructional challenge. Very little environmental education is required of preservice teachers (i.e., those in training who have not yet begun their teaching careers), and there is limited organizational infrastructure for it at the state level. Fewer than 15 percent of preservice teachers take a formal EE course, and state-level data are equally slim. Kirk et al. (1993) offer perhaps the most recent state-level overview. The current teaching force lacks training in environmental education, and there is no provision for it in the preservice training of new teachers or in ongoing in-service training.

Most EE curricular materials were designed as supplemental lessons to be infused episodically into a given curriculum (for instance, Project WILD, Project Learning Tree). A plethora of print and video materials of highly variable quality is offered by many private and public curriculum developers. Some of these materials have been described as biased, inaccurate, incomplete, or propagandizing by both critics and supporters of environmental education. In the face of this criticism, NAAEE has developed a set of guidelines for developing, selecting, and evaluating materials (NAAEE, 1999b). The guidelines address fairness and accuracy, balanced viewpoints, depth of understanding, critical and creative thinking, and civic responsibility, as well as other instructional criteria. Major initiatives are needed to evaluate existing curricula to ensure that the highest quality products are recommended. Despite attempts to upgrade the quality of EE materials, conservative factions in the United States continue to criticize materials that are related to specific issues (e.g., the greenhouse effect). Instead they promote a version of environmental science that is "fact based." For example, the study of eutrophication as a concept—meaning the process of a body of water's becoming rich in nutrients but deficient in oxygen—is acceptable to them, but the examination of eutrophication as a function of nonpoint pollution in Galveston Bay, Texas, and of its sources, is not acceptable.

A MODEL CURRICULUM

One environmental education curriculum program, called Investigating and Evaluating Environmental Issues and Actions (IEEIA), has been developed over time, accumulating an extensive research and evaluation base (Hungerford et al., 1996; Ramsey, 2000; Winther, 2000). It meets the NAAEE guidelines and augments many of the outcomes identified by other discipline standards, such as national science standards. It is structured for insertion (as opposed to supplemental infusion) into the curriculum. And it has been the target of numerous research and evaluation publications.

Its initial development was as a one-semester curriculum designed for use at the middle school level. Subsequently, it was published in two themes, environmental and science-technology-society, and in two formats, modular and case study. These changes expanded the initial program's use from Grade 5 through high school. The case study programs use the same instructional structure as the initial IEEIA program but are built around specific topics, including coastal marine issues, endangered species issues, and solid waste issues.

The following discussion focuses on the environmental theme and the modular format. The model grew out of one teacher's desire to allow his junior high school students both to investigate environmental issues of interest to them and to enable them to develop the skills needed to conduct such an issue-based investigation.

Over the years the model has been refined as more and more teachers and students have provided input and as more research information has become available. In addition, it became apparent early in the development process that a component of citizenship participation (i.e., citizen action) was needed because students often wished to do something about the issues they investigated after completing their research. Today, the published versions of the curriculum reflect generally accepted instructional goals beginning with background information that leads to issue awareness, issue investigation and evaluation, and citizenship participation/issue resolution.

The curriculum is organized into a series of six modules or chapters. The modules are interdisciplinary in nature and introduce students to the characteristics of issues, the skills needed for obtaining and processing information, the skills needed for analyzing and investigating issues, and the skills needed by responsible citizens for issue resolution. The following description provides a brief overview of each module.

Module I: Environmental Problem Solving: This module contains lessons using actual environmental issues to develop the skills necessary to understand and analyze issues independently. These skills include discriminating among the interrelationships of events, problems, and issues, as well as understanding the role of beliefs and values in issues. Issue analysis, the skill of unpacking the critical components of an issue, is introduced and practiced. The concept of interaction, that is, the interrelatedness of human activities and the natural world, is also introduced, demonstrated, and applied. Rather than focus on a particular body of information or ideas, these lessons focus on the skills necessary for students to analyze the complexity of environmental issues.

Module II: Getting Started on Issue Investigation: These lessons begin the skills necessary to start an issue investigation. Students identify issues, write research questions, and learn how to obtain information from secondary sources and how to compare and evaluate information sources. These lessons focus on finding, analyzing, and evaluating secondary source information about issues.

Module III: Using Surveys, Questionnaires, and Opinionnaires in Environmental Investigations: Students learn how to obtain information using primary methods of investigation. Initially, they learn how to develop surveys, question-naires, and "opinionnaires." Subsequently, they learn sampling techniques, how to administer data collection instruments, and how to record collected data. These lessons focus on social science inquiry skills in the context of environmental beliefs, attitudes, and behaviors.

Module IV: Interpreting Data From Investigations: Students learn how to draw conclusions, make inferences, and formulate recommendations. They also learn how to produce and interpret graphs. These lessons prepare students to interpret and communicate findings using data related to environmental issues.

Module V: Investigating an Environmental Issue: Students autonomously select and investigate an issue. This process involves the application and synthesis of skills learned thus far. The model's developers recommend that students' investigations be reported back to their peers in formal classroom presentations. In this section of the program, students "take over," undertaking an inquiry into an authentic environmental issue approved and facilitated by the teacher.

Module VI: Environmental Action Strategies: Students learn the major methods of citizenship action, analyze the effectiveness of individual versus group action, and develop issue resolution action plans. This action plan is evaluated against a set of predetermined criteria designed to assess the social, cultural, and ecological implications of citizenship actions. Finally, the action plan may be implemented if the students wish. In this section students use their investigation data to formulate a plan for possible participation as a citizen in the solution of the issue under investigation.

The recommended outcomes of the program are to enable students to

- inquire successfully into ill-defined problems,
- demonstrate responsible citizenship in the community,
- interact successfully with environmental issues,
- use higher-order thinking skills, and
- think reflectively in terms of alternative positions related to issues.

The foundation of the program is the preparation for and undertaking of an authentic environmental investigation on the part of a student or a small group of students. Its structure provides a framework for teachers and students to manage complex intellectual activities. It is important to note that the most powerful educational experiences for students result from projects for investigation that they choose in their local community or region.

IEEIA has its roots in a variety of philosophical perspectives, beginning with John Dewey, who wrote at length on instructional models that reflect the democratic process and the scientific method. A number of eminent educators who followed Dewey either supported the same notion or independently arrived at a similar philosophy of education. Among these were Kilpatrick (progressive education), Counts (social problem solving and reconstructionism), and Hullfish and Smith (reflectivity).

Curricular approaches such as IEEIA are structured to help learners understand that democracy for a social group involves the investigation of problems and the development of solutions. Furthermore, this model provides for an attempt at issue resolution by having learners choose a desired method for helping resolve the issue (i.e., an action plan) and subsequently evaluate that method. In these ways, this model appears to reflect progressivism quite well. And, given that students' action plans often call for some form of social reform, the model carries with it characteristics associated with reconstructionism as well.

RESEARCH ABOUT ENVIRONMENTAL BEHAVIOR

The previous discussion noted that problem solving in terms of personal and social environmental decision making is a critical goal of environmental education. Given this, let's look at what is known about responsible environmental behavior. A number of studies of adults have been done from an environmental education perspective that offer insight into the relevant psychological attributes (Hines et al., 1986/1987; Sia et al., 1985/1986; Sivek, 1989/1990; Lierman, 1995; Marcinkowski, 2000; Volk and McBeth, 2000; Zelezny, 2000). (Studies from other perspectives are discussed elsewhere in this volume: Schultz, Chapter 4; Lutzenhiser, Chapter 3; Thøgerson, Chapter 5; Stern, Chapter 12.) Hines et al. (1986/1987) conducted a meta-analysis of research on responsible environmental behavior, reviewing studies from a variety of fields and using statistical procedures to determine the strength of the relationship between responsible environmental behavior and associated variables. Positive correlations were found for verbal commitment, locus of control, attitude, personal responsibility, knowledge, education level, income, and economic orientation. Using Hines' findings, Sia et al. (1985/1986) studied the predictors of environmental behavior in two populations of adults, one environmentally active and the other environmentally inactive. Sia's prediction model was based on eight variables, six of which were determined to be significant using regression analysis procedures and which accounted for 52 percent of the variance. The findings indicated that skill in using action strategies, environmental sensitivity, and knowledge of environmental action strategies accounted for the majority of the variance. Sia's findings were replicated by Sivek (1989/1990) and extended by Marcinkowski (2000) and Lierman (1995). Thus, the research indicates that responsible environmental behavior is associated with the following variables:

- Environmental sensitivity (i.e., feelings of comfort in and empathy toward natural areas),
- Knowledge of ecological concepts,
- Knowledge of environmental problems and issues,

- Skill in identifying, analyzing, investigating, and evaluating environmental problems and solutions,
- Beliefs and values (i.e., beliefs are what individuals hold to be true, and values are what they hold to be important regarding problems/issues and alternative solution/action strategies),
- Knowledge of environmental action strategies (i.e., consumerism, political action, persuasion, legal action, and physical actions),
- Skill in using environmental action strategies, and
- Internal locus of control (i.e., the belief that by working alone or with others an individual can influence or bring about the desired outcomes).

Hungerford and Volk (1990) used these variables to generate a model of responsible environmental behavior for environmental educators. Their model contains all the variables identified in the previous research, but the terms "ownership," "empowerment," and "entry-level" were added as category descriptors indicating the relationship of the variables to IEEIA instruction. Ownership refers to a construct of factors associated with personal knowledge and affect about environmental issues. Empowerment refers to a construct of factors associated with a sense of efficacy about issue solutions. Entry-level refers to factors that could be thought of as prior knowledge and dispositions (see Figure 9-1).

This discussion reflects the attempts of environmental educators and researchers to understand psychological and other factors associated with responsible environmental behavior. These findings were used as a reference framework in the design of the IEEIA curriculum. Additional research was then undertaken to determine the extent to which the key variables associated with responsible environmental behavior were affected by IEEIA instruction. The following section presents these studies.

RESEARCH ABOUT IEEIA

Eleven studies have examined the effects of IEEIA instruction in middle-grade settings: Ramsey et al. (1981), Klingler (1982), Volk and Hungerford (1981), Ramsey (1987), Ramsey (1993), Holt (1988), Bluhm et al. (1995), Bluhm and McBeth (1996), Withrow (1988), Simpson (1991), and Culen and Volk (2000). All these studies reported statistically significant, positive differences in responsible environmental behavior as a result of instruction, and many reported positive increases in the associated variables. For example, Ramsey et al. (1981) compared IEEIA-based instruction with environmental awareness and control treatments in Grade 8. He reported positive results on two outcome variables, knowledge of action and responsible environmental behavior. Three years later, Ramsey conducted a followup study of the students involved in the original study. Graduate students conducted double-blind interviews with students involved in all three groups. The graduate students identified all the subjects

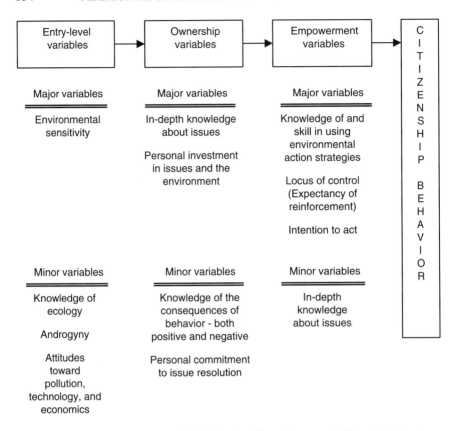

FIGURE 9-1   Major and minor variables involved in environmental citizenship behavior.

participating in the IEEIA treatment and found higher levels of responsible environmental behavior in the IEEIA group, despite the absence of subsequent instructional reinforcement during the ensuing three-year period.

One variable, environmental sensitivity, was not found to be affected by IEEIA treatment in any of the studies. Environmental sensitivity focuses on attributes that provide an individual with an empathetic view of the environment. Sensitivity research (e.g., Peterson, 1982; Sward and Marcinkowski, 2000; Tanner, 1980) strongly indicates that environmental sensitivity is one of the major precursors to environmental behavior. It seems to develop at an early age, when individuals experience pristine outdoor settings with adults who are important to them. Thus, it would be surprising if IEEIA, a formal classroom instructional treatment, could influence environmental sensitivity. What is important for environmental educators are the findings that IEEIA can foster responsible environmental behavior as well as gains in many of the allied factors.

In summary, research on IEEIA shows that for instruction to be effective, five elements are necessary. Students should have

1. Sound problem identification skills: They should be able to identify problems that are important to them in the communities or regions in which they live (Volk and Hungerford, 1981).
2. A degree of environmental sensitivity: Sensitivity is critical as a precursor to behavior. And although it may be possible, it is not easy for the formal classroom to accomplish this (Peterson, 1982; Sward and Marcinkowski, 2000; Tanner, 1980).
3. Issue investigation and evaluation skills: The ability to investigate and subsequently to evaluate issues runs throughout much of the research discussed in this chapter. It would be hard to tease out the precise components (in a research sense), but we know that students must be able to effectively evaluate important issues before they can make intelligent decisions about what to do about them. It also appears that a key element in the concept of ownership is personal involvement by students in issues under investigation (Sia et al., 1985/1986; Hines, 1986/1987; Marcinkowski, 2000; Ramsey, 1987, 2000; Ramsey et al., 1981).
4. Knowledge of and perceived skill in the use of citizenship action strategies: These skills include persuasion, political action, consumerism and the variables show up over and over again in one form or another in a great preponderance of the research discussed here. Also, these variables may well be the easiest to deal with in the classroom. How valuable they would be in and of themselves, however, without the framework of issue investigation, is unclear. It is hypothesized that there is a synergistic effect here, and the Klingler (1982) research indicates rather strongly that both are needed (Sia et al., 1985/1986; Hines et al., 1986/1987; Marcinkowski, 2000; Ramsey, 1993, 2000; Ramsey et al., 1981).
5. An internal locus of control: Locus of control is a key element in the concept of empowerment. Knowledge of action strategies without a concomitant feeling that the action will result in something positive probably won't get the job done. So opportunities must be provided that give students a feeling of success (even though we know that success is not met at every turn in citizenship roles). The teacher is a powerful force in helping students make good citizenship decisions, helping them find success on one hand and salving their defeats on the other (Sia et al., 1985/1986; Hines et al., 1986/1987; Marcinkowski, 2000).

In the future, other researchers may find that variables left out of the list above should have been included. It may be that other variables that show significant implications may operate only with certain populations under certain conditions. And some may be related to the ones listed above, such as knowledge of issues, verbal commitment to take action, beliefs about and attitudes toward pollution and technology, a sense of personal commitment, and attitudes toward economics.

SOME SUGGESTIONS

It is important to remember that environmental sensitivity needs to be initiated at an early age. Because this is a difficult attribute for the formal classroom situation to influence, it may be the most difficult to achieve. Classroom teachers can't turn learners into family campers, trappers, hunters, fishers, hikers, and people associated with other sensitivity-building avocations. Of course, the school can sponsor outdoor activities, but can it provide them in a dimension designed specifically to promote sensitivity? Remember, the outdoor activities reported by sensitive individuals focus largely on long-term experiences in relatively pristine environments. And these activities are done either on an individual basis or with one or two close associates. The class field trip may not be an appropriate vehicle. However, it may be possible for the school to accommodate some of these experiences by planning activities that take place in relatively small groups and in relatively pristine environments at times that can maximize at least a modicum of awe and wonder, that is, sincere appreciation. And there must be many such experiences.

Perhaps the best opportunity that the school has for achieving sensitivity is to combine high-quality outdoor activities with high-quality role models. Teachers should themselves demonstrate a high level of sensitivity, be able to communicate this sensitivity to learners, and be willing to lead students to aesthetic environmental experiences via books, television, and other media, along with outdoor experiences.

Beyond sensitivity, a number of behavior-related attributes can be influenced by planning for instruction that eventually involves learners in the investigation and resolution of issues. Young children can receive instruction on environmental issues through what is called the extended case study. The traditional case study deals with issues at a basic level of awareness. The extended case study is divided into five components:

1. A carefully selected *issue topic* around which a case study can be developed, such as municipal solid waste disposal, a locally endangered species, land use management in the community/region, air/water/aesthetic pollution, loss of wetlands, forest fire management, preservation of ecologically important plant/animal communities, and population growth;
2. *Science content,* which serves as prerequisite knowledge to understanding the scientific nature of a chosen issue;
3. *Issue awareness,* which focuses on the anatomy of that issue (the players

involved and their positions, beliefs, and values), the history of the issue, and possible solutions and impediments to them;

4. Some aspect of *issue investigation,* which gets learners involved in data collection regarding that issue (e.g., surveys, questionnaires, opinion-naires, interviews with key players), and

5. *Citizenship skills* (strategies such as political action and consumerism) that can be used to help resolve the issue coupled with an action plan that is developed cooperatively by the students and teachers and implemented if desired.

Older students, middle school and higher, should receive both case study instruction (at a more sophisticated level) and IEEIA instruction. With this strategy, teachers guide students through an introduction to issues, identifying problems, analyzing issues, using primary and secondary sources to obtain infor-mation about issues, recording and interpreting collected data, and demonstrat-ing citizenship strategies used in society for the remediation of issues. Major activities in this strategy include allowing the students to choose an issue of interest, guiding them in investigating and evaluating it, and reporting the find-ings to peers. The issue investigation is followed by the development of an action plan for helping to remediate that issue; it can be implemented or not, depending on the attitude of the student and judgment of the teacher.

It should be stressed that behavior-directed instruction needs to be articulat-ed across grade levels. There is some evidence (not reported in this chapter) that the behaviors sought will tend to erode unless there is periodic reinforcement across grade levels. This erosion is not complete, but students, as they grow older and receive no reinforcement, tend to back away from citizenship behavior as they lose teacher support and a social support system. Similarly, the skills associated with responsible citizenship behavior should be developed across sub-ject areas with a number of content specialists (such as science, social studies, language arts, and home economics) working cooperatively using a team-teach-ing/infusion approach.

Whether the school should fulfill the role of change agent in society depends entirely on the perspectives held by those making instructional decisions. How-ever, many educators firmly believe that "teaching about something" will influ-ence behavior. If this were absolutely true, then everyone would vote; no one would contract a venereal disease; everyone would be scientifically literate; the average citizen would love classical literature; man's inhumanity to man would be diminished or absent; no teenager would have an unwanted pregnancy; all laws would be respected; no animals or plants would be endangered; and people would not smoke. The same is probably true for citizenship responsibility re-garding the environment. Environmental educators have long argued for the importance of making people aware of environmental issues. But researchers have known for a long time that this assumption is faulty (see Schultz, this volume, Chapter 4).

Needless to say, what people know is important. Yet knowing will not provide the learner with what we refer to as ownership and empowerment. For learners to become actively involved in issue investigation and evaluation as well as citizenship behavior outside school, it is rather clear that they must own the issues on which they focus and be empowered to do something about them.

Instruction for the elements of ownership and empowerment is not traditionally part of the teacher's repertoire for instruction. This is noted to point out that teacher training has failed in its responsibility to give teachers the skills and motivation they need to make necessary instructional changes. And these instructional skills are not easy ones to come by. Our experiences in training teachers suggest that acquiring them takes time. At the very least, training necessitates a modification of philosophy, the acquisition of a wide variety of skills (many of which are foreign to most teachers), practice in the use of these skills and the methods associated with them, and help in learning how to evaluate students for grading purposes.

Changing a pattern of inadequate teacher education is beyond the purview of this chapter. Those who wish to see changes take place must consider how to make changes happen, both in the classroom and in students' lives. In a sense, we need to consider dual dimensions: attitude changes and skill acquisition on the parts of both teachers and students. The question then becomes how to give teachers and students both ownership and empowerment, its dual dimensions.

REFERENCES

Bluhm, W., H. Hungerford, W. McBeth, and T. Volk
    1995   The middle school report: A final report on the development and pilot assessment of the Middle School Environmental Literacy Assessment Instrument. In *Environmental Literacy/Needs Assessment Project: Assessing Environmental Literacy of Students and Environmental Education Needs of Teachers: Final Reports for 1993-1995*, R. Wilkie, ed. Stevens Point: University of Wisconsin-Stevens Point.

Bluhm, W., and W. McBeth
    1996   *Evaluation Report for Investigating and Evaluating Environmental Issues and Actions.* Washington, DC: U.S. Department of Education.

Culen, G., and T. Volk
    2000   Effects of an extended case study on environmental behavior and associated variables in seventh and eighth grade students. *Journal of Environmental Education* 31:9-15.

Disinger, J.F.
    1983   *Environmental Education's Definitional Problems.* No. 2. Columbus, OH: ERIC Clearinghouse for Science, Mathematics, and Environmental Education.

Disinger, J.F., and M. Monroe
    1994   *Defining Environmental Education.* (EE Toolbox). Ann Arbor, MI: Consortium for Environmental Education and Training.

Harvey, G.D.
    1977   Environmental education: A delineation of substantive structure. *Dissertation Abstracts International* 38:611-A.
Hines, J.M., H.R. Hungerford, and A.N. Tomera
    1986   Analysis and synthesis of research on responsible environmental behavior: A meta-
    /87   analysis. *Journal of Environmental Education* 18:1-8.
Holt, J.G.
    1988   Study of the Effects of Characteristics Associated with Environmental Behavior in Non-gifted Eighth Grade Students Issue Investigation and Action Training. Unpublished research paper, Southern Illinois University at Carbondale.
Hungerford, H.R., R.B. Peyton, J. Ramsey, and T.V. Volk
    1996   *Investigating and Evaluating Environmental Issues and Actions: Skill Development Modules.* Champaign, IL: Stipes.
Hungerford, H.R., and T. Volk
    1990   Changing learner behavior through environmental education. *The Journal of Environmental Education,* 21:8-21.
Kirk, M., R. Wilkie, and A. Ruskey
    1993   Survey of the status of state-level environmental education in the U.S. *Journal of Environmental Education,* 29:9-16.
Klingler, G.
    1982   Effect of an Instructional Sequence on the Environmental Action Skills of a Sample of Southern Illinois Eighth Graders. Unpublished research paper, Southern Illinois University at Carbondale.
Lierman, R.
    1995   Predicting Responsible Environmental Behavior Through the Secondary Environmental Literacy Instrument: A Secondary Analysis. Unpublished master's report, Florida Institute of Technology.
Marcinkowski, T.J.
    2000   Predictors of environmental behavior: A review of three dissertation studies. In *Essential Readings in Environmental Education.* H.R. Hungerford et al., eds. Champaign, IL: Stipes.
North American Association for Environmental Education
    1999a   *Excellence in Environmental Education: Guidelines for Learning (K-12).* Washington, DC: North American Association for Environmental Education.
    1999b   *Environmental Education Materials: Guidelines For Excellence.* Washington, DC: North American Association for Environmental Education.
Peterson, N.J.
    1982   Developmental Variables Affecting Environmental Sensitivity in Professional Environmental Educators. Unpublished master's thesis, Southern Illinois University at Carbondale.
Ramsey, J.
    1987   Study of the Effects of Issue Investigation and Action Training on Characteristics Associated with Environmental Behavior in Seventh Grade Students. Unpublished doctoral dissertation, Southern Illinois University at Carbondale.
    1993   Effects of issue investigation and action training on eighth-grade students' environmental behavior. *Journal of Environmental Education* 24:31-36.
    2000   Comparing four environmental problem solving models. In *Essential Readings in Environmental Education,* H.R. Hungerford et al., eds. Champaign, IL: Stipes.
Ramsey, J., H.R. Hungerford, and A.N. Tomera
    1981   Effects of environmental action and environmental case study instruction on the overt environmental behavior of eighth grade students. *Journal of Environmental Education* 13:24-29.

Sia, A.P., H.R. Hungerford, and A.N. Tomera
  1985/   Selected predictors of responsible environmental behavior: An analysis. *Journal of*
  1986   *Environmental Education* 17:31-40.
Simmons, D.
  2000   Are we meeting the goal of responsible environmental behavior? In *Essential Readings in Environmental Education,* H.R. Hungerford et al., eds. Champaign, IL: Stipes.
Simpson, P.
  1991   Effects of Extended Case Study Instruction on Citizenship Behavior and Associated Variables in Fifth and Sixth Grade Students. Unpublished dissertation, Southern Illinois University at Carbondale.
Sivek, D., and H. Hungerford
  1989/   Predictors of responsible behavior in members of three Wisconsin conservation organi-
  1990   zations. *The Journal of Environmental Education* 21(2):35-40
Stapp, W.B.
  1969   Concept of environmental education. *Environmental Education* 1:30-31.
Sward, D., and T. Marcinkowski
  2000   Environmental sensitivity: A review of the literature, 1980-1998. In *Essential Readings in Environmental Education,* H.R. Hungerford et al., eds. Champaign, IL: Stipes.
Tanner, T.
  1980   Significant life experiences: A new research area in environmental education. *Journal of Environmental Education* 11:20-24.
Volk, T.L., and H.R. Hungerford
  1981   Effects of process instruction on problem identification skills in environmental education. *Journal of Environmental Education* 12:36-40.
Volk, T., and W. McBeth
  2000   *Environmental literacy in the United States.* Pp. 73-86 in Essential Readings in Environmental Education. H. Hungerford, ed. Champaign, IL: Stipes.
Winther, A.
  2000   Investigating and evaluating environmental issues and actions. In *Essential Readings in Environmental Education,* H.R. Hungerford et al., eds. Champaign, IL: Stipes.
Withrow, V.
  1988   Effects of an Issue-oriented Case Study on Fifth and Sixth Students' Knowledge and Citizen Action. Unpublished master's research paper, Southern Illinois University at Carbondale.
Zelezny, L.
  2000   Educational interventions that improve environmental behaviors: A meta-analysis. In *Essential Readings in Environmental Education,* H.R. Hungerford et al., eds. Champaign, IL: Stipes.

10

A Model of Community-Based Environmental Education

Elaine Andrews, Mark Stevens, and Greg Wise

This chapter focuses on one model for achieving community flexibility and responsiveness to environmental issues. The model, termed community-based environmental education, differs from traditional education in that the educational activities not only build individual knowledge and skills, but also help to build an infrastructure for change that is sustainable, equitable, and empowering.

When the "classroom" is the community, an education strategy can take the form of employee training, media marketing, "point of purchase" information, workshops, study circles, one-on-one demonstrations, or a group initiative to gather data about a local problem. Typically, the educator chooses the education or diffusion strategy and bases the choice on considerations of the topic; audience skills; and personal skills and resources. But in community-based environmental education, the educator has an unconventional role. The community-based model presented in this chapter emphasizes selection of the education strategy in a way that also builds local skills and supports voluntary actions. Practitioners work in collaboration with the community to choose a strategy; to consider how and when the strategy could be used; and to guide whether the strategy is applied alone or in combination with others.

The "community" of the community-based environmental education model may be a community of *place*; a community of *identity*; or a community of *interest*.[1] In each situation, the intent is to build the skills of citizens to gather, analyze, and apply information for the purpose of making environmental management decisions. Successful application of the model contributes to the "environmental policy capacity" of the community, as described by Press and Balch (this volume, Chapter 11).

To ensure that education activities will support long-term and/or structural change, this collaborative strategy invites those involved to ask questions such as:

- Are the goals of the activity determined by a bottom-up process, or a top-down process?
- Is the intervention targeted narrowly to a specific audience or broadly to whole populations?
- Is the locus of control generated by individuals or community groups, or by marketing agents? ("Locus of control" is a term that refers to the source of personal empowerment. Does the person's sense of power to act come from within, or from the group, or is the person affected by an external agent?)
- Is the interest group actively involved in creating information and targeting research, or is the interest group a passive consumer of information?
- Does the intervention build sustainability for its impacts by engaging people at different levels of responsibility within the community (such as property owners, political leaders, and the agency that has jurisdiction)?

Community-based environmental education incorporates public participation, social marketing, environmental education, and right-to-know strategies. Measures that contribute to the effectiveness of volunteer activities also are encompassed in this model. The community-based model, however, contrasts with Ramsey's definition of environmental education, in that community-based environmental education goals incorporate a behavior change or policy change objective. Community education goals are designed to be responsive to the reality of the community economic, political, and social contexts. Application of specific education and dissemination elements is described in other chapters in this volume (Lutzenhiser, Chapter 3; Schultz, Chapter 4; Thøgersen, Chapter 5; Mileti and Peek, Chapter 7; Valente and Schuster, Chapter 6; Ramsey and Hungerford, Chapter 9; Nash, Chapter 14; Herb et al., Chapter 15; Harrison, Chapter 16).

DEVELOPMENT OF THE COMMUNITY-BASED
ENVIRONMENTAL EDUCATION MODEL

The U.S. Environmental Protection Agency (EPA) and the U.S. Department of Agriculture (USDA) both recognize that managing the environment requires investment in the community for two powerful reasons: (1) local activities affect the quality of the local environment, and (2) community members have a common interest in protecting and improving their community's quality of life. Consequently, these agencies have promoted environmental management via local decision-making and voluntary compliance with regulations and have considered ways to support these situation-specific processes and offer more effective environmental education.

Guided by research that describes how community members work together to make change (Wise, 1998) and how individuals make decisions about what they will do (in this volume, see Lutzenhiser, Chapter 3; Schultz, Chapter 4; Thøgersen, Chapter 5; Stern, Chapter 12), the EPA and USDA Cooperative Extension worked in partnership to investigate potential qualities of community-based environmental education. The resulting Community-Based Environmental Education (CBEE) model was defined through a four-part process: (1) by examining community efforts that had a common goal of improving local environmental management; (2) by consulting theoretical writings along with empirical studies of "what works"; (3) by identifying what appeared to be the critical elements of a common model; and (4) by then presenting the written model to practitioners for review (Andrews, 1998).[2, 3]

What we learned from the EPA/USDA Partnership project is that effective community-based environmental education builds on community development processes (including problem solving, community building, and systems interaction) and focuses on generating positive actions, rather than criticism or protest of current policies (see Figure 10-1). In a community-based education model, a community:[4]

- Has or establishes a vision and goals,
- Inspires an instigator who, stimulated by these goals, enlists or gathers a group or coalition to start an initiative and to keep it going,
- Supports group activities to gather and analyze information, and finally
- Through the group, engages the larger community in carrying out what it has learned through policy changes, new regulations, and/or education.

For example, property owners around Lake Example have a recognized or implicit vision for clean and healthy water. Inspired by this vision, the president of the property owners' association initiates a project to establish a wastewater collection system. To implement the project, property association members and other interested people would need to learn what technology is needed, how much it would cost, who would pay for it, what benefits would result, and what other ways are available to solve the same problem. Once the information is collected and analyzed, the owners' association might develop an information campaign to reduce local use of lawn and garden pesticides and lobby a government representative to propose an ordinance that requires a wastewater collection system to be installed around all local lakes. Feedback from these new activities influences community vision and goals, and the process begins again.

Each of these actions, viewed separately, can be seen as similar to a great deal of everyday community activity. What is distinctive about the CBEE model is that it integrates the elements as a linked chain. With such an inherently complex structure, it is difficult to estimate potential outcomes and impacts for the CBEE model as a whole. A number of studies relevant to the model, howev-

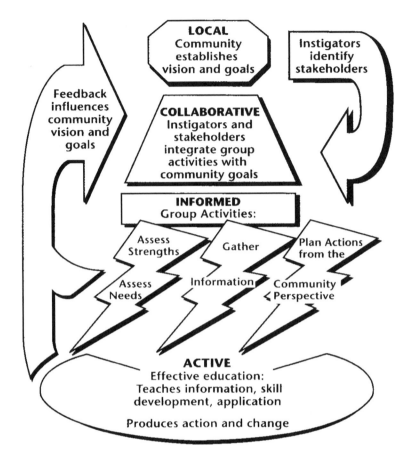

FIGURE 10-1   Building capacity: Applying the principles of community-based education.

er, have been published since the original project was completed. Their implications are discussed later in this chapter.

WHAT IS COMMUNITY-BASED ENVIRONMENTAL EDUCATION?

Community-based education means more than "education based in the community." It implies an education plan created as a result of community involvement and designed to match community interests.[5] "Community interests" refer to standard community issues, such as affordable housing or workforce development, as well as to activities with a recognizable environmental component such as road building, stormwater management, "permitting" a new development, or addressing environmental health concerns in an urban neighborhood.

Ideally, the education plan helps strengthen citizens' skills to plan or act with the environment in mind. Goals of community-based environmental education are to:

- Expand the community's ability to improve environmental quality,
- Integrate environmental management goals with other community development activities,
- Lead to actual environmental improvement, and
- Increase involvement of more community interests (both groups and points of view) in community environmental management activities.

CBEE activities have four key qualities. Activities are *community based, collaborative, information based,* and *action oriented.* The choice and sequence of activities relies on community development strategies for determining environmental goals;[6] a modified action research process for identifying information about the environmental problem and engaging stakeholders in the development of that information base;[7] and a combination of communication, environmental education, innovation diffusion, and social marketing to involve the broader public or "community of interest" in carrying out selected goals. Details of each of the elements are provided in Box 10-1.

EXAMPLES OF COMMUNITY-BASED
ENVIRONMENTAL EDUCATION

Applying a community-based approach is both an art and a science. The *art* is in the educator's ability to notice and take advantage of community links and opportunities. The *science* involves applying skills needed for working with a coalition or group. How the approach is applied depends on the characteristics of the community and of the groups or agencies involved.

Consider, for example, the activities at the Sea Change Resource Center, a community-based organization in Philadelphia.[8] Challenged by urban problems, Penn State Extension educators could have tried to improve the local economy by offering their own education program. Instead, educators worked in collaboration with Sea Change, which works to enhance economic development in selected Philadelphia neighborhoods by developing entrepreneurial solutions to local environmental problems. Sea Change activities are effective because they are well connected to neighborhood and city political structures. In the Sea Change/ Penn State Extension partnership, Sea Change identifies training needs for local groups in consultation with Penn State, and invites Penn State specialists, such as horticulture and urban forestry professionals, to provide technical assistance and training. Penn State has the potential to make a real difference in people's

BOX 10-1
Community-Based Environmental Education

LOCAL—Education Is Locally Based
- Responds to a locally identified/initiated issue or concern.
- Takes advantage of opportunities (such as a new law or current event) and community assets.
- Works in or with representative groups, including targeted audience (i.e., the people who collaborate represent all the interests associated with the issue).
- Works towards a *positive outcome* to a specific concern.

COLLABORATIVE—Education Works with a Coalition or Group
- Identifies someone who takes responsibility for managing or leading the process.
- Attends to *process* objectives and *product* objectives.
  — Process objectives = group building, leadership development, capacity building, conflict management
  — Product objective = successfully addressing a substantive issue
- Relies on systematic planning procedures.
- Uses expert facilitation.
- Uses consensus decision making.
- Develops linkages to enhance the group's effectiveness.
  — To other communities or regions
  — To other partners
  — To resources—technology, experts, agencies, funds
- Communicates broadly using multiple venues (e.g., newsletters, town meetings, TV, festivals).
- Provides recognition and rewards.
- Is flexible both to process and conditions; adopts a "learning organization" perspective.

INFORMED—Education Takes Action Based on Information
- Relates actions to long-term community vision and goals.
- Considers the community as a whole.
  — Evaluates context
  — Considers sociopolitical, economic, historical, cultural influences
  — Looks to the future

lives due to adaptation of resources to meet community needs, and can deliver programs as part of a well-established and respected community organization.

In the CBEE model, leadership is not a fixed status, but involves roles that shift back and forth over time. The educator is both working with the instigator and is influenced by the instigator's efforts. Education activities range from providing training in group process and planning, to providing information and resources for investigating the environmental problem. With this foundation, the

- Generates and makes use of data about the local condition.
- Involves citizens in gathering and analyzing data.
- Builds on locally existing skills and resources and scales actions appropriately to community resources and skills.
- Respects, encourages, and rewards local initiative.
- Evaluates and reports accomplishments.

ACTIVE—Education Practices Quality Education with Broader Groups
- Uses social marketing techniques.
  - Identifies and addresses *individual barriers* to preferred behavior (e.g., a tag on an outside faucet helps residents to remember when to water)
  - Identifies and addresses *social or structural barriers* to preferred behavior (e.g., encourage recycling by providing curbside pickup)
- Uses training to support a community-based initiative, for example, provides training to:
  - Improve planning process skills
  - Generate and refine implementation ideas
  - Improve data gathering and analysis by citizens
  - Increase access to resources by group/coalition
  - Teach skills that group has identified as needed to accomplish goals
- Implements an education strategy that:
  - Presents all points of view
  - Relates to a specific audience and its needs
  - Takes place close to the targeted behavior
  - Presents behaviors that
    - provide immediate, observable consequences
    - are similar to what people already do
    - do not require a lot of steps or training
    - are relatively low cost in terms of time, energy, money, materials
  - Provides details on how to do the exact behavior
  - Provides target audience with opportunities for:
    - self-assessment
    - a personal discussion about the new behavior
    - verbalizing a commitment to the change
    - practicing or applying new skills
  - Uses creative approaches
  - Reaches people in multiple ways

leadership group, along with additional members of the affected public or interest group, then engage in problem investigation and planning.

A case in point is the story of the Horicon Marsh Area Coalition (HMAC). Horicon Marsh is the largest freshwater cattail marsh in the United States and is a designated wetland of international importance (Thoms and Andrews, 2000). Recognizing the diversity of potentially conflicting interests and the increasing demand on the marsh and its surrounding areas, a local conservation group began thinking about how to protect the marsh before any major conflicts arose. Eventually the group contacted the University of Wisconsin Extension in Dodge County (i.e., a local outreach office) for assistance. Together, they planned a one-day Horicon Marsh Forum, convened and facilitated by the extension educator. This forum attracted 80 people representing 23 interest groups. Using group facilitation processes, the group identified eight priority issues. Work groups formed around each issue.

Forum organizers and a representative from each work group convened a steering committee, the HMAC, including representatives from diverse stakeholders, local government, and agencies. This group agreed to a set of "Organizational Principles, Policies, and Guidelines" based on a collaborative approach introduced by the extension educator. As HMAC continued to meet, the county extension educator introduced new process skills based on what participants were interested in learning. Experts from the university and other agencies occasionally provided content information and shared analytical skills when asked to explain research findings.

The CBEE model emphasizes qualities of equity, empowerment, and sustainability as part of environmental management decision processes. Case studies also indicate that while each of the four elements of the CBEE model are significant, the dynamic or interplay of CBEE elements is as important as successful implementation of any one element. Box 10-2 summarizes four other models that integrate education with community planning and have similar goals.

THE CBEE MODEL AND RELATED APPROACHES

CBEE integrates information dissemination, traditional education, participatory decision making, and other tools used in communication/diffusion approaches. We call this community-based model an education model for several reasons. First, CBEE's community context and process approach exemplifies the ideal application of learning theory, which maintains that individuals are not motivated to learn unless the information is *relevant to their lives* and they have a sense of control about the learning process (Carlson and Maxa, 1997; Heimlich and Norland, 1984). The CBEE model also provides educators with guidelines for developing education activities that are relevant to society's needs, and it provides a context for quality education practices because it requires higher order learning skills and integrates education into real-life experiences (Bloom, 1956; Horton and Hutchinson, 1997; Joplin, 1995; Knox, 1993; Westwater and Wolfe, 2000).

Education relies on the existence of a body of knowledge, but its power is in the fact that the knowledge is not only transferred to the individual, but is instrumental in transforming the individual. For education to take place, the individual has to actively receive the knowledge and know what to do with it (Bloom, 1956; Whitehead, 1929; Weintraub, 1995). The educator's job is to provide the education in a way or at a time when the individual is receptive and to assure that

BOX 10-2
Community-Based Education Models

U.S. EPA's Urban Environmental Initiative (UEI)—UEI applies a modified version of the CBEE model to guide EPA professionals and city managers in their work with the community. Professionals focus on listening to community concerns and leveraging resources (e.g., providing grants) to address them with meaningful improvements. Stakeholders identify information needs, participate in developing and analyzing relevant information, and develop outreach plans for communicating results. The goal of UEI is similar to that for CBEE—to build a community infrastructure that allows people to work effectively on environmental issues, but clearly places the professional in the role as "instigator." (This information was provided by Kristi Rea, UEI City Manager, Providence, Rhode Island.)

Take Charge—This is a national Cooperative Extension program for economic development in small communities. This education program engages community members in a "community visioning" model. Community stakeholders participate in a variety of information gathering and analysis activities that enable them to answer three questions: (1) Where Are We Now? (2) Where Do We Want to Be? and (3) How Do We Get There? This workbook is the basis for many popular community development efforts of the 1990s, not only those limited to economic development (Ayres et al., 1990).

The Local Agenda 21 Planning Guide—This is a United Nations Environment Program (UNEP) for helping communities to organize systems for devising appropriate solutions to local environment and development issues. According to the authors, planning should include five components: partnerships, community-based issue analysis, action planning, implementation and monitoring, and evaluation and feedback. Each section explains procedures, provides work sheets or resources, and illustrates concepts with community-based case studies. References and explanations of many valuable planning tools are included such as: Rapid Urban Environmental Assessment; setting targets and triggers for action planning; creating effective structures for accomplishing actions; and the UN Conference on Human Settlements Indicators Project (ICLEI, 1996).

Starting with Behavior—This strategy engages community members in self-help education initiatives. For example, a local team in Ecuador worked to design, test, and implement a methodology for monitoring and measuring observable changes in behavior related to sustainable use of land in buffer zones surrounding an ecological reserve. They involved community members in selecting target behaviors that include: defining the ideal behavior, conducting research with "doers" and "non-doers," selecting and negotiating target behaviors, and developing strategies which reflect findings (Booth, 1996).

the individual knows what to do with specific knowledge. That is, "the individual can find appropriate information and techniques in his previous experience to bring to bear on new problems and situations. This requires some analysis or understanding of the new situation; it requires a background of knowledge or methods which can be readily utilized; and it also requires some facility in discerning the appropriate relations between previous experience and the new situation" (Bloom, 1956:38).

Environmental education integrates basic learning skills with innovation diffusion approaches to create an education process focused on natural and sociocultural environments. The four themes of environmental education literacy have been incorporated into the CBEE model. They are (1) knowledge of environmental processes and systems; (2) questioning and analysis skills; (3) skills for understanding and addressing environmental issues; and (4) personal and civic responsibility (Simmons et al., 1999).

An education program, if it is going to accomplish transformation, or even if it is merely to result in the adoption of a target behavior, must include communication, skill development, and application. The CBEE model stresses the importance of a careful match between the person who will learn and the choice of education process. For practical purposes, it is less important to clearly distinguish among communication, diffusion, social marketing, and education concepts than it is to identify how to use each to create sustainable processes for supporting voluntary measures in environmental protection.

Education programs developed based on the CBEE model rely, primarily, on informal learning—learning through activities that occur outside formal educational settings and that are characterized as voluntary, as opposed to required for school credit. Just as in formal education, however, informal learning experiences can be structured to meet a stated set of objectives and can be designed to influence attitudes, convey information, and/or change behavior (Crane et al., 1994). CBEE activities may also be supported by formal education opportunities. For example, drinking water quality described by the Consumer Confidence Reports found in homeowner water bills might be studied in the high school chemistry class.

Informal learning may include any of the information and diffusion strategies discussed in Chapters 3 through 7 of this volume. For example:

- **Information dissemination and communication** efforts use various media to provide information to specific target audiences or to the public. Effectiveness of information campaigns has been studied relative to a variety of audiences and purposes (see Chapters 3-7 and 12 of this volume).
- **Behavior change** efforts involve teaching an ideal behavior or an environmental practice (a series of several related behaviors that, together, could affect the environmental problem, Booth, 1996). An ideal behavior or practice is usually defined by experts. Behavior change efforts also may in-

volve encouragement for personal commitment, use of external prompts, and changing social norms (McKenzie-Mohr, 1996; Stern, 2000).

* **Diffusion** approaches emphasize the spread of innovations by communication among the members of a social system. In diffusion theory, innovators, diffusers, and potential adopters communicate to understand the innovation; how and why it works; and what its advantages, disadvantages, and consequences are in specific situations. Research about innovation diffusion usually refers to how citizens adopt new technology, but the concepts can apply equally to new information (Rogers, 1995).

There is extensive research about techniques used in informal and adult education and with public participation. Educators can learn numerous details about effectiveness of workshops, types of signs to use, visitor attention span, benefits of linking television programs with local support groups, and other information. For example, see studies summarized in Crane et al. (1994), Chess and Purcell (1999), and the President's Commission on Americans Outdoors (1986). Our challenge is to figure out how to use communication, diffusion, and education strategies to infuse environmental management considerations into the mix of everyday discussion and decision making. The CBEE model provides numerous avenues to use these strategies for increasing environmental management capacity among many audiences.

EFFECTIVENESS OF THE CBEE MODEL

There is a rich set of resources about what makes community-based involvement and outreach effective—too many resources to describe here, except in the most general sense. Details have been captured in the CBEE Model (see Box 10-1). Yet across this wide variety of publications, there is a consistent emphasis on application of community development techniques to solving community problems. In itself, this commonality of theme indicates something about the value of this approach.

Finding definitive research about the effectiveness of the CBEE elements when applied to environmental management, however, was difficult. It was easy to identify guides, literature reviews, and descriptive materials, but difficult to find information that summarizes impacts of specific program strategies. Reports and newsletter articles provide periodic summaries for some community-based programs, such as Farm*A*Syst (Jackson, 1990), Groundwater Guardian (Kreifels, 1997), the River Network (Wallin and Haberman, 1992), and Save Our Streams (Firehock, 1994). Otherwise, impact information is available primarily through a small number of studies of individual local programs or studies of program elements.

Some reports involve collecting and summarizing case studies and highlighting commonalities. These studies attempt to build theories of community-based efforts or to provide a list of keys to success. Some were useful in building the CBEE model. These studies of groups of cases include studies of: 9 (out of 618) federally funded watershed-based projects (U.S. General Accounting Office, 1995); 30 community-based management initiatives (U.S. Environmental Protection Agency, 1997); various watershed management plans and related education initiatives (Ficks, 1997); 5 river case studies (Wallin and Haberman, 1992); annual summaries of state progress in adapting Farm*A*Syst resources for local outreach education needs (Jackson et al., 1997); case analysis of public involvement through Great Lakes Remedial Action Plan citizen advisory committees (Landre and Knuth, 1993); surveys of Rouge River neighborhood programs (Powell et al., 2000); stormwater pollution case studies (Aponte Clarke et al., 2000); investigation of impacts from a homeowner nutrient management program; and local management of "common-pool" resources (for example, Ostrom, 1990; Singh and Ballabh, 1994).

In addition, there are major text books and literature reviews based on examination of the mainly case-based literatures about public involvement and collaboration in natural resources management (MacKenzie, 1996; National Research Council, 1999; Renn et al., 1995a; Wondolleck and Yaffee, 2000).

IMPLICATIONS OF CBEE FINDINGS

CBEE could be described as a process of changing the community's idea of acceptable environmental management behavior, as a result of direct involvement of citizens in the management process. In spite of the difficulty of describing and studying such a complex process, this participatory, engaged approach provides a community involvement and outreach model that can be responsive to political as well as ecological necessity. For example, studies show that the new science of ecosystem-based management depends on application of community development problem-solving processes, as described by the CBEE model (Kellogg, 1999; MacKenzie, 1996; Wondolleck and Yaffee, 2000; National Research Council, 1999).

Community interests work together to find and implement solutions to common problems. The question is how and when to apply the CBEE model to address environmental protection needs. When is education an important element of environmental decision making? What types of education needs are best supported through this model? Who are the people who can assure that this complicated process can be carried out? How can the effectiveness of the process be evaluated? How can it be applied to larger scale problems?

Some of these questions will be answered as researchers study how citizen participation models[9] or development of social infrastructures required to manage common-pool resources[10] could be applied in the CBEE model. (Common-pool resources usually refer to an *economic* resource, such as animal grazing land, which is collectively owned by an identifiable community.) For example, more discussion is needed about the role of education in managing common-pool resources and how findings apply to their social uses (health, well-being, beauty, recreation).

The role of education seems clear. At a minimum, it is important to help people develop the capacity to make decisions and take responsibility (Horton and Freire, 1990; Ostrom, 1994). In managing common-pool resources, users need knowledge of resource conservation and use to help in correct and timely diagnosis of problems and to assure they have the best knowledge they can have, because resource decisions are usually made based on "best available" knowledge (e.g., nutrient best management practices). Policymakers need education so they can understand the nature and causes of problems and the tools for management (Singh, 1994). Public participation in policy development requires equal access to information (Lynn and Kartez, 1995; Dienel and Renn, 1995). Communities need a source of leadership in environmental management (Kellogg, 1999), and natural resources are more likely to be managed sustainably when decision making is decentralized (Wondolleck and Yaffee, 2000).

The question of when to use the CBEE model refers to the type of decisions needed. If individual behaviors are the primary management elements, then application of CBEE can provide peer support and motivation, but if transfer of relevant information is the only goal, CBEE is only one of many workable approaches. If a policy or infrastructure change is needed, then application of CBEE is one of few ways to accomplish the goal sustainably.

Who can assure that the CBEE model is properly applied is a very significant question; its answer also helps to answer the question of how CBEE could apply to larger scale problems. *Government* can enhance the skills of its own staff and ensure that policies provide the time and perspective necessary for community flexibility and responsiveness to environmental issues. *Institutions* that provide community outreach also can assure that educators build skills for facilitating or supporting different steps of the CBEE model. Leaders of *community organizations* can commit to supporting the comprehensive CBEE process.

In CBEE, government agency personnel, in particular, need to commit to authentic efforts with communities. Citizen advisory committee studies show, for example, that success depends on the citizen perception that the underlying purpose of the sponsoring institution is sincere and legitimate (Lynn and Kartez, 1995; MacKenzie, 1996). Goals must be established through genuine collaboration and with all participants committing to them—even when they differ from the initial ideas, plans, or missions of some participants. Application of the CBEE model also depends on availability of resources that enable communities to respond effectively, and on agency personnel who are ready to support community assumption of responsibility.

Remedial Action Plans (RAP) for the Great Lakes ecosystem, and ecosystem-based water resource management schemes in several areas, serve as tests for how to apply the CBEE process while addressing larger scale problems. RAPs were mandated by the federal government but were written by state and local governments with input from citizens, business, and industry (Renn, 1991). Centralizing goals, but not mechanisms, provides an opportunity for maximizing success at the local level through application of the CBEE model.

Several authors have suggested that a collaborative, sequential, or "nested" administrative structure, such as that found in the RAP process, is needed to enhance successful implementation of public participation in larger scale problems (Born and Genskow, 2001; National Research Council, 1999; Ostrom, 1990; Renn and Finson, 1991). For example, a study of watershed strategies found that organizations for watershed management are most likely to be effective if their structure matches the scale of the problem (National Research Council, 1999). In this example, local issues are handled by local self-organized watershed councils, where the CBEE process could be applied, while larger organizations should deal with broader issues.

Other examples where CBEE could be effective in the application of a vertical decision-making strategy include an effort like the Dutch government's initiative to develop a national policy on energy (Midden, 1995) and efforts to improve effectiveness of Citizen Advisory Committees (CACs). CAC impacts could be increased by combining their activities with techniques providing more representation, such as surveys or referenda (Vari, 1995) *or* the CBEE approach.

Finally, evaluation tools have been developed to help practitioners determine whether their community-based education efforts have been effective or applied appropriately. It's one thing to provide a citizen education or participation model, but another to know whether its application accomplished the goal of increased citizen ownership for the product. Questionnaires can help practitioners evaluate community involvement for competence and fairness (Renn et al., 1995b) or for appropriate choice of steps toward the involvement process (University of Wisconsin Extension, 1998).

CONCLUSION

When educators, business/industry administrators, politicians, or government agency representatives suggest public education as one way to meet an environmental management goal, the education strategy must go beyond simplistic solutions to be effective. The usual suggestions—hold a meeting, write a manual, develop a curriculum, provide training—will not support long-term or structural change on their own. Coupling these standard education resources with the CBEE process sets the stage for meaningful education; that is, education designed to provide the context and relevance recognized by the learner and to generate the opportunity for the learner to apply knowledge to the environmental problem.

If CBEE's collaborative and participatory processes are complemented by an authentic commitment to participate in and use its key qualities of being community based, collaborative, information based, and action oriented, we can achieve community flexibility and responsiveness to environmental issues. Further study of the elements of community-based education and representative programs would enhance our ability to determine when CBEE should be emphasized and how to train and support practitioners to facilitate successful participation in this dynamic process.

NOTES

1 A "community of interest" is that form of community whose commonality lies in the benefits received from a resource or the costs imposed on it (Wondolleck and Yaffee, 2000).

2 A 1996-98 project investigated ways to strengthen partnerships among the USDA Cooperative Extension, EPA, and communities in the service of these environmental management and education efforts. The Steering Committee included representatives from two EPA regions (Region 3, Philadelphia, and Region 10, Seattle) and the University of Wisconsin Project staff. More information is available online at http://www.reeusda.gov/nre/figs/usdaepa.pdf or http://www.wisc.edu/erc/.

3 Based on steering committee recommendations, project staff reviewed published case studies, U.S. EPA and USDA agency activities, and exemplary local programs that considered the whole community (i.e., programs which linked environmental education to management of local ecosystem components and community sustainability goals as defined by the President's Council on Sustainable Development [1996]). Staff also identified literature reviews, monographs, manuals, conference proceedings, and studies that provided further information about community development models, social marketing experiences, outstanding models of community-based education (as identified by peers), and community-based environmental education strategies. In addition to community development references, cited by Wise (1998), published references that influenced development of the model included Andrews et al. (1995), Andrews et al. (1996), Ayres et al. (1990), Beckenstein et al. (1996), Berger and Corbin (1992), Booth (1996), Butler et al. (1995), Byers (1996), Cairn et al. (1996), Chavis and Paul (1990), Cole-Misch et al. (1996), De Young (1993), Domack (1995), Drohan et al. (1997), Dwyer et al. (1993), Environmental Defense Fund Pollution Prevention Alliance Staff (1996), Ficks (1997), Firehock (1994), Fishbein and Gelb (1992), Flora (1997), Gigliotti (1990), Harker and Natter (1995), Himmelman (1992), Howe and Disinger (1988), Hungerford and Volk (1990), Hustedde et al. (1984), Israel and Ilvento (1996), Jackson et al. (1990), Jansen (1995), Johnson et al. (1996), Kreifels (1997), Kretzman and McKnight (1993), Lewis et al. (1993), McKenzie-Mohr (1996), National Association of Service and Conservation Corps (1996), Nuzum (1996), Olden and Poje (1996), Rocha (1997), Rogers (1995), Rusky and Wilke (1996), Sargent et al. (1991), Selin and Chavez (1995), Sexton (1996), Sidel et al. (1996), Sorenson (1985), U.S. Environmental Protection Agency (1997), U.S. General Accounting Office (1995), Walzer et al. (1995), Wallin and Haberman (1992), Wise and Kenworthy (1993).

4 As explained in the introduction, *community* refers to the topic or situation under discussion. *Community of interest* is a useful characterization because *community,* as used here, implies more than merely a physical place, although it can and often does include a geographic element. It may reference a discrete collection of persons who have a common interest, yet they may be located in different places and may not be aware of their shared interest. The community of interest also need not be made up of similar perspectives. Indeed, often it is made up of diverse perspectives surrounding a common issue (Wise, 1998).

5 Although this definition was developed by the EPA/USDA Partnership (Andrews, 1998), it has its origins in several other traditions, that are closely related to each other. Knox (1993) describes community problem-solving education as education that aims at community and organizational de-

velopment and social change, in contrast to traditional education, which is aimed at development and change of the individual. Based on this extensive study of national and international programs, Knox defines this type of education as "the process and result of an effort to include a broad cross section of people in educational activities to enable them to work together to solve organizational or community problems that have usually entailed consciousness raising, empowerment, and structural transformation." John Dewey, Myles Horton, and Paulo Freire are leaders in this tradition. Knox cites examples that include citizenship schools, county board workshops, participatory literacy, and workplace programs.

6   The purpose of community development is to satisfy local needs and welfare of people. Empowerment is emphasized as a means of identifying issues, managing change, and facilitating community-based solutions. Community development has been described as having four parts: a *process* moving by stages from one condition to the next; a *method*, a way of working toward the attainment of a goal; a *program*, whereby if activities are carried out, goals will be accomplished; and a *movement*, a cause to which people become committed. Emphasis is on what happens to people, and accomplishing a goal through activities and inciting people to take action (Wise, 1998).

7   Action research involves the student in generating new information to improve understanding of how knowledge content is developed, using critical thinking skills, and creating a sense of ownership of the knowledge. Action research has been used extensively in training and development in corporations, and in adult education in environmental, agricultural, and health settings (Quigley, 1997).

8   Information was obtained through personal communication with Roz Johnson, Director, Sea Change Resource Center, as part of the EPA/USDA community-based education investigation (Andrews, 1998).

9   Citizen participation models include citizen advisory committees, citizen panels (also known as planning cells), citizen juries, citizen initiatives, negotiated rule making, mediation, compensation and benefit sharing, and Dutch study groups (Renn et al., 1995a).

10   See Indiana University's materials for the Workshop in Political Theory and Policy Analysis for an extensive bibliography of studies and research about common-pool resources (Hess, 1996). For a recent summary of the field, see National Research Council (2002).

REFERENCES

Andrews, E.
    1998   An EPA/USDA Partnership to Support Community-Based Education. 910-R-98-008. [Online]. Available: http://www.reeusda.gov/nre/gifs/usdaepa.pdf [Accessed April 2000].
Andrews, E., E. Farrell, J. Heimlich, R. Ponzio, and K. Warren
    1995   *Educating Young People About Water - A Guide to Program Planning and Evaluation.* Columbus: ERIC Clearinghouse for Science, Mathematics, and Environmental Education, The Ohio State University.
Andrews, E., J. Hawthorne, and K. Pickering
    1996   Watershed Education - Goals and Strategies for Training, Communication, and Partnerships. Watershed '96 Pre-conference Symposium. National Fish and Wildlife Foundation. Baltimore, MD, June.
Aponte Clarke, G.P., P.H. Lehner, D.M. Cameron, and A.G. Frank
    2000   Community responses to stormwater pollution: Case study findings with examples from the Midwest. Pp. 124-131 in *Proceedings of the National Conference on Tools for Urban Water Resource Management and Protection.* Cincinnati, OH: U.S. Environmental Protection Agency, Office of Research and Development.
Ayres, J., R. Cole, C. Hein, S. Huntington, W. Kobberdahl, W. Leonard, and D. Zetocha
    1990   *Take Charge: Economic Development in Small Communities.* North Central Regional Center for Rural Development. Ames: Iowa State University.

Beckenstein, A.R., F. Long, T. Gladwin, B. Marcus, and the Management Institute for Environment and Business

1996   *Stakeholder Negotiations: Exercises in Sustainable Development.* Chicago, IL: Richard D. Irwin.

Berger, I.E., and R.M. Corbin

1992   Perceived consumer effectiveness and faith in others as moderators of environmentally responsible behaviors. *Journal of Public Policy and Marketing* 11(2):79-89.

Bloom, B.S., ed.

1956   *Taxonomy of Educational Objectives, Handbook 1: Cognitive Domain.* New York: David McKay Company.

Booth, E.M.

1996   *Starting with Behavior—A Participatory Process for Selecting Target Behaviors in Environmental Programs.* Washington, DC: GreenCOM, Academy for Educational Development.

Born, S.M., and K.D. Genskow

2001   *Toward Understanding New Watershed Initiatives: A Report from the Madison Watershed Workshop.* Madison: University of Wisconsin-Madison. [Online]. Available: http://www.tu.org/newsstand/library_pdfs/watershed.pdf [Accessed February 19, 2002].

Butler, L.M., C. Dephelps, and K. Gray

1995   *Community Ventures: Partnerships in Education and Research Circular Series.* Pullman: Western Regional Extension Publications, Washington State University.

Byers, B.A.

1996   *Understanding and Influencing Behaviors in Conservation and Natural Resources Management.* African Biodiversity Series, No. 4. Washington, DC: Biodiversity Support Program (U.S. Agency for International Development-funded consortium of World Wildlife Fund, The Nature Conservancy, and World Resources Institute).

Cairn, R., S. Cairn, E. Row, and E. Andrews

1996   *Give Water A Hand - Action Guide.* Environmental Resources Center. Madison: University of Wisconsin. [Online]. Available: http://www1.uwex.edu/ces/erc/gwah/gwahform.cfm [Accessed March 6, 2002].

Carlson, S., and S. Maxa

1997   *Science Guidelines for Nonformal Education.* Center for 4-H Youth Development, College of Education and Human Development, University of Minnesota. Summarized from Bartlett, F.C. (1932), *Remembering: A Study in Experimental and Social Psychology.* London: Cambridge University Press.

Chavis, D.M., and F. Paul

1990   *Community Development, Community Participation, and Substance Abuse Prevention: Rationale, Concepts and Mechanisms.* Santa Clara, CA: Department of Health, County of Santa Clara.

Chess, C., and K. Purcell

1999   Public participation and the environment: Do we know what works? *Environmental Science and Technology* 33(16):2685-2692.

Cole-Misch, S., L. Price, and D. Schmidt

1996   *Sourcebook for Watershed Education.* Arlington, VA: Global Rivers Environmental Education Network at Earth Force.

Crane, V., M. Chen, S. Bitgood, B. Serrell, D. Thompson, H. Nicholson, F. Weiss, and P. Campbell

1994   *Informal Science Learning: What the Research Says About Television, Science Museums, and Community-Based Projects.* Dedham, MA: Research Communications, Ltd.

De Young, R.

1993   Changing behavior and making it stick: The conceptualization and management of conservation behavior. *Environment and Behavior* 25(4):485-505.

Dienel, P.C., and O. Renn
	1995	Planning cells: A gate to "fractal" mediation. In *Fairness and Competence in Citizen Participation,* O. Renn, T. Webler, and P. Wiedemann, eds. Boston: Kluwer Academic.
Domack, D.R.
	1995	*Creating a Vision for Your Community.* Madison: University of Wisconsin Extension - Cooperative Extension (G3617).
Drohan, J., C. Abdalla, B. Marshall, and E. Stevens
	1997	*Lessons from Successful Project Leaders.* 1996 conference summary. Pennsylvania Groundwater Policy Education Project. State College: Pennsylvania State University Cooperative Extension and the League of Women Voters of Pennsylvania.
Dwyer, W.O., F.C. Leeming, M.K. Cobern, B.E. Porter, and J.M. Jackson
	1993	Critical review of behavioral interventions to preserve the environment: Research since 1980. *Environment and Behavior* 25(3):275-321.
Environmental Defense Fund Pollution Prevention Alliance Staff
	1996	*Environmental Sustainability Kit.* Washington, DC: Environmental Defense Fund.
Ficks, B.
	1997	*Top 10 Watershed Lessons Learned.* Office of Water and Office of Wetlands, Oceans, and Watersheds. EPA840-F-97-001. Washington, DC: U.S. Environmental Protection Agency.
Firehock, K.
	1994	*Save Our Streams—Volunteer Trainer's Handbook.* Gaithersburg, MD: Izaak Walton League of America.
Fishbein, B., and C. Gelb
	1992	*Making Less Garbage—A Planning Guide for Communities.* New York: INFORM.
Flora, C.B.
	1997	Innovations in community development. *Rural Development News* 21(3):1-3,12.
Gigliotti, L.M.
	1990	Environmental education: What went wrong? What can be done? *Journal of Environmental Education* 22(1):9-12.
Harker, D., and E. Natter
	1995	*Where We Live - A Citizen's Guide to Conducting a Community Environmental Inventory.* Washington, DC: Island Press.
Heimlich, J.E., and E. Norland
	1984	*Developing Teaching Style in Adult Education.* San Francisco: Jossey-Bass Publishers.
Hess, C.
	1996	*Common Pool Resources and Collective Action: A Bibliography* (v. 3). Bloomington: Workshop in Political Theory and Policy Analysis, Indiana University.
Himmelman, A.T.
	1992	*Communities Working Collaboratively for a Change.* Monograph, 1996 revised ed., IA No. 4. Minneapolis, MN: Hubert H. Humphrey Institute of Public Affairs, University of Minnesota.
Horton, M., and P. Freire
	1990	*We Make the Road by Walking: Conversations on Education and Social Change.* B. Bell, J. Gaventa, and J. Peters, eds. Philadelphia: Temple University Press.
Horton, R.L., and S. Hutchinson
	1997	*Nurturing Scientific Literacy Among Youth Through Experientially Based Curriculum Materials.* Center for 4-H Youth Development, College of Food, Agricultural and Environmental Sciences. Columbus: The Ohio State University.
Howe, R.W., and J. Disinger
	1988	Environmental education that makes a difference—Knowledge to behavior changes. *ERIC/SMEAC Environmental Education Digest* ED320761(4):1-5. [Online]. Available: http://www.ed.gov/databases/ERIC_Digests/ed320761.html [Accessed March 6, 2002].

Hungerford, H.R., and T.L. Volk
  1990   Changing learner behavior through environmental education. *Journal of Environmental Education* 21(3):8-21.

Hustedde, R., R. Shaffer, and G. Pulver
  1984   *Community Economic Analysis: A How To Manual*. The North Central Regional Center for Rural Development. Ames: Iowa State University.

International Council for Local Environmental Initiatives, International Development Research Centre, United Nations Environment Program
  1996   *The Local Agenda 21 Planning Guide*. Toronto: Local Agenda 21 Initiative.

Israel, G.D., and T. Ilvento
  1996   *Building a Foundation for Community Leadership, Involving Youth in Community Development Projects*. Mississippi State, MS: Southern Rural Development Center.

Jackson, G.
  1990   Farm*A*Syst, Farmstead Assessment System. Adapted by states throughout the decade. Available from state Cooperative Extension offices. Madison: University of Wisconsin - Extension. [Online]. Available: http://www.uwex.edu/farmasyst [Accessed March 6, 2002].

Jackson, G., and Farm*A*Syst/Home*A*Syst Staff
  1997   *National Directory - Farm*A*Syst/Home*A*Syst*. Madison: University of Wisconsin at Madison, Environmental Resources Center.

Jansen, L.
  1995   Citizen activism in the foundations of adult environmental education in the United States. *Convergence* 28(4):89-97.

Johnson, A.W., J.R. Denworth, and D.R. Trotzer
  1996   *The EAC Handbook - A Guide for Pennsylvania's Municipal Environmental Advisory Councils*. Phiadelphia: Pennsylvania Environmental Council.

Joplin, L.
  1995   On defining experiential education. In *The Theory of Experiential Education*, K. Warren, M. Sakofs and J.S. Hunt, Jr., eds. Dubuque, IA: Kendall/Hunt Publishing Co.

Kellogg, W.A.
  1999   Community-based organizations and neighborhood environmental problem solving: A framework for adoption of information technologies. *Journal of Environmental Planning and Management* 42(4):445-469.

Knox, A.
  1993   *Strengthening Adult and Continuing Education: A Global Perspective on Synergistic Leadership*. San Francisco: Jossey-Bass.

Kreifels, C., ed.
  1997   *A Community Guide to Groundwater Guardian*. Lincoln, NE: The Groundwater Foundation.

Kretzman, J., and J. McKnight
  1993   *Building Communities From the Inside Out: A Path Toward Finding and Mobilizing A Community's Assets*. Chicago: ACTA Publications.

Landre, B., and B. Knuth
  1993   Success of citizen advisory committees in consensus-based water resources public involvement programs. *Society and Natural Resources* 6:229-257.

Lewis, S.J.
  1993   *The Good Neighbor Handbook: A Community-Based Strategy for Sustainable Industry*. 2nd. ed. Waverly, MA: The Good Neighbor Project for Sustainable Industries.

Lynn, F.M., and J.D. Kartez
  1995   The redemption of Citizen Advisory Committees: A perspective from critical theory. In *Fairness and Competence in Citizen Participation*, O. Renn, T. Webler, and P. Wiedemann, eds. Boston: Kluwer Academic.

MacKenzie, S.H.
    1996   *Integrated Resource Planning and Management*. Washington, DC: Island Press.
McKenzie-Mohr, D.
    1996   *Promoting a Sustainable Future: An Introduction to Community-Based Social Market-ing*. Ottawa, Ontario, Can.: National Round Table on the Environment and the Economy.
Midden, C.
    1995   Direct participation in macro-issues: A multiple group approach. An analysis and critique of the Dutch national debate on energy policy, fairness, competence, and beyond. In *Fairness and Competence in Citizen Participation*, O. Renn, T. Webler, and P. Wiedemann, eds. Boston: Kluwer Academic.
National Association of Service and Conservation Corps
    1996   *Tools for Environmental Service: An Inventory of Project and Environmental Education Resources for Conservation and Service Corps and Other Stewards of the Earth*. Washington, DC: National Association of Service and Conservation Corps.
National Research Council
    1999   *New Strategies for America's Watersheds*. Committee on Watershed Management. Washington, DC: National Academy Press.
    2002   *The Drama of the Commons*. Committee on the Human Dimensions of Global Change. E. Ostrom, T. Dietz, N. Dolšak, P.C. Stern, S. Stonich, and E. U. Weber, eds. Washington, DC: National Academy Press.
Nuzum, R.
    1996   Know Your Watershed Program. Conservation Technology Information Center. [Online]. Available: http://www.ctic.purdue.edu [Accessed March 6, 2002].
Olden, K., and G. Poje
    1996   The emergence of environmental justice as a national issue. *Health & Environment Digest* 9(9):77-79.
Ostrom, E.
    1990   *Governing the Commons: The Evolution of Institutions for Collective Action*. Cambridge, Eng.: Cambridge University Press.
    1994   *Covenants, Collective Action, and Common-Pool Resources*. Workshop in Political Theory and Policy Analysis. Bloomington: Indiana University.
Powell, J., Z. Ball, and K. Reaume
    2000   Public involvement programs that support water quality management. Pp. 214-221 in *Proceedings of the Conference on Tools for Urban Water Resource Management and Protection*. Cincinnati, OH: U.S. Environmental Protection Agency, Office of Research and Development.
President's Commission on Americans Outdoors
    1986   *A Literature Review*. Washington, DC: President's Commission on Americans Outdoors.
President's Council on Sustainable Development
    1996   *Sustainable America: A New Consensus for Prosperity, Opportunity, and a Healthy Environment for the Future*. Washington, DC: U.S. Government Printing Office.
Quigley, B.A.
    1997   The role of research in the practice of adult education. In *Creating Practical Knowledge Through Action Research: Posing Problems, Solving Problems, and Improving Daily Practice*, B.A. Quigley and G.W. Kuhne, eds. San Francisco: Jossey-Bass.
Renn, O., and R. Finson
    1991   *The Great Lakes Clean-up Program: A Role Model for International Cooperation?* Florence: European University Institute.

Renn, O., T. Webler, and P. Wiedemann

1995a *Fairness and Competence in Citizen Participation*. Boston: Kluwer Academic.

1995b The pursuit of fair and competent citizen participation. In *Fairness and Competence in Citizen Participation*. O. Renn, T. Webler, and P. Wiedemann, eds. Boston: Kluwer Academic.

Rocha, E.M.

1997 A ladder of empowerment. *Journal of Planning Education and Research* 17:31-44.

Rogers, E.M.

1995 *Diffusion of Innovations*, 4th ed. New York: Free Press.

Ruskey, A., and R. Wilke

1996 *Promoting Environmental Education: An Action Handbook for Strengthening EE in Your State and Community*. National Environmental Education Advancement Project (NEEAP). Stevens Point: University of Wisconsin.

Sargent, F.E.O., P. Lusk, J.A. Rivera, and M. Varela

1991 *Rural Environmental Planning for Sustainable Communities*. Washington, DC: Island Press.

Selin, S., and D. Chavez

1995 Developing a collaborative model for environmental planning and management. *Environmental Management* 19(2):189-195.

Sexton, K.

1996 Environmental justice: Are pollution risks higher for disadvantaged communities? *Health & Environment Digest* 9(9):73-77.

Sidel, V., B. Levy, B. Johnson

1996 Environmental injustice: What must be done? *Health & Environment Digest* 9(9):79-89.

Simmons, D., et al.

1999 *Excellence in Environmental Education–Guidelines for Learning (K-12)*. Rock Spring, GA: North American Association for Environmental Education.

Singh, K., and V. Ballabh

1994 *Role of Leadership in Cooperative Management of Natural Common Pool Resources: A Collective Goods Theoretic Perspective*. Working Paper No. 50. Anand, Gujarat, India: Institute of Rural Management.

Sorenson, D.

1985 Organizing an information program for nonpoint pollution control. *Journal of Soil and Water Conservation* 40(1):82-83.

Stern, P.C.

2000 Toward a coherent theory of environmentally significant behavior. *Journal of Social Issues* 56(3):407-424.

Thoms, C., and E. Andrews

2000 Building Capacity, Community-Based Environmental Education in Practice. In US Environmental Protection Agency/Cooperative Extension Partnerships, No. 8. [Online]. Available: http://www.uwex.edu/erc/pdf/EPA8.pdf [Accessed: March 6, 2002].

U.S. Environmental Protection Agency

1997 *Community-Based Environmental Protection: A Resource Book for Protecting Ecosystems and Communities*. EPA 230-B-96-003. Washington, DC: U.S. Environmental Protection Agency.

U.S. General Accounting Office

1995 *Agriculture and the Environment: Information on and Characteristics of Selected Watershed Projects*. Report to the Committee on Agriculture, Nutrition, and Forestry, U.S. Senate. GAO/RCED-95-218. Washington, DC: General Accounting Office.

University of Wisconsin Extension
    1998   *Community Group Member Survey* G3658-9 and *Evaluating Collaboratives: Reaching the Potential,* G3658-8. Madison, WI: Cooperative Extension Publications.
Vari, A.
    1995   Citizens' Advisory Committee as a model for public participation: A multiple-criteria evaluation. In *Fairness and Competence in Citizen Participation.* O. Renn, T. Webler, and P. Wiedemann, eds. Boston: Kluwer Academic.
Wallin, P., and R. Haberman
    1992   *People Protecting Rivers: A Collection of Lessons from Successful Grassroots Activists.* Portland, OR: River Network.
Walzer, N., S.C. Deller, H. Fossum, G. Green, J. Fruidl, S. Johnson, S. Kline, and D. Patton
    1995   *Community Visioning/Strategic Planning Programs: State of the Art.* Ames: Iowa State University, North Central Regional Center for Rural Development.
Weintraub, B.A.
    1995   Defining a fulfilling and relevant environmental education. *Urban Education* 30(3):337-366.
Westwater, A., and P. Wolfe
    2000   The brain-compatible curriculum. *Educational Leadership - The Science of Learning* 58(3):49-52.
Whitehead, A.N.
    1929   *The Aims of Education.* New York: New American Library.
Wise, G.
    1998   Applying U.S. community development process lessons. Appendix A in An EPA/USDA Partnership to Support Community-Based Education. 910-R-98-008. Unpublished manuscript. University of Wisconsin Environmental Resources Center, Madison.
Wise, M., and L. Kenworthy
    1993   *Preventing Industrial Toxic Hazards - A Guide for Communities.* New York: INFORM.
Wondolleck, J.M., and S.L. Yaffee
    2000   *Making Collaboration Work: Lessons from Innovation in Natural Resource Management.* Washington, DC: Island Press.

11

Community Environmental Policy Capacity and Effective Environmental Protection

Daniel Press and Alan Balch

Many of our current public policy debates are variations on these age-old questions: Is it better to regulate . . . through mandatory standards or through voluntary guidelines and individual discretion? Should social welfare programs be centralized, with uniform standards applying to all the states, or would decentralization allow local officials to apply their knowledge of local circumstances in ways that would make for better policy? (Stone, 1997:238)

For the past 30 years, federally based command-and-control regulation and, to a lesser degree, market-based incentive approaches have been the primary focus of U.S. environmental policy. Extensive experience with, and analyses of, such efforts reveal strengths and weaknesses in both. Scholars and policy analysts are giving new attention to different policy paths such as the devolution of authority and/or responsibility from federal and state authorities to local communities (Sabel et al., 2000; Vig and Kraft, 2000). The political result is a growing effort to shift away from a federal command-and-control paradigm toward more community-specific approaches that are based on local decision making and that create opportunities for collaboration among agencies, local governments, industry, nongovernmental organizations (NGOs), and citizens.

In this chapter, we focus on such community-based environmental protection measures, beginning with some working definitions, then moving on to a framework for understanding the factors in and around a community that shape both its responses to environmental problems and the effectiveness of those responses. We illustrate this framework with research on local open space preservation and recycling activity in California.

WHAT ARE COMMUNITY-BASED ENVIRONMENTAL
PROTECTION MEASURES?

"Community" includes actors inside and outside of local government, and thereby encompasses private citizens and companies, NGOs, and local government agencies. We use the term "community-based" to focus on local environmental protection activities and decisions that are driven *primarily* by local actors and institutions, although they may be reacting to or receiving support from wider regional, state, provincial, federal, or even international spheres. In some cases, the community (or some part thereof) decides to take up an issue voluntarily and determines what action, if any, to take. For example, no state or federal mandate requires local California communities to purchase open space, although a significant amount of local open space is purchased via community action. In other cases, local protection efforts may respond to an external governing body telling the community that it must act; however, the consequent actions can be considered locally based only if the community is given discretion to determine what type of action to take. In California, for example, the state required local jurisdictions to reduce solid waste disposal by 50 percent, but gave localities significant latitude to determine how to achieve those reductions.

To summarize, community-based efforts arise when communities are provided the option—or take the initiative—to fashion place-specific remedies to problems. What forms might such remedies take? Localities may take steps that are command-and-control oriented. However, many community-based attempts to address environmental issues are largely nonregulatory, often relying on extensive voluntarism.

Most community-based voluntary environmental measures can be grouped into one of four categories: (1) information gathering, (2) resource restoration or protection, (3) persuasion/endorsement, and (4) personal or lifestyle changes. Information activities span a range from applied research to monitoring and data collection on environmental health and quality (including biotic and abiotic assessments). Resource restoration and protection activities range from the fee-simple purchase of open space lands to one-time beach, creek, or park cleanups to long-term, multiyear revegetation and invasive exotic species removals.

Persuasion and endorsement efforts include political lobbying, campaigning or canvassing on local issues, and brokering collaborations or consensus on controversial environmental issues that arise between local actors. Personal and/or lifestyle changes encompass a variety of efforts such as water conservation, carpooling, recycling, composting, and energy conservation that require alterations in behavior and habits. Of course, any of these four actions can be promoted at the local level by local, state, provincial, and federal governments, and they may even be required rather than voluntary. They become community-based voluntary measures when they are discretionary in nature and the impetus for such action comes primarily from within the community

TABLE 11-1   Examples of Community-Based Voluntary Activities for Environmental Protection

Community Action	Program Details and Foci	Examples
Restoration and protection	Restoration of local areas important for wildlife habitat and natural systems	Save the Bay's Habitat Restoration Program works to restore critical Narragansett Bay habitats, beach vegetation (Save the Bay, 2002)
Information gathering	Monitoring of natural systems or agency processes to track change or flag problems	Sacramento Tree Foundation monitors the spread of Dutch Elm disease (Sacramento Tree Foundation, 2002)
Persuasion/ endorsement	Door-to-door advocacy on environmental issues and candidates; phone banks	Sonoma County Conservation Action: door-to-door grassroots organizing to mobilize letters to elected officials and familiarize voters with candidates' voting records (Sonoma County Conservation Action, (2002)
Lifestyle changes	Promotion of environmentally beneficial behaviors and actions	Citizens in Los Angeles can attend free "Smart Gardening Classes" offered by the city to promote backyard composting (City of Los Angeles, 2002)

The types of community-based groups that may promote these activities are as varied as the activities themselves. These may include local government agencies, environmental groups, schools, neighborhood associations, and local businesses. Table 11-1 provides examples of both voluntary activities and the local groups supporting them.

Most community-based groups actually engage in more than one of the activity types listed in Table 11-1. Some groups, especially those organized to protect large communities or regions, may engage in most or even all of the activities, or form a coalition of groups that combine tasks. For example, the Lake Michigan Federation "works to restore fish and wildlife habitat, conserve land and water, and eliminate toxics in the watershed of America's largest lake. We achieve these through education, research, law, science, economics, and strategic partnerships" (Lake Michigan Federation, 2002).

Even small local groups who focus primarily on one task, such as water quality monitoring, sometimes engage in other related tasks, as data from the U.S. Environmental Protection Agency's (EPA's) Office of Water suggest. This office collects data on volunteers who monitor water quality throughout the country. Of the 778 volunteer organizations the EPA surveyed, nearly a third engaged in just one activity (e.g., biological water quality monitoring or physical-chemical analysis) (U.S. Environmental Protection Agency, 2000). Only half of these groups added two or more major activities to their responsibilities (such as debris cleanup and restoration, storm-drain stenciling, and land use surveys).[1]

States varied with respect to how many volunteer watershed groups were active within their borders (from 1 to 58 groups per state). Moreover, groups varied in the number of activities they assumed beyond their primary water quality assessment tasks (from no additional tasks to five). We divided the 50 states into those with fewer than 11 groups statewide ("Low Group States") and those with more ("High Group States"). Volunteer organizations in Low Group States were, on average, no more likely to take on additional activities than organizations in High Group States ($t = 1.14$, $p < 0.05$, df $= 28$). At this stage in the development of water quality NGOs, it is thus unlikely that the relative absence of volunteer groups spurs existing groups to take on a wider range of activities.

POLICY CAPACITY FOR COMMUNITY-BASED ENVIRONMENTAL MEASURES

One way to study the transformation from past practices to more sustainable, scientific, ecosystem-based management practices is to compare systematically the communities that are implementing sustainable land use with those that are not and try to isolate the key variables that account for the differences (Mazmanian and Kraft, 1999:297).

Do community-based efforts, especially those that are voluntary, result in positive environmental outcomes? What explains the variations in community response and performance on key environmental issues across California? We will offer insights into these questions through examples and results from separate studies on open space preservation and solid waste diversion in California, both of which are largely grounded in community-based efforts. Despite a wealth of efforts across the state, some cities and counties have been significantly more successful in these areas than others.

To guide research into the conditions shaping community willingness and ability to implement effective environmental measures on a local level, we rely on a policy capacity model (Press, 1998; Boyne, 1985; Ringquist, 1993). A successful policy capacity model for explanatory and heuristic purposes should identify all the theoretically plausible independent variables, then explain the mechanisms by which each variable potentially could affect environmental outcomes. The model we present is based on the oft-repeated observation that some communities are more capable of mounting environmental protection activities than others. In places with more environmental policy success, our model suggests a positive feedback loop whereby a community strongly supports certain environmental protection measures, which translates into further support from local leaders and generates the political, economic, and technical resources necessary for sustaining and implementing environmental programs. Organized efforts to enact and implement such programs in turn may rely on members of the general community for support, cooperation, and participation. If the community is supportive of environmental protection in the first place, then its members are likely to respond positively to environmental protection efforts and take voluntary action if called on to facilitate the success of such efforts.

We are not suggesting that environmental attitudes automatically translate into certain environmental behaviors in all cases (see Schultz, this volume, Chapter 4). We are suggesting, however, that when a community organization or political entity decides to enact and implement a local environmental program (e.g., establishing a citywide greenbelt), that program is likely to be well received in communities with relatively high concentrations of environmentally concerned citizens. Indeed, it is probable that the presence of such citizens is part of the reason such action is being taken in the first place (i.e., the actions of elected officials or community organizations are often reflections of popular demand). Moreover, if the effectiveness of such environmental programs hinges on widespread, voluntary citizen participation (e.g., recycling), then communities with environmental predilections among the general populous are likely to encounter relatively high levels of program participation. Among the many possible mitigating factors in this attitude to action equation is the amount of effort required (Schultz and Oskamp, 1996; McKenzie-Mohr et al., 1995). Thus, community groups can provide citizens with capacity tools that facilitate the effort required to translate attitudes into action (e.g., providing curbside pickup or circulating a petition). In principle, the easier it is to participate and take action, the more likely it is that people will express their preferences through action (McKenzie-Mohr et al., 1995).

We refer to "environmental policy capacity" as a community's ability to engage in collective action that secures environmental public goods and services. Much like Putnam's (1993) conception of institutional performance, we envision a relatively simple model of policy capacity and performance. The model is integrative, relying on four general components that contribute to a community's environmental problem-solving ability. First, we consider the resources and constraints on local policy responses that are (a) *internal* to the community in question. These consist largely of a locality's sociopolitical, demographic, and economic characteristics (e.g., local revenues, demographics, income, political ideology, party identification, development pressure). Second, a community's policy response may be facilitated by or in response to (b) *external* factors, such as development pressure in neighboring jurisdictions or the nature of state, provincial, or federal mandates and funding. Third, a community's (c) *policy network* consists of the political mechanisms by which the external and internal factors translate into policy mobilization, formulation, and implementation regarding a particular issue (Ringquist, 1993). This category consists of the various public and private institutions and actions that potentially could play a role in crafting and/or implementing the social and political choices made by a community. Finally, the types of (d) *policy outputs* generated by these three factors will dramatically affect a community's ability to achieve the desired levels of environmental protection.

Each of these four components may affect environmental outcomes through a number of venues. Internal influences, for example, may come from the government in the form of tax revenues, from civil society in the form of political ideology, from markets in the form of development pressure, or from the environment in the form of geographic features. Thus, each of the four categories can be divided further into at least two or three subcategories based on the question of who is responsible for the action within that category. Is it the government, in its execution of official duties? Is it civil society, in its pursuit of collective action? Is it the free market, in its pursuit of profits and wages? These four general categories (and their subcategories) combine to determine a community's capacity to address environmental issues. Table 11-2 provides a breakdown of these categories and subcategories in addition to examples of each.

How do these four components come together to create a model useful for the study of politics and society? Figure 11-1 presents the four categories and their potential relationships to one another in the context of local policy choices. Because policymaking is an evolving and dynamic process, we explore the role played by each of these factors and how they may combine to determine local environmental policy capacity.

Because local policy capacity exists in a particular setting (city, county, region) during a given time period (i.e., a particular decade), it is subject to both (a) internal and (b) external constraints. These internal and external factors not only influence each other (e.g., the state or provincial economy affects the local economy), but they can separately or simultaneously shape the makeup and response of (c) local policy networks, especially by molding or changing the relationship between local desires and local expectations. As an example of the impact of internal factors, consider a wealthy city with a high degree of environmentalism.

Local civic environmentalism reflects public expectations concerning the provision of environmental goods. Collective norms strongly influence factors within the policy network, such as the political ideology of elected officials and the focus of local interest groups. In the model of a community with strong environmental policy capacity, the social norm is to expect a high level of environmentally sound individual behavior and institutional performance. Such a city likely will have in place the elected officials and organized interest groups (i.e., entities found in the policy network) to address environmental issues.

TABLE 11-2 Policy Capacity Categories

Category	Subcategories	Examples
(a) Internal influences	Government	Tax revenues
		Current regulations
		Government type and/or size
	Civil society	Demographics
		Political ideology
		Party identification
		Environmental values/support/knowledge
		Civic environmentalism/voluntarism
	Environmental	Landscape features
	Market	Development pressure
		Taxable sales
		Employment
(b) External influences	Government	State/provincial and/or federal grants
		State/provincial and/or federal mandates
		State/provincial and/or federal fines
	Civil society	State/provincial and federal nongovernmental organizations
		Private foundation grants
	Environmental	Droughts
		Floods
	Market	Regional and state/provincial economic activity
		Regional development pressure
(c) Policy network	Government	Bureaucratic commitment
		Attention from elected officials
		Policy entrepreneurialism
		Administrative and technical expertise
	Civil society	Interest group activity and mobilization
		Local foundation funding
		Grassroots activism
	Market	Business advocacy groups
(d) Policy outputs	Government	Regulations and laws
		Programs
		Program staff and spending
		Grant requests
		Voluntary activity
		Resource restoration/protection
		Persuasion/endorsement
		Personal or lifestyle changes

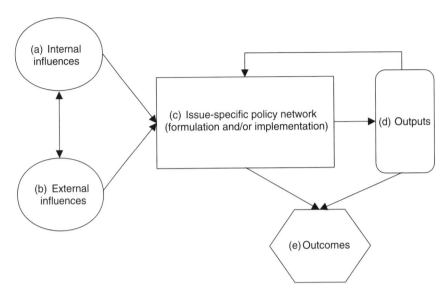

FIGURE 11-1 Local environmental policy capacity model.

Whereas local environmentalism may provide the political will for environmental action, local wealth can provide the fiscal resources required for effective community action. The community policy networks attempting to address environmental problems vary quite a bit in their ability to raise local funds, either because they attract a different tax base or because they vary in public support for ballot-box financing of bonds, taxes, and fees.

That same policy network also may be influenced by external factors, such as funding for environmental programs or environmental mandates from state, provincial, and federal governments, or perhaps by state, provincial, and national interest groups with local chapters. Moreover, local desire for environmental protection may be tempered by low or heavily encumbered tax revenues or by state, provincial, and federal limitations on taxation (proposition 13)[2] or restrictions on land use regulation (such as those shaped by federal takings cases).[3] External environmental and market factors also may play a role, such as pollution or development pressure from a nearby city that prompts local concern among the public and/or within the local policy network.

The factors included in (a) and (b) provide constraints and opportunities, which set the context for political action within the (c) policy network. Thus, internal factors such as political ideology, income, and party identification all can affect local policy responses, but only through the influence they have on the "intervening political mechanisms" that shape policy choices (Ringquist, 1993;

Boyne, 1985). Such intervening can include interest group activity (Dahl, 1956; Sabatier and Jenkins-Smith, 1993; Ringquist, 1993), policy entrepreneurialism (Schneider et al., 1995; Mintrom, 2000; Kingdon, 1995), and stakeholder partic- ipation (Mazmanian and Sabatier, 1989).

In the local policy network, civic and government attention to an issue can translate into social choices and action targeted at that issue. Such efforts com- monly are referred to as (d) policy outputs and can include spending, regulations, hearings, new programs, and new laws (see Figure 11-1). The four types of voluntary action described previously also fall under the policy output umbrella. In some cases, outputs can lead directly to collective environmental goods and services ("on the ground" [e] outcomes). For example, a local land trust may partner with a city agency to purchase a particular parcel of local open space. However, turning policy outputs into successful outcomes, especially when rely- ing on voluntary actions, can be far more complicated. Many outcomes require sustained attention from a (c) policy network, which may be similar to the one that generated the output, or may be a completely different policy network, or may be a mixture of both. A city's recycling program requires citizen participa- tion; a county's carpool program may rely on support from local businesses; an environmental group's habitat restoration program relies on membership partici- pation; a bond measure for open space acquisition funds may hinge on voter approval. These examples are just a small sample of the various types of outputs that rely on a network of private and/or public stakeholders and target groups that are responsible for turning environmental outputs into successful outcomes (Mazmanian and Sabatier, 1989). This observation is especially true for volun- tary programs.

To summarize, internal and external factors influence each other and the policy network; the policy network translates social attention into effort. De- pending on the type of action taken, outputs can result in immediate environ- mental protection, or such protection may hinge on the support, participation, and cooperation of community members. Based on this model, one would ex- pect policy capacity in the environmental context to be highest where:

- Environmental conditions and problems are locally visible.
- Local budgetary, technical, and administrative resources are relatively high.
- Community expectations of, and desire for, institutional performance in environmental protection are high.[4]
- Political leaders sustain a commitment to environmental policy and pro- grams.

Results from our California local open space preservation and solid waste diversion studies confirm many of these expectations.

LOCAL OPEN SPACE PRESERVATION IN CALIFORNIA

Even though slowing growth now enlists nationwide voter support, sprawl has been left for local communities to address, and these are under tremendous pressure to develop open space in order to secure tax revenues. With ever-increasing populations and a political economy dependent on growth, how can people in some places in California manage to preserve open space? More specifically, what are the conditions for creating innovative, effective land preservation institutions at the local level?

The Community and Conservation in California study[5] investigates these questions by first assessing the extent of local acreage acquired by cities, counties, special districts, and land trusts in California. Doing so reveals that communities have acquired a little over a million acres of valuable open space, mostly since 1950. These acreages are very unevenly distributed. The study explores this variation by analyzing county-level policy capacity. Accordingly, we gathered data on civic environmentalism (through a telephone survey of 4,100 residents), voting on statewide environmental policy measures (through state records), local fiscal and administrative resources (city, county, and special district revenues; planning and geographic information system [GIS] resources), and development pressure and landscape features (using new housing starts, topography, and river density).

As the policy capacity model suggests, internal community factors may influence local policy outputs and environmental outcomes. Development pressure and compelling landscapes stimulate concern for preservation (Figures 11-2 and 11-3); fiscal resources enable land acquisition (Figure 11-4), and civic engage-

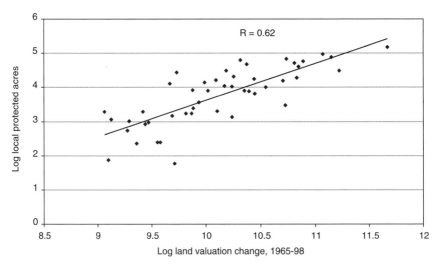

FIGURE 11-2 Land valuation change versus local protected acres.
Source: California State Controllor's Office (1965-1998).

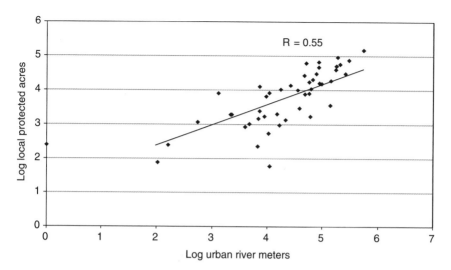

FIGURE 11-3 Urban rivers versus local protected acres.
Source: California Spatial Information Library (2000).

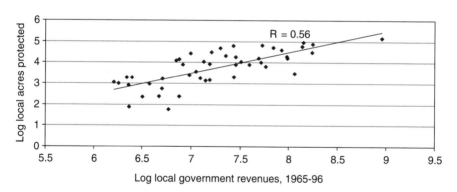

FIGURE 11-4 Local government revenues versus local acres protected.
Source: California Department of Finance (1965/1996).

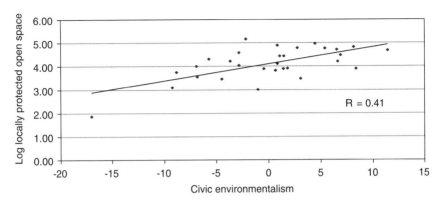

FIGURE 11-5 Civic environmentalism versus local acres protected.
Source: Press (2002).

ment[6] provides the values and political support necessary to mobilize local action (Figure 11-5; for the full study, see Press, 2002).

SOLID WASTE DIVERSION

In 1989, California passed legislation designed to promote a dramatic shift in local solid waste management. The Integrated Waste Management Act (AB 939) mandated that every city and county across the state achieve a 50-percent reduction in landfilled solid waste by 2000. Communities were given significant latitude in determining the most appropriate paths for achieving the required diversion levels. Cities and counties across California were (and continue to be) under intense political pressure to divert waste from landfills, and they all implemented various programs in response. Not surprisingly, communities have varied in their abilities to actually translate diversion efforts into outcomes by recovering materials from the waste stream. The scientific advantage of investigating these variations in California is that relatively equal pressure was applied to communities across the state to achieve the same 50-percent diversion goal. One way of measuring these variations in success at the local level is by looking at the amount of recyclable material diverted from landfills. We used 1999 county-level recycling tonnage data provided by the state's Department of Conservation as an environmental outcome measure. Unfortunately, the state only collects recycling data for a handful of items: glass containers, aluminum containers, plastic containers, and some paper fibers. Communities can and do collect a wider variety of materials for recycling. However, because the items included in this data set are among the most common materials collected by curbside and other local recycling programs, they should provide a reliable (albeit limited) indication of recycling levels.

To control for variations in population size across counties, we used recycling tonnage per person as the dependent variable.   For the independent variables, we considered a variety of internal factors that could explain variations in local recycling capacity.   For example, we measured community interest in and support for environmental protection (i.e., local civic environmentalism) using county percentage of registered Green Party voters in 1999 and county average vote on statewide environmental measures from 1998-2000 (see Figure 11-6).[7]

Both voting and party registration are acts of civic responsibility. However, although voting for an environmental ballot measure is undoubtedly an act of environmental support, registering as a Green Party member is more a statement of affiliation than an environmental act.   Our choice to use this variable as a surrogate for local civic environmentalism was based on an assumption that such voters perceive environmental issues as among the most important to them politically, and such voters are likely to engage in various forms of environmental activism.   Thus, a higher percentage of registered Green Party voters in a county could translate into a higher level of support for, participation in, or emphasis on environmental issues.

A multiple regression of these variables on per capita recycling data for 57 of California's 58 counties suggests that those counties with a high degree of environmental support and interest also have high levels of per capita recycling ($R^2 = 0.55$; $p < 0.0001$). An index of broader internal civic engagement based on these measurements in addition to survey data for 30 California counties (collected through the Community and Conservation Study) also proved a reason-

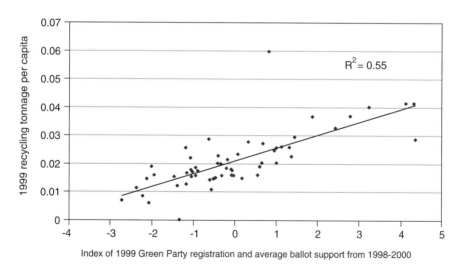

FIGURE 11-6 Environmental support versus recycling per capita.
Sources: California Department of Conservation (1999); California Secretary of State (1999).

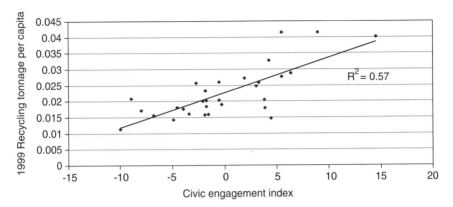

FIGURE 11-7 Civic engagement versus recycling.
Sources: California Department of Conservation (1999); California Secretary of State (1999); and Press (2002).

ably strong predictor of per capita recycling levels (R^2= 0.57, p< 0.0001).[8] Figure 11-7 contains a scatterplot of these results.[9]

SOLID WASTE, OPEN SPACE, AND THE
POLICY CAPACITY MODEL

Most of the independent variables explored in the two cases we provided would be classified as internal civil society, government, and environmental variables in the policy capacity model. How do such internal factors shape open space protection and recycling activity? One plausible explanation is that these factors enable a policy network of public and private institutions that reflect and pursue the community's interest in environmental issues. In other words, internal factors shape the policy network by supporting and in some cases creating important institutions and infrastructure capable of producing environmental outcomes. However, open space and recycling efforts may produce decidedly different roles for the community in terms of implementation. In the case of open space, preservation often requires minimal effort from the community at large beyond providing latent support for the institutions and actors pursuing such ends. Occasionally, these actors and institutions may turn back to the community for political and/or financial support at critical times (e.g., when voters must pass a bond measure). Community recycling efforts, on the other hand, often rely on large segments of the community on a daily basis in order to achieve success. Whether community members participate in recycling programs may hinge on many of the same internal factors that either limit or empower expansive community recycling programs in the first place, such as environmental attitudes and values and the amount of effort required for participation (Schultz and Oskamp, 1996; Stern and Dietz, 1994; Stern et al., 1993; Hopper and Nielsen, 1991; McKenzie-Mohr et al., 1995; Derksen and Gartrell, 1993). Thus in the case of recycling, civil society can have a significant impact on both policy formulation and implementation.

CONCLUSION

Logic and evidence concerning community-based voluntary measures suggest that this "third way" can make viable, important contributions to environmental protection under certain circumstances and in particular places. Over time, some communities develop expectations about collective environmental goods and the capacity to provide these goods through largely local efforts. Voluntary or discretionary community-based efforts often complement, extend, or leverage regulatory or incentive-based environmental policies. Indeed, vigorous community assistance can vastly enhance the programs implementing these policies. A good example comes from the many coastal zone watchdog organizations. Such groups often conduct water quality tests far more comprehensively and frequently than government officials could ever hope to mount on their own; they also extend government enforcement and patrolling of coastal waterways. Thus it would be a mistake to view community-based measures as somehow standing apart from command-and-control regulation or market incentives mandated by governments at all levels.

State, provincial, and federal agencies and policymakers can—and do— enhance community-based environmental voluntarism. Government can do so first, by enhancing the capacities of communities to translate local willingness into action. The returns on a few dollars of capacity-building can be huge. For example, a little time and effort on the part of some water district staff results in miles of stream cleanups on many weekends throughout the country. Second, agency officials who actively encourage and respect participation by volunteers and community groups benefit from not only from local activities that relieve their management burdens, but also from the wide, sustained political support that may follow. Finally, government can design traditional regulatory or incentive-based environmental policies with an eye to a role for community-based activities. For example, municipal waste diversion incentives would be far less attractive to urban residents, businesses, and industry in the absence of the many NGOs who routinely provide public education programs on recycling and reuse or perform free commercial waste audits.

Community-based voluntary activities not only get the work of environmental protection and restoration done, they extend governance over this important area to a much wider sphere than is possible when only two agents—say, polluters and regulators—are involved. Because government officials always will be underequipped to provide entirely adequate governance over environmental issues, community participation spreads the burden widely and provides insurance against, or compensation for, the shortcomings of traditional environmental management.

NOTES

1 For this analysis, we used the EPA's counts of watershed groups, which list the number of groups per state as well as counts of their activities. We included construction site inspections, pipe surveys, and human use and land use surveys in the category "land use surveys." We included debris monitoring and photographic surveys in the category "other surveys." Our thanks to Betsy Herbert for her assistance with the watershed data.

2 Proposition 13, passed in 1978, was a constitutional amendment passed by initiative. "Proposition 13 rolled back property tax assessments to 1975 levels, permitted an annual increase in assessment of only 2 percent except in the event of a sale, and, for all practical purposes, capped property-tax rates at 1 percent per year. (A higher rate requires a two-thirds vote, which is very difficult to obtain.) Since property tax rates at the time were approaching 2 percent in many parts of the state, Proposition 13 cut local government revenues dramatically"(Fulton, 1991:209).

3 See Lucas v. South Carolina Coastal Council, 505 U.S. 1003 (1992), Dolan v. City of Tigard, 512 U.S. 374 (1994).

4 Institutions here include administrative agencies, elected policymakers, and voluntary civic associations.

5 The Community and Conservation in California study was led by Daniel Press, Principal Investigator, University of California, Santa Cruz, with support from the EPA and the John Randolph Haynes and Dora Haynes Foundation (for the full study see Press, 2002).

6 We assessed civic environmentalism using survey and voting data, constructing an index measuring: (1) informational resources (knowledge of development or land use problems and conflicts, familiarity with land trusts), (2) financial resources in the form of willingness to pay for collective environmental goods (either as property taxes or indirectly as income tax for park bond issues), (3) participation in a wide variety of face-to-face activities, (4) NGO resources in the form of volunteer activity for civic and environmental causes, and (5) a county's average vote on statewide environmental measures, 1924-2000.

7 Both the Green Party data and the environmental ballot approval data were downloaded from the California Secretary of State's Web site. The Green Party variable was created by taking total registered Green Party voters in a county for 1999 and dividing that figure by the total number of registered voters in the county for the same year. The ballot measure variable is an average percentage of yes votes in the county on the four statewide ballot measures dealing with environmental issues between 1998 and 2000: 1998 Prop 4 (Animal Trap Ban), 1998 Prop 7 (Air Emissions Credits), 2000 Prop 12 (Parks and Water), 2000 Prop 13 (Water Conservation and Supply).

8 The civic engagement index was created by standardizing and then combining results from several different categories: (1) environmental values (1999 Green Party registration), (2) environmental liberalism (stated preference for increased governmental services and regulation to address environmental issues), (3) political mobilization (contacting public officials, volunteering time to political organization/candidate, and attending public meeting attendance), (4) environmental volunteerism, such as Adopt-Creek, and (5) environmental support (county's average vote on statewide environmental measures, 1998-2000).

9 It is important to note that these results are preliminary and part of a larger research analysis that will include a variety of demographic, economic, and political variables and their possible relationships to disposal and recycling levels. Thus, the results presented here must be viewed with caution because a variety of factors that could skew the results, many of which will be taken into consideration in the complete analysis, were not explored fully in these preliminary findings. For

example, certain national, state, provincial, and even regional policies and characteristics could be responsible for some of the variations in per capita recycling at the county level. How a county is structured economically also could affect recycling levels.

REFERENCES

Boyne, G.A.
    1985   Theory, methodology and results in political science - The case of output studies. *British Journal of Political Science* 15:473-515.

California Department of Conservation
    1999   Unpublished data provided to authors. Sacramento: State of California Department of Conservation. [Online]. Available: http://www.consrv.ca.gov [Accessed 3/5/02].

California Department of Finance
    1965/   *California Statistical Abstract.* Sacramento: State of California State Controller's
    1996   Office.

California Secretary of State
    1999   *California Voter Registration Statistics as of October 5, 1999.* Updated Nov. 29, 1999. Sacramento: State of California Secretary of State. [Online]. Available: http://www.ss.ca.gov/elections/ror/ror_1099.htm [Accessed 3/5/02].

California Spatial Information Library
    2000   Unpublished data provided to authors. Sacramento: State of California Spatial Information Library. [Online]. Available: http://www.gis.ca.gov [Accessed: 3/5/02].

California State Controller's Office
    1965-   *Assessed Valuation Annual Report.* Sacramento: State of California State Controller's
    1998   Office.

City of Los Angeles
    2002   *Smart Gardening Classes.* [Online]. Available: http://www.lacity.org/san/sancpost.htm [Accessed 3/5/02].

Dahl, R.
    1956   *A Preface to Democratic Theory.* Chicago: University of Chicago Press.

Derksen, L., and J. Gartrell
    1993   The social context of recycling. *American Sociological Review* 58(June):434-442.

Fulton, W.
    1991   *Guide to California Planning.* Point Arena, CA: Solano Press.

Hopper, J., and J. Nielsen
    1991   Recycling as altruistic behavior. *Environment and Behavior* 23(2):195-220.

Kingdon, J.W.
    1995   *Agendas, Alternatives, and Public Policies.* 2nd ed. New York: HarperCollins College Publishers.

Lake Michigan Federation
    2002   *About LMF.* [Online]. Available: http://www.lakemichigan.org/about/default.asp [Accessed 3/5/02].

Mazmanian, D.A., and M.E. Kraft, eds.
    1999   *Toward Sustainable Communities: Transition and Transformations in Environmental Policy.* Cambridge, MA: MIT Press.

Mazmanian, D.A., and P.A. Sabatier
    1989   *Implementation and Public Policy.* Lanham, MD: University Press of America.

McKenzie-Mohr, D., L.S. Nemiroff, L. Beers, and S. Desmarais
    1995   Determinants of responsible environmental behavior. *Journal of Social Issues* 51(4):139-156.

Mintrom, M.
  2000 *Policy Entrepreneurs and School Choice.* Washington, DC: Georgetown University Press.
Press, D.
  1998 Local environmental capacity: A framework for research. *Natural Resources Journal* 38(Winter):29-52.
  2002 *Saving Open Space: The Politics of Local Preservation in California.* Berkeley: University of California Press.
Putnam, R.D.
  1993 *Making Democracy Work.* Princeton, NJ: Princeton University Press.
Ringquist, E.J.
  1993 *Environmental Protection at the State Level: Politics and Progress in Controlling Pollution.* Armonk, NY: M.E. Sharpe.
Sabatier, P.A., and H.C. Jenkins-Smith
  1993 *Policy Change and Learning: An Advocacy Coalition Approach.* Boulder, CO: Westview.
Sabel, C.F., A. Fung, and B. Karkkainen
  2000 *Beyond Backyard Environmentalism.* Boston: Beacon Press.
Sacramento Tree Foundation
  2002 *Save the Elms Program.* [Online]. Available: http://www.sactree.com/about/#step [Accessed 3/5/02].
Save the Bay
  2002 *Save the Bay's Community-Based Restoration Program.* [Online]. Available: http://www.savesfbay.org/cbrmain.html [Accessed 3/5/02].
Schneider, M., P. Teske, and M. Mintrom
  1995 *Public Entrepreneurs.* Princeton, NJ: Princeton University Press.
Schultz, P.W., and S. Oskamp
  1996 Effort as a moderator of the attitude-behavior relationship: General environmental concern and recycling. *Social Psychology Quarterly* 59(4):375-383.
Sonoma County Conservation Action
  2002 *Effective Organizing for Sonoma County's Environment.* [Online]. Available: http//www.conservationaction.org/effect.htm [Accessed 3/5/02].
Stern, P.C., and T. Dietz
  1994 The value basis of environmental concern. *Journal of Social Issues* 50(3)65-84.
Stern, P.C., T. Dietz, and L. Kalof
  1993 Value orientations, gender, and environmental concern. *Environment and Behavior,* 25(3):322-348.
Stone, D.
  1997 *Policy Paradox: The Art of Political Decision Making.* New York: W.W. Norton.
U.S. Environmental Protection Agency, Office of Water
  2001 *National Directory of Volunteer Monitoring Programs.* Updated October 12, 2001. [Online]. Available: http://yosemite.epa.gov/water/volmon.nsf [Accessed 3/5/02].
Vig, N., and M. Kraft
  2000 *Environmental Policy in the 1990s.* 4th ed. Washington, DC: CQ Press.

12

Changing Behavior in Households and Communities:
What Have We Learned?

Paul C. Stern

C hapters 3 to 11 examine the use of what have been called communication and diffusion instruments (Kaufmann-Hayoz et al., 2001) to change environmentally significant behavior in households and communities. These instruments include information, education, the use of models, other informal social influences, and other interventions that rely primarily on language and visual symbols. Communication and diffusion instruments are used to supplement the traditional policy instruments of regulation (command and control), economic influence, and the provision of infrastructure and services to make desired behaviors more feasible. They are the centerpiece of social marketing efforts in environmental policy (McKenzie-Mohr and Smith, 1999).[1]

The key policy questions about these instruments are how much they can contribute to environmental protection objectives and how best to use them to achieve this potential. As chapters 3 to 11 indicate, much has been learned about how to design these instruments for greatest effectiveness and about what they can be expected to accomplish, both on their own and combined with other policy instruments.

THE POTENTIAL OF COMMUNICATION AND
DIFFUSION INSTRUMENTS

The potential of any policy instrument depends on its fit with the policy objective. An instrument has the greatest potential when it can provide just what is needed to overcome the barriers to attaining the objective. For example, regulations distinctively provide assurance of fairly equal compliance across target firms. Thus, they have great potential value when the firms would comply voluntarily, except for the concern that they might be put at a competitive disadvantage. When there are major barriers to the desired behavior that a policy instrument cannot remove, that instrument has very limited potential. Thus, regulations have limited effect when they call for changes that are technologically or economically infeasible.

Figure 12-1 identifies causal links between the types of policy instruments identified by Kaufmann-Hayoz et al. (2001) and a range of factors that in turn influence environmentally significant behavior. Although informed observers will disagree on which of these links are most important, available knowledge strongly supports the key point: Each type of policy instrument has particular capabilities and thus can influence only a subset of the many factors that drive behavior. Depending on what is standing in the way of a target behavior, a particular instrument may be highly successful or nearly useless. Communication and diffusion instruments, as shown in Figure 12-1, can influence some aspects of the target individuals and their immediate social contexts, but cannot directly affect the broader social, economic, or technological contexts.[2] They cannot make inconvenient behaviors convenient, make expensive behaviors inexpensive, or remove institutional or legal barriers to behavioral change. They often cannot even get people to put environmental actions high enough on their personal to-do lists to get them done, even if they are convinced to act. Environment-related actions must compete with other demands on a person's time and energy. It follows that when such contextual factors stand in the way of a target behavior, communication and diffusion measures by themselves will have little effect. Similarly, when the target behavior is seriously impeded by lack of information, social support, behavioral models, and the like, regulatory and economic instruments by themselves may have little effect.

These points may seem self-evident, but they have not always been reflected in the design of environmental policies and programs. Many documented failures of environmental and energy information programs in the household sector can be attributed in part to a failure to address significant noninformational barriers to behavioral change (see, e.g., National Research Council, 1984; Gardner and Stern, 1996; Lutzenhiser, this volume, Chapter 3; Schultz, this volume, Chapter 4). Similarly, the disappointing performance of many financial incentive programs targeting these behaviors can be attributed in part to a failure to diffuse the programs adequately (e.g., Stern et al., 1986). The most effective interventions tend to combine various types of communication and diffusion instruments with each other and with other policy instruments (Gardner and Stern, 1996; McKenzie-Mohr and Smith, 1999).

The implication for communication and diffusion instruments is that they have their greatest potential under two sets of conditions. In the first, the factors that communication and diffusion can influence (see Figure 12-1) are the only important barriers to the desired behavioral change. Under these conditions, well-designed communication and diffusion programs can bring about important

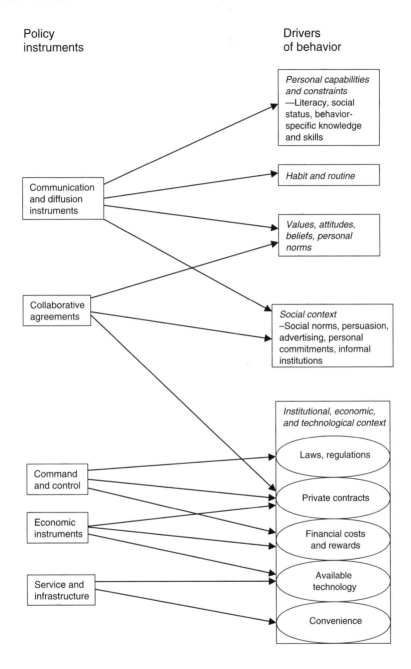

FIGURE 12-1 Paths of influence of five types of environmental policy instruments on five factors that affect environmentally significant behavior.

behavioral change without the aid of other policy instruments. In the second set of conditions, the barriers include both factors that communication and diffusion can influence and factors they cannot—but other policy instruments are available to remove the other barriers. Under these conditions, combining communication and diffusion instruments with the other policy instruments can bring about important behavioral changes that neither policy type alone could achieve. Communication and diffusion instruments are thus important as adjuncts to or partners with other policy tools.

The next section summarizes knowledge about how to design communication and diffusion instruments. Applying this knowledge is essential for the instruments to work well under either of the sets of conditions already described, although under the second set of conditions, communication and diffusion tools are not enough, no matter how well designed. The subsequent section discusses the application of communication and diffusion tools in situations where they are not sufficient for attaining policy objectives.

DESIGNING COMMUNICATION AND DIFFUSION
FOR GREATEST EFFECT

Chapters 3 to 7 summarize current knowledge about how to design communication and diffusion instruments to be as effective as possible. Chapters 3 to 5 cover the most carefully studied applications in environmental policy. Chapters 6 and 7 summarize knowledge from well-studied domains outside environmental policy where there is a long history of research on communication and diffusion instruments. They arrive at conclusions quite consistent with those mentioned in Chapters 3 to 5. The generalizations that receive the most consistent support across domains are described in the following subsections.

Design the Intervention from the Behaver's Perspective

Environmentally significant behavior is a product of the individual and the situation; more specifically, it is a product of the individual's values and attitudes, personal capabilities and constraints, and habits and routines, as well as of contextual factors that provide incentives, possibilities, and constraints (Stern, 2000). Because these things vary with the individual, successful efforts to change behavior are those that are matched to the individual's needs. This does not mean that effective communications must be individualized. Programs can succeed by targeting types of people whose situations are similar with regard to a target behavior or by being so multifaceted that they can be effective across a variety of people and behavioral contexts. To pursue either approach effectively, however, program designers must make explicit efforts to understand the behaver's perspective. This can be done by employing social research techniques (e.g., surveys, focus group techniques, ethnographic methods) and by involving members of the target group—or other people who have detailed experience-based understanding of the target audiences—in the design of the program (e.g., Gardner and Stern, 1996; Werner and Adams, 2001).

Build on Interpersonal Communication

Impersonal communication efforts, such as mass mailings and mass media advertising, are easy for policy makers to organize at large scales, but have much less influence on any individual than personal communication that comes from people the target individual cares about or trusts. Personal communication is especially important when elements of the message are controversial or when the original source of the message (e.g., a government agency) has limited credibility with portions of the audience (National Research Council, 1984).

Devising ways to gain the benefits that can come from interpersonal communication can take ingenuity, especially with large-scale policy objectives. One useful strategy is to induce respected leaders or people central to communication networks to adopt a desired behavior and thus act as *models* whose behavior may be readily adopted by others (Rogers, 1995; Valente and Schuster, this volume, Chapter 6). Another strategy is to partner with community groups and voluntary associations that can act as *intermediaries* who convey messages between policy makers and target individuals. Such groups often can make personal contact and can command a level of attention and trust from their constituencies that mass appeals rarely achieve. These groups are not simply channels for transmitting messages. They are most effective when they adopt the intervention as their own, perhaps adapting the message in the process to make it meaningful to their constituencies. A third strategy is to make existing *social norms* more visible, as Schultz (this volume, Chapter 4) did in an experimental manipulation with curbside recycling. This approach is most promising in situations in which, as with curbside recycling, the expectations and opinions of others matter and those others can monitor the relevant behavior.

Use Multiple Channels to Communicate the Message

As a rule, messages are most influential when they reach audiences in many forms and from many sources (Mileti and Peek, this volume, Chapter 7; Valente and Schuster, this volume, Chapter 6). This is probably the case because different people attend to and trust different sources, because different channels may have advantages for conveying different parts of the message, and because multiple channels provide an effective way to repeat and reinforce messages.

Apply Psychological Principles for Message Design

Messages are most effective when presented in terms, metaphors, and images the audience understands and finds attractive. When they involve calls for action, they can be made more effective by emphasizing the costs or dangers of inaction—but only when they also provide clear advice on what to do to avoid those hazards, thus giving audience members a sense that they can control their fates rather than creating fear and anxiety (Rogers, 1983; Weinstein and Sandman, 1992; Gardner and Stern, 1996). Useful summaries of the research on message design can be found in Chapters 6 and 7 of this volume and elsewhere (e.g., McKenzie-Mohr and Smith, 1999; Morgan et al., 2002). [3] As already noted, effective message design depends on understanding the target audiences and how they perceive the target behavior.

Maintain a Program's Momentum

Experience with communication and diffusion efforts indicates that programs maintained over long periods can be much more effective than one-shot or short-term programs (disaster preparedness and public health provide good examples, as noted in Chapters 6 and 7). Repetition helps messages sink in and increases the likelihood that a message will be received when the recipient is receptive, such as during a crisis or near the time of a noncrisis decision.

Set Realistic Expectations

Communication and diffusion instruments take time to be effective. A message must get into awareness and penetrate into a decision process in order to bring about behavioral change (Thøgerson, this volume, Chapter 5; the Knowledge-Attitude-Practice curves described by Valente and Schuster, this volume, Chapter 6). In addition, action may be delayed even when someone is psychologically prepared to change. For example, a household will acquire a more energy-efficient motor vehicle only when the time comes to change vehicles. Someone may not change an old habit until the right occasion arises (e.g., reconsidering the use of mass transit when one's work location changes). Because of such predictable delays, communication and diffusion instruments should be evaluated against a behaviorally defensible timetable for progress. Expectations should also consider contextual factors that may limit the effect of understanding on behavior change. This point is discussed further in the next section.

Continually Evaluate and Modify Programs

Policy interventions should not be expected to be at their best the first time they are tried, nor to maintain a constant level of performance in a changing environment. They need to be evaluated and adjusted if they are to achieve their full potential.

USING COMMUNICATION AND DIFFUSION WITH OTHER TOOLS

As already noted, communication and diffusion instruments can have little direct effect on changing the institutional, economic, or technological contexts of environmentally significant behavior. Where these contexts are unfavorable, the best use of communication and diffusion is in conjunction with other policy tools that address the relevant contextual issues. Therefore, it is important to understand the context in order to find the best use of communication and diffusion tools.

One important kind of contextual influence is discussed by Press and Balch (this volume, Chapter 11). They argue that the effectiveness of all locally implemented environmental policy instruments, including communication and diffusion instruments, is contingent on the policy capacity of local institutions. Put more provocatively, their argument implies that no matter how well designed a community-based communication program may be, it will only be effective in certain kinds of communities. If a community is lacking in local finances, administrative expertise, civic involvement, and some other qualities, Press and Balch argue, implementation likely will fail. Communication and diffusion instruments need to be supplemented with, or to follow after, efforts at community capacity building.

In some contexts, communication and diffusion can be combined with other policy instruments for synergistic effect. A good historical example was the financial incentives used to promote energy efficiency in homes in the aftermath of the 1970s energy crises. Several U.S. electric utility companies offered such incentives, but the rate of acceptance was fairly low—apparently due in part to inadequate communication and diffusion efforts. Some programs, however, were 10 or more times as effective as others that offered identical incentives, but marketed them in different ways (Stern et al., 1986). Apparently, communication and incentives had complementary functions: communication drew attention to the programs (as indicated by requests for energy audits), and once consumers noticed, larger incentives increased acceptance of the financial incentives (see Figure 12-2). When incentives were large enough, communication and diffusion had a very large practical effect by getting consumers to consider the incentives (Stern, 1999).

Communication and diffusion may have similar synergistic effects with service and infrastructure instruments such as the provision of new public transit lines or curbside recycling services: These new services may not be well used unless they are well marketed. Available evidence suggests that the standard marketing strategy—simple information dissemination—is usually not enough. What is needed is to combine new services and infrastructure with communication programs designed according to the principles described in the previous section.

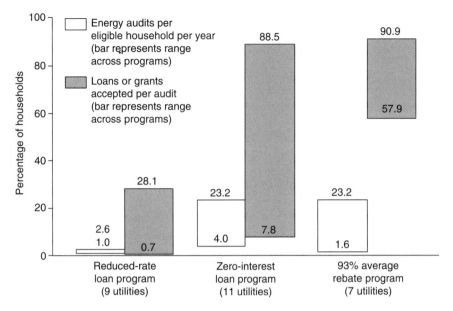

FIGURE 12-2   Households requesting energy audits (white bars) and accepting incentives once they have received audits (shaded bars) in three home energy conservation incentive programs.
Source: Stern et al. (1986). Reprinted with permission of Allyn and Bacon.

THE ROLE OF EDUCATION

Although education relies primarily on communication, it is different in its objectives from the kinds of social marketing efforts already described. When communication is used to achieve an environmental policy goal (i.e., for social marketing), its objective is to change the prevalence or frequency of a target behavior that directly affects environmental quality. In environmental education as defined by Ramsey and Hungerford (this volume, Chapter 9), communication is used to improve understanding of environmental and related phenomena and to enable and encourage environmental citizenship, but not normally to change specific behaviors that directly alter environmental conditions.[4] If students from a single environmental education class all became active in environmental lobbying, but took opposite sides on an issue, the class might be properly counted an educational success.

The effects of environmental education, defined in this way, on environmental quality are hard to assess. For one thing, they are mainly indirect, operating through public policy. For another, the effects on policy may not all be in the same direction. What good environmental education does for environmental policy is to raise public discussion to a higher level. Disagreements based on misinformation give way to those based on alternative interpretations of correct but ambiguous information or different judgments about what to do under environmental uncertainty. The policies that result from better informed debates are not predictable, but if democracy works well, they tend toward results that citizens want—including environmental results.

When environmental education is the self-education of adults, it can target a wider range of behaviors, as Andrews and colleagues point out (Chapter 10). A community that devises its own adult environmental educational program may begin with consensus on environmental objectives. For example, it may decide that water conservation is imperative and devise a program aimed at explaining local water supply conditions, showing why water conservation is necessary, and teaching people how to conserve water. Such a program may include elements of both education and social marketing and, if well designed, may greatly influence environmental outcomes. However, combining education and social marketing in this way can be highly controversial when it is proposed as a public policy strategy because of objections to government attempting proactively to influence the publics it is supposed to represent. This objection can be overcome if a legitimate public decision is made to adopt a social influence policy, as has been done in the United States for combating the use of illegal drugs and for driving under the influence of alcohol. A useful guideline for when it is appropriate to use education for social influence has been stated in another context by the National Research Council (1989:90): It is justifiable "only to the extent that some legitimate public process has culminated in a decision that using [education] to influence behavior serves an important public purpose." Community-based social marketing has the potential to meet this test and thus achieve wide acceptance.

CONCLUSION

Research has shown that communication and diffusion instruments can, under certain conditions, make significant contributions to meeting environmental policy objectives. It has identified a number of robust principles for designing these instruments to reach their potential. It has shown that these principles must be implemented in ways that are sensitive to the situation and that systematic evaluation is needed to achieve and maintain the fit of programs to settings. Research also has helped distinguish three kinds of situations: those in which communication and diffusion instruments can yield significant environmental effects on their own, those in which they have potential only when supplemented by other instruments, and those in which they are unlikely to be successful even when combined with other policies. Thus, it has helped to clarify the functions of communication and diffusion in the environmental policy toolbox and to show how they can be used to greatest effect.

The potential of communication and diffusion can be quantified only in relation to particular situations. In favorable situations for communication and diffusion alone, such as the 1970s energy crises, well-conducted communication programs have reduced household resource consumption by 10 to 20 percent for short periods, beyond what could be achieved by conventional information dissemination (Stern, 1992). In situations appropriate for combining communication and diffusion with other policies, the instruments have even greater potential. For example, an integrated residential energy-efficiency program in Hood River, Oregon, in the 1980s achieved nearly complete adoption of recommended energy-efficiency improvements throughout the community, a result never approached by other programs, even when very strong financial incentives were offered (Hirst, 1987).

NOTES

1 Communication and diffusion are also at the heart of commercial advertising, much but not all of which runs counter to the goals of proenvironmental social marketing. This tension between environmental policy goals and those expressed in commercial "countermarketing" is discussed by Lutzenhiser (this volume, Chapter 3). Countermarketing may focus on specific behaviors (usually purchases); it also may promote general values and attitudes that support a range of environmentally consumptive behaviors.

2 Communication can change institutional and other contexts indirectly by influencing societal ways of thinking. The classic example is the effect of books like *Silent Spring* (Carson, 1962) on the U.S. environmental movement and public support for environmental regulation in the 1960s and early 1970s.

3 Chapters 6 and 7 report on research and practice in the public health and disaster preparedness communities, which operate from a philosophy very friendly to social marketing. They deal with hazards that are widely accepted as important and for which there is broad public support for using government to influence people to act in ways that promote both personal and social interests. Social consensus is harder to find in environmental policy. Consequently, practitioners of environmental "risk communication" often operate on a philosophy that favors providing balanced information that people can use to make informed decisions (National Research Council, 1989). Energy conservation, recycling, and "green" purchasing are among environmental policy goals for which various communication philosophies may operate in different communities or countries and for which the community philosophy may change with the times.

4 A great many environmentally significant behaviors can be classified as environmental activism, nonactivist behaviors in the public sphere (e.g., contributing to organizations that work on environmental issues, attending public meetings, expressing opinions about environmental policies), or private-sphere environmentalism (e.g., purchasing "green" products, composting household waste, maintaining automobile engines to reduce pollution) (Stern, 2000). Environmental citizenship consists mainly of the second class of behaviors; only the last class of behaviors directly affects environmental quality.

REFERENCES

Carson, R.
    1962 *Silent Spring.* Boston: Houghton-Mifflin.

Gardner, G.T., and P.C. Stern
 1996 *Environmental Problems and Human Behavior.* Needham Heights, MA: Allyn and
  Bacon.
Hirst, E.
 1987 *Cooperation and Community Conservation.* Report 1300Y 1987. Portland, OR: Bon-
  neville Power Administration..
Kaufmann-Hayoz, R., C. Bättig, S. Bruppacher, R. Defila, A. Di Giolio, U. Friederich, M. Garbely,
H. Gutscher, C. Jäggi, M. Jegen, A. Müller, and N. North
 2001 A typology of instruments for the promotion of sustainable development. In *Changing
  Things—Moving People: Strategies for Promoting Sustainable Development at the
  Local Level,* R. Kaufmann-Hayoz and H. Gutscher, eds. Basel, Switz.: Birkhäuser.
McKenzie-Mohr, D., and W. Smith
 1999 *Fostering Sustainable Behavior: An Introduction to Community-Based Social Market-
  ing.* Gabriola Island, British Columbia, Can.: New Society Publishers.
Morgan, M.G., B. Fischhoff, A. Bostrom, and C.J. Atman
 2002 *Risk Communication: A Mental Model Approach.* New York: Cambridge University
  Press.
National Research Council
 1984 *Energy Use: The Human Dimension.* Committee on the Behavioral and Social Aspects
  of Energy Consumption and Production. P.C. Stern and E. Aronson, eds. New York:
  W.H. Freeman.
 1989 *Improving Risk Communication.* Committee on Risk Perception and Communication.
  Washington, DC: National Academy Press.
Rogers, E.
 1995 *The Diffusion of Innovations.* New York: The Free Press.
Rogers, R.
 1983 Cognitive and physiological processes in fear appeals and attitude change: A revised
  theory of protection motivation. In *Social Psychology: A Sourcebook,* J. Cacioppo and
  R. Petty, eds. New York: Guilford Press.
Stern, P.C.
 1992 What psychology knows about energy conservation. *American Psychologist* 47:1224-
  1232.
 1999 Information, incentives, and proenvironmental consumer behavior. *Journal of Consumer Policy* 22:461-478.
 2000 Toward a coherent theory of environmentally significant behavior. *Journal of Social
  Issues* 56:407-424.
Stern, P.C., E. Aronson, J.M. Darley, D.H. Hill, E. Hirst, W. Kempton, and T.J. Wilbanks
 1986 The effectiveness of incentives for residential energy conservation. *Evaluation Review*
  10:147-176.
Weinstein, N. and P. Sandman
 1992 A model of the precaution adoption process: Evidence from home radon testing. *Health
  Psychology* 11:170-180.
Werner, C.M., and D. Adams
 2002 Changing homeowners' behaviors involving toxic household chemicals: A psychological, multilevel approach. *Analyses of Social Issues and Public Policy* 1:1-32.

PART III

VOLUNTARY MEASURES IN THE PRIVATE SECTOR

Introduction

Chapters 13-20 focus on the private sector and the use of new tools to influence the behavior of corporations. The most prominent of these new approaches, and the focus of seven of the eight chapters, are voluntary agreements undertaken either between firms and governments or among firms without direct government participation. These voluntary agreements take many forms, varying in who participates, how they are organized, and what is expected of participants. Indeed, one contribution of the chapters in this section is that they map the diverse terrain that falls under the general heading of voluntary agreements, contrasting different forms with each other and examining the relationship between voluntary measures and command-and-control regulation.

We should note at the outset that the scope of our examination is focused on the United States. Although several chapters draw on the experience of other nations and Harrison reviews some Canadian programs similar to the most visible U.S. programs, the majority of the discussion is based on the U.S. experience. As we argue in Chapter 20, the time has come to develop more comparative analyses of voluntary measures. But we limit our scope to the United States in part because there is more research on U.S. initiatives than on those in any other nation (though this is changing quickly) and because we hope that the analyses presented here will be of value to those designing and evaluating voluntary programs in the United States. In addition, because of the unusual legal and policy context in the United States, particularly the major role of adversarial processes in environmental policy, the relevance of other countries' experiences to the U.S. case should not be taken for granted. A robust comparative literature can provide useful guidance to policy, but until that literature emerges and this issue of transferability across political systems is thought through, the U.S. experience provides the best guidance to the design of U.S. programs. The introduction of a comparative literature on voluntary agreements is beginning to appear (tenBrink, 2002).

Three themes emerge across the chapters that follow. One is the potential for firms to "free-ride." A firm that does not participate in a voluntary program when many of its peers in the same industry do might receive many of the public relations and goodwill benefits of the program without the costs of reducing environmental impact. Or a firm may sign on, but do little to change its behavior. The problem of free-riders, or more generally the contrast between altruism and narrowly self-interested behavior, is a central topic for the social sciences and is frequently engaged in the analyses that follow. A second common theme is that firms are embedded in networks of suppliers, customers, investors, and competitors, and also embedded in the communities where they are located. These networks are very consequential for a firm's decisions with regard to environmental protection, as will be noted repeatedly in this section. The third theme is that voluntary programs have indirect effects on organizational culture and practices as well as more direct effects on targeted environmental impacts. In the long term, these changes in firms may be even more important than short-term reductions in environmental impact.

The section begins with a survey by Mazurek (Chapter 13) of the kinds of voluntary agreements that have been brokered among individual firms, industry associations and the U.S. federal government. She reviews what has been learned from the U.S. Environmental Protection Agency's (EPA's) Green Lights, 33/50, and Project XL programs. These are the largest voluntary programs in the United States. Some suggest they are also among the most successful. They are certainly the most carefully examined. But as will be evident throughout this section, careful research on who participates in voluntary agreements, why they participate, what participants do, and what the overall effects are is just beginning to emerge. There are more questions at this point than careful scholarship to answer the questions.

Nash (Chapter 14) examines the causes and effects of voluntary codes of environmental conduct that are established by networks and formal associations of private firms. These include the American Chemistry Council's Responsible Care initiative and related efforts by other parts of the chemical industry, as well as programs by the National Paint and Coatings Association, the American Petroleum Institute, the American Forest & Paper Association, and the American Textile Manufacturers Association. Her analysis considers who adopts such codes, how effective they have been, and what factors might enhance or retard their impact.

Herb, Helms, and Jensen (Chapter 15) focus on another form of "new tool"—community "right-to-know" (RTK) policies. In the United States the most important RTK effort is the Toxics Release Inventory, which requires manufacturing plants to report to the federal government their environmental releases of toxic chemicals. The EPA in turn, makes this information available to the public. The TRI is not a voluntary program—compliance is mandatory. But, as Herb and colleagues argue, it can lead to changes in organizational behavior as a part of increased awareness of the firms themselves, the public, and even the investment community.

Harrison (Chapter 16) considers the problem of evaluating voluntary programs, drawing on both U.S. and Canadian examples. She notes that the first assessments offered of programs may not adequately control for factors other than the voluntary program that might have led to firms reducing their environmental impact. Accurate assessment of programmatic impacts will require careful thinking about the proper measures of program effects, about the appropriate basis for comparison, and about the many factors that can influence corporate environmental behavior. Fortunately, problems of evaluation methodology are not unique to voluntary programs. A substantial and sophisticated literature on program evaluation can provide guidance to future research on this topic.

As noted, voluntary codes are subject to the problem of free-riders and require coordinated action among a network of actors. Furger (Chapter 17) draws on the theories of collective action and networks as well as on case studies to develop hypotheses regarding why firms might participate in voluntary codes. He also argues that some aspects of voluntary agreements that have been subject to criticism—that they are too vague or generic and reduce accountability—may be features essential for their long-term effectiveness.

Prakash (Chapter 18) examines both external and internal factors that may facilitate or retard effective participation in voluntary measures. The theory of collective goods provides a basis for hypothesizing about the incentive structure faced by firms. Prakash reminds us that whatever other motivations may weigh in corporate decision making, profits and the balancing of economic costs and benefits always will be important.

Randall (Chapter 19) adds to this discussion by considering the role of government that influences participation in voluntary measures. As rational actors, firms will anticipate government actions, and voluntary measures may be adopted to avoid regulations that are perceived as more onerous. He reviews some recent theoretical analyses that elucidate the effects that monitoring, enforcement, and other factors can have on compliance with voluntary and regulatory measures and offers some tentative conclusions.

In the final chapter of this section, Dietz (Chapter 20) extracts some themes and lessons from the other chapters. It is clear that care must be taken in evaluating voluntary measures and that sophisticated tools exist to aid in this task. Understanding participation in voluntary measures will require an integrated approach to corporate behavior that takes account of strategic rational action, the networks in which firms are embedded, and the multiple actors and concerns involved in making organizational decisions. Finally, although this volume focuses on the United States, it is clear that further work must become increasingly comparative both because such comparisons facilitate understanding of voluntary measures within the United States and because corporate action and voluntary measures are increasingly global in scope.

Voluntary measures are seen by many as an exciting new approach to environmental policy. The chapters that follow review the best research available on voluntary measures. To some who support voluntary measures, these chapters may be sobering. The experiments to date may not have accomplished as much as is sometimes claimed. But there is a hopeful message that is even more important. Both practical experience and theory provide a much broader repertoire of ideas for designing environmental policy than was the case a decade ago. Furthermore, the chapters in this section demonstrate that a community of researchers is emerging who are providing the scientific basis for understanding voluntary policies so that new initiatives can build on both practical experience and rigorous analysis.

REFERENCE

tenBrink, P.
  2002   *Voluntary Environmental Agreement: Process, Practice, and Future Use.* Shefield, Eng.: Greenleaf Publishing.

13

Government-Sponsored Voluntary Programs for Firms: An Initial Survey

Janice Mazurek

U.S. regulatory agencies and industry view voluntary agreements (VAs) as an increasingly popular alternative to conventional air, water, waste, and toxic control laws. Some observers view VAs as potentially more effective, efficient, and less adversarial than traditional command-and-control approaches. Since 1988, 42 voluntary initiatives have been developed at the federal level by the U.S. Environmental Protection Agency (EPA) and industrial trade organizations such as the American Chemistry Council.[1,2] As of 1998, more than 7,000 corporations, small businesses, local governments, and nongovernmental organizations participated in public voluntary and negotiated programs administered by the EPA, according to the agency's most recent estimates (U.S. Environmental Protection Agency, 1998). The EPA had projected the number would increase to approximately 13,000 in 2000 (See Figure 13-1).

In contrast, more than 350 such agreements are in place in Organization for Economic Co-operation and Development (OECD) member countries (Dowd and Boyd, 1998). European examples include the Dutch covenant system and the Danish CO_2 Agreements. The OECD (1999) compared VAs in Europe, Japan, and the United States. Although it is difficult to generalize across political economies and cultures, the report concludes that VAs in the United States appear to represent a special case because they must operate within the context of a stringent, complicated, and often adversarial legal context.

Despite their growing popularity, a 1997 study commissioned by the U.S. Congress found EPA voluntary initiatives to be "marginal" to the agency's regulatory activities (National Academy of Public Administration [NAPA], 1997). Similarly, a 1996 study commissioned by 21 U.S. companies found the EPA's major voluntary programs "peripheral, both to business and to society" (Davies

Number
of firms

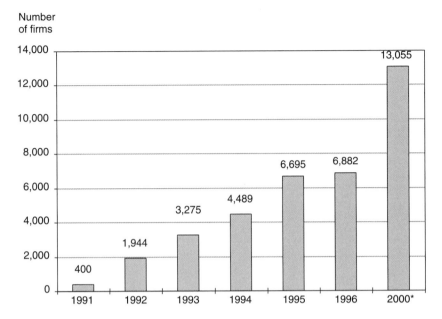

FIGURE 13-1 Participation in EPA voluntary programs, 1991-2000.
Sources: EPA (1997a, 1998, 2002).
NOTE: * Projected. Year 2000 participation number extrapolated from trends between 1991 and 1996. Note that actual data for years 1997 through 2000 were not developed by EPA. EPA estimates the total number of partners in 2002 to be about 11,000.

et al., 1996). Observers agree that the existing legislative framework limits the EPA's ability to use voluntary efforts to improve environmental regulation (NAPA, 1995, 1997; U.S. General Accounting Office [GAO], 1997a). In a study of the EPA's most prominent voluntary agreements, Davies et al. (1996) conclude: "There is no way around the difficult task of trying to legislate a better system."

To illustrate how laws limit VA effectiveness and efficiency, this chapter is organized into three sections. The first section uses Lévêque's (1996) typology to illustrate how and under what circumstances the EPA and industry apply VAs. In the United States, most VAs are between the EPA and individual firms. In general, the EPA and industry use voluntary agreements to (1) address risks that U.S. laws and regulations fail to adequately target, and (2) integrate individual air, water, waste, and toxics laws (NAPA, 1995, 1997; Davies and Mazurek, 1998). The subsequent section shows how federal pollution control laws and regulations impede effective VA implementation. Another section pairs assessment data developed by implementing regulatory agencies with independent studies to describe the performance of three prominent voluntary initiatives: Green Lights, the 33/50 program, and Project XL.

In most cases, poorly designed program evaluation methods make it difficult to attribute environmental changes exclusively to voluntary programs (NAPA, 1997; GAO, 1997b). Because little data exist to demonstrate environmental effectiveness, it is virtually impossible to assess whether or to what degree voluntary programs affect abatement cost. To supplement what is known about VA effectiveness, this chapter draws from a small but growing literature to examine three of the most prominent U.S. public voluntary programs. They include Green Lights; the 33/50 program, and Project XL (GAO, 1994, 1997a, 1997b; INFORM, 1995; Arora and Cason, 1995; NAPA, 1995, 1997; Davies et al., 1996; Storey et al., 1996; Kappas, 1997; Dowd and Boyd, 1998; Boyd et al., 1998). Appendix 13-A provides an overview of each program.

The literature underscores the degree to which the lack of data and evaluation methods complicates assessment. Some data exist with which to assess the administrative cost of voluntary programs. This chapter concludes that to promote transparency and acceptance of voluntary programs, public agencies must develop better evaluation methods.

PUBLIC VOLUNTARY PROGRAMS PREVAIL

VAs in the United States consist primarily of what Lévêque (1996) defines as "public voluntary" programs, the focus of this chapter.[3] The EPA independently or in tandem with other federal agencies administers 33 of the 42 voluntary federal initiatives (see Table 13-1). Of these, 31 are purely public voluntary programs. Two, Project XL and the Common Sense Initiative (CSI), are hybrids. Project XL involves negotiation between the EPA and individual industrial facilities. The EPA uses CSI to develop voluntary agreements with industry sectors.

Voluntary programs, including 33/50 and Green Lights, were first popularized under the former Bush Administration to promote more market-oriented incentives for environmental performance. Today, most of the EPA's voluntary programs are designed to reduce greenhouse gas emissions or to adopt voluntary goals established under the Pollution Prevention Act of 1990. Voluntary climate change programs are designed to provide participants with technical information in order to promote energy conservation. For example, the EPA designed the Green Lights program in 1991 to encourage the installation of energy-efficient lighting technologies in commercial and industrial buildings.

Voluntary pollution prevention programs are designed to reduce a subset of toxic chemicals released and transferred by manufacturers. For example, the 33/50 program, initiated under the Bush Administration in 1991 and concluded in 1995, encouraged manufacturers to voluntarily reduce emissions of 17 target chemicals by 50 percent. A primary goal of negotiated strategies is to improve efficiency by reducing regulatory burden. In practice, Project XL aims to reduce administrative costs associated with reporting, monitoring, and permitting.

TABLE 13-1   Voluntary Agreement Categories

PUBLIC VOLUNTARY

Climate Change	Pollution Prevention	Negotiated Agreements
1. AgStar Program (1993)	**1. 33/50 (1991)**	**1. Project XL (1995**
2. Climate Wise (1993)	2. Design for the Environment (1991)	2. Common Sense Initiative (1994)
3. Chlorofluorocarbon Substitutes (post-1993)	3. Environmental Accounting Project (1992)	
4. Coalbed Methane Outreach	4. Environmental Leadership Program (1994)	
5. Commuter Choice (post-1993)	5. Green Chemistry (1992)	
6. Energy Star Buildings (1994)	6. Indoor Environments Program (1995)	
7. Energy Star Homes (1995)	7. Pesticide Environmental Stewardship Program (1993)	
8. Energy Star Office Equipment (1993)	8. Waste Minimization National Plan (1994)	
9. Energy Star Transformer Program (1995)	9. Water Alliances for Voluntary Efficiency (WAVE) (1992)	
10. Environmental Stewardship Initiative (1997)	10. Voluntary Standards Network (1993)	
11. Green Lights (1991)		
12. HFC-23 Reductions (post-1993)		
13. Landfill Methane Outreach Program (1994)		
14. Natural Gas Star (1993, 1995)		
15. Ruminant Livestock Methane Efficiency Program (1993)		
16. Seasonal Gas Use for the Control of Nitrous Oxide (post-1993)		
17. State and Local Climate Change Outreach Program (1993)		
18. Transportation Partners (1995)		
19. The U.S. Initiative on Joint Implementation (1993)		
20. Voluntary Aluminum Industrial Partnership (1995)		
21. WasteWise (1992)		

Sources: EPA (1996a, 1997b); Dowd and Boyd (1998).
Note: Programs listed in boldface are discussed in this chapter.

Public voluntary programs in the United States use information subsidies, technical assistance, and/or public recognition to encourage participants to voluntarily reduce pollution. Public recognition may be provided through awards, press announcements, and the use of product logos.

Among U.S. VAs, only Project XL contains legally binding provisions. This is because only Project XL promises to provide firms with relief from existing laws and regulations. In exchange, participants must be able to demonstrate environmental performance superior to status quo standards. Typically, the legally binding portions of an XL agreement are contained as a separate document, such as a permit to ensure the agreement's enforceability. Nonbinding provisions appear in what is known as a "Final Project Agreement (FPA)." Enforceable provisions carry sanctions such as compliance actions and fines. Failure to meet nonbinding commitments results in FPA termination.

Most EPA voluntary initiatives such as Green Lights require participants to sign nonbinding letters of agreement such as a Memorandum of Understanding (MOU), which imposes no sanction for program withdrawal. Failure to meet the MOU terms means that the company can no longer claim the benefits of participation, which usually includes public recognition. The threshold for participation in 33/50 was even lower. The program simply asked potential participants to send the EPA a letter indicating their willingness to reduce the 17 targeted chemicals. In the case of 33/50, firms were free to reduce as much or as little as they saw fit.

POLLUTION CONTROL LAWS IMPEDE IMPLEMENTATION

In theory, the primary disadvantage of VAs arises from the collective nature of their benefits—participants have a strong incentive to act as free-riders. Voluntary agreements also may act to exclude competitors and restrain trade. Such practices may privately benefit participants, but not society in general, by reducing supply and increasing cost. Another potential problem is that industry may use VAs to influence and capture the details of environmental policy. In practice, such problems have not yet been observed because laws impede VA implementation. In particular, negotiated strategies such as Project XL, which are designed to provide participants with regulatory relief, are problematic.

For public voluntary programs in particular, environmental laws impede implementation because Congress and the courts require the EPA to focus attention and resources on meeting legal requirements and judicially imposed deadlines (NAPA, 1995). The persistence of pollution control laws also makes it difficult for groups that traditionally act as adversaries to effectively harness cooperative strategies (Davies et al., 1996). Although cooperative strategies tend to be more inclusive than status quo approaches, they are also less transparent.

Negotiated Agreements

Implementation of Project XL is hampered because Congress has not given the EPA the authority to provide firms with relief from existing laws and regulations and because potential participants run the risk of civil lawsuits (NAPA, 1995, 1997; Davies et al., 1996). The results are twofold: First, the lack of regulatory flexibility has led to suboptimal outcomes, with projects that are largely possible under existing regulations. For example, the EPA cannot authorize companies to reduce abatement costs via plant-level pollutant trades. The second problem is procedural. When government or trade associations fail to possess legal authority, they can act only by achieving some degree of consensus. This situation gives each participant a potential veto power and leads to large, sometimes intractable transaction costs. Reliance on consensus-based methods also fails to maximize outcomes. Instead, they tend to result in goals that represent the lowest common denominator on which all parties agree. In the extremely adversarial context of U.S. environmental regulations, consensus is typically difficult to achieve (Davies et al., 1996).

Project XL's limitations stem more directly from uncertainties regarding its legality. As long as the EPA lacks the authority to grant firms relief from laws, firms face the risk of civil lawsuits. However, the time and resources required to negotiate the first three XL agreements were higher than forecast due to procedural problems. Under Project XL, the EPA agreed to give up "letter of the law" compliance with all applicable regulations in return for environmental performance exceeding what traditional regulation could bring. Because the experiment involves negotiation, it was understood that initial transaction costs to industry, to regulators, and to public participants would be high for all parties. It was hoped that the benefits in cost reductions accorded by increased compliance flexibility would more than make up for delays and costs of negotiations.

Concerns regarding the legality of Project XL resulted in participation rates lower than EPA originally envisioned. Although the EPA originally had hoped to admit 50 firms to Project XL, since 1995 the agency has approved 46—all of which are underway (EPA, 2000). Fourteen additional projects are in various stages of development or negotiation, and 30 proposals have been withdrawn or rejected. As mentioned, questions regarding the legality of XL projects also have resulted in proposals that fall largely within the scope of existing laws. As a result, environmental benefits are likely to be lower than originally envisioned.

Although XL projects largely are possible under current laws, environmental groups nonetheless worry that XL projects could set precedents that would weaken existing laws and regulations. For example, at Intel's XL effort in Arizona, local participants agreed to provide Intel with relief from air permitting requirements in exchange for a set of binding and voluntary environmental commitments. The local community supported the plan. However, 130 nonlocal environmental organizations and individuals signed a petition in protest of the agreement. It has been suggested that national environmental groups protested the Intel XL plan because they were not invited to participate in the formal 6-month project negotiation (NAPA, 1997). The EPA reasoned that only parties directly affected by the project outcome should participate in the bargaining process (EPA, 1996c).

To summarize, the EPA and U.S. industry employ VAs to address the shortcomings of pollution control laws. However, the persistence of pollution control laws impedes VA implementation, particularly of industry-led efforts and public projects that employ negotiation (Boyd et al., 1998). As a result, voluntary approaches remain largely "marginal" to federally mandated air, water, waste, and toxic control programs. Implementation of voluntary agreements may be improved by taking into account more fully the legal uncertainties associated with attempts to circumvent laws. However, it is likely that the effectiveness of VAs in the United States will require legislative remedy.

Implementation problems have led to lower-than-expected environmental results for all VA categories. Among the different types of VAs employed in the United States, programs designed to reduce greenhouse gas emissions and a subset of toxic chemicals have contributed to emissions declines. However, poor evaluation methods likely caused the EPA to overstate the environmental effectiveness of both climate change and prevention programs.

VA PERFORMANCE

This section draws from a small but growing literature that examines the three most prominent U.S. voluntary programs: Green Lights, the 33/50 program, and Project XL. Although the EPA reports that each of these programs was a success, independent studies report otherwise (GAO, 1994, 1997b; INFORM, 1995; Arora and Cason, 1995; NAPA, 1995, 1997; Davies et al., 1996; Storey et al., 1996; Dowd and Boyd, 1998; Boyd et al., 1998). This section briefly reviews the results of independent studies to illustrate that poor program evaluation methodology makes it difficult to assess the success of voluntary programs.

Green Lights

The experience of the Green Lights program illustrates best uncertainties surrounding VA program measurement and performance. The EPA reports that 2,300 Green Lights participants have experienced rates of return of up to 50 percent (see Table 13-2). The agency also reports that total energy savings translate into $100 million dollars per year (EPA, 1996a). However, the U.S. General Accounting Office (GAO) finds that poor assessment methods caused the EPA to initially overestimate the effectiveness of Green Lights (GAO, 1997b).

TABLE 13-2   Participants, Funding, and Other Details About Green
Lights

Targeted Gas(es)	Carbon Dioxide
Type of participants	Business and government
Number of participants	2,308
FY 1996 funding (million $)	$20.1
Greenhouse Gas (GHG) reductions through FY 1995 Million Metric Tons of Carbon Equivalent (MMTCE)	0.6
GHG reductions in 2000 MMTCE	3.9

Source: U.S. General Accounting Office (1997b).

GAO questioned the basis of the EPA's emission reduction estimates for Green Lights. GAO also questioned the extent to which Green Lights was responsible for adoption decisions among a quarter of the program's participants. GAO found that 593 of the 2,308 Green Lights participants represented companies that were likely to install energy-efficient lighting even in the absence of Green Lights. The subset consisted of companies that manufacture, sell, and install lighting products. Combined, the 593 companies contributed to about 6 percent of total emissions reductions attributed by the EPA to Green Lights.

Finally, GAO found evidence to suggest that a substantial amount of floor space was upgraded before the Green Lights program was well established. GAO based its findings on a national survey of commercial buildings conducted by the Department of Energy's Energy Information Administration (EIA). The EIA survey found that 43 percent of commercial floor space had lighting conservation features in the years prior to EPA implementation of Green Lights.

The 33/50 Program

GAO faulted the EPA's 33/50 assessment methods for many of the same reasons it faulted the EPA's climate change assessments: Poor assessment methods caused the EPA to initially overestimate program effectiveness (GAO, 1997b). The EPA points to the 33/50 program as one of the agency's most successful voluntary initiatives. The initiative did reduce toxic emissions. However, GAO (1994) and two nonprofit organizations (Citizen Fund, 1994; INFORM, 1995) found that the EPA overstated the success of the 33/50 program. Moreover, there is no evidence to suggest that participants used prevention rather than control methods to achieve 33/50 program goals.

According to the EPA, the 33/50 program's interim and final emissions reduction goals were both met a year ahead of schedule (EPA, 1996c). The EPA calculated 33/50 program reductions for the 17 chemicals by aggregating reduc-

TABLE 13-3   Emissions and Transfer Declines, Participants and Nonparticipants

Years	33/50 Program Participants	Nonparticipants
1991-1994	−49%	−30%

Source: EPA (1996c).

tions from all reporting firms. By this method, the EPA found that participants reduced the targeted chemicals by 590 million pounds in 1991 and by 757 million pounds in 1994. In 1995, releases and transfers for the 17 target chemicals totaled 664 million pounds, a 55.6-percent reduction from the program's 1988 baseline. The EPA reports that between 1991 and 1994, reductions in releases and transfers among program participants outpaced nonparticipant reductions by 19 percent (see Table 13-3), but the actual figure may be even lower (EPA, 1996c).[4,5]

GAO, however, found that the EPA incorrectly attributed emissions reductions to the 33/50 program. GAO researchers found evidence to suggest that companies had made substantial reductions prior to 33/50's implementation. GAO also faulted the EPA's decision to use 1988 emissions as a baseline against which to compare performance under 33/50. GAO's findings were reinforced by another study which found that prior to 33/50's implementation, about 83 percent of all facilities had started to make reductions in 33/50 program chemicals (Citizen Fund, 1994).

Researchers from INFORM, a nonprofit environmental research organization, similarly found that 31 percent of 33/50 program participants already had initiated reduction activities prior to 1991. Based on these findings, GAO recommended that EPA only consider reductions achieved between 1991 and 1994. Table 13-4 shows how the 33/50 program's results change when the baseline is modified from 1988 to 1991. From this perspective, 33/50 program chemical emissions fell by only 204 million tons (as opposed to 757) by 1994—a 27-percent, rather than a 51-percent, decline (Davies et al., 1996).

GAO estimated that about 38 percent of 33/50 program reductions were made by nonparticipating companies (GAO, 1994). EPA's (1996c) estimates

TABLE 13-4   Baseline Selection and 33/50 Program Results

Reduction Goal/Year	1994 Result (1988 baseline)	1994 Result (1991 baseline)
33% by 1992	40%	12%
50% by 1995	51%	28%

Source: Adapted from Davies et al. (1996).

TABLE 13-5   Production Related Waste Declines, 33/50 and Nonprogram
Chemicals

Years	33/50 Program Chemicals	Nonprogram Chemicals
1991-94	-1%	9%

Source: EPA (1996c).

are slightly lower. The EPA found that about 26 percent (196 million pounds) of reductions attributed to 33/50 were made by nonparticipants. The EPA nonetheless concludes that 33/50 influenced nonparticipants to make such reductions. Moreover, between 1991 and 1994, the waste declines achieved were of chemicals not covered by the program (see Table 13-5).

The 33/50 program's ancillary goal was to promote pollution prevention. However, GAO found no evidence to suggest that 33/50 promoted prevention measures, as opposed to less favorable strategies such as abatement (GAO, 1994). INFORM (1995) similarly found 33/50's impact on prevention to be questionable. INFORM found 33/50's prevention goals ineffectual because the EPA failed to require participants to link reported reductions to the use of prevention methods.

Project XL

As in the case of 33/50, the calculation of environmental benefits under Project XL is complicated by poor baseline measures. The best example is Intel Corporation's XL project in Arizona. The world's largest microprocessor manufacturer negotiated with the EPA for a specialized air permit to facilitate frequent production changes. To achieve refinements and optimize its production process, Intel must modify process chemistries up to 35 times a year and equipment 5 times a year. However, the manufacturer's ability to make refinements in a timely manner is threatened by air permitting provisions. The facility must obtain air permit approval each time it makes a manufacturing change.

To address these issues, Intel under Project XL sought a 5-year air permit that approved chemical and equipment changes in advance. The binding, enforceable air permit is part of a larger package of Project XL commitments to reduce water use and waste generation at the company's newest manufacturing facility in Phoenix. The air permit covers emissions of conventional and hazardous air pollutants at Intel's new facility, and gives the manufacturer the ability to construct an additional manufacturing facility without having to secure a new permit. In exchange, Intel pledged to accept air pollution caps for the Phoenix facility set lower than the requirements of the Clean Air Act Amendments of 1990 (see Table 13-6).

TABLE 13-6   Intel's Project XL Emissions and Two Baselines

Pollutant (Tons/Per Year)	Federal Requirements for Minor Sources	1994 Plant Air Permit	Project XL Site Permit[a]
Carbon monoxide	<100	59	49
Nitrogen oxide	<100	53	49
Sulfur dioxide	<250	10	5
Particulate matter^{-10}	<70	7.8	5
Total volatile organic compounds	<100	25	40
Hazardous air pollutants (HAPs)[b]	<25 aggregate; for any individual HAP	5.5	10 total organic^{-10} 10 total inorganic[c]

Source: U.S. EPA (1996b).

[a] The emissions levels under the XL permit column are for two plants, or the entire site.

[b] HAPs are those listed in Section 112(b) of the federal Clean Air Act, as amended; the 10-ton-per-year limits for total organic HAPs and total inorganic HAPs assume that more than one HAP will be emitted from the site.

[c] If a single HAP is emitted from the site, the emissions limit is 9.9 tons per year; based on Intel's modeling exercise and Arizona Ambient Air Quality Guidelines, the permit establishes a separate limit for phosphene, at 4 tons per year, and sulfuric acid at 9 tons per year, to be included in the aggregated combined inorganic HAP emissions plant site emissions limit.

Boyd et al. (1998) considered various types of costs and benefits registered as part of Intel's XL air permit. Cost and benefit categories include environmental benefits, abatement costs, and transaction costs. They also examined potential benefits associated with reducing permit-based delays in production. Overall, they found that the air permit could raise abatement costs for Intel and increase environmental benefits over a standard air permit baseline, although there is ample room for debate over both of these conclusions.

At the time, the Intel XL air permit applied to a new plant and to a second manufacturing facility that only existed on paper. Because one facility was new, and the other remained in blueprint form, the site lacked an emissions history with which to craft a baseline, or base case to determine what emissions would have been in the absence of Project XL.

EPA and Intel XL project stakeholders decided that absent historic data, the theoretical maximum under the Clean Air Act Amendments of 1990 constituted an appropriate baseline. Nonlocal environmental groups objected to the use of this baseline measure, claiming that it failed to constitute environmental performance "superior" to the status quo. National environmental groups encouraged the EPA to develop an industry air emissions benchmark against which to compare how well the Intel facility actually performed. The EPA attempted such a calculation, but concluded that the exercise was complicated because of the lack of industrywide air emissions data (EPA, 1996c). In response to criticism from environmental groups, Intel and the EPA adopted as binding a set of voluntary ambient air pollution guidelines issued by the state of Arizona.

Boyd and colleagues (1998) conclude that precise measurement of the XL air permit's incremental effects is likely to be difficult, if not impossible. This is because of the site-specific nature of the XL air permit. It is also because of the complex effects that any form of regulation—including command and control—can have on the private sector. Although it is difficult to isolate the effects of the XL air permit with precision, analyses of such effects can provide suggestive evidence on the social welfare effects of the XL agreement.

CONCLUSIONS

VA assessment is hampered by program novelty, lack of data, and weak metering and evaluation methods. As Harrison shows in Chapter 16 of this volume, evaluating voluntary programs is challenging. In most cases, it is difficult to attribute environmental changes exclusively to voluntary programs. Partially because of the lack of environmental data, virtually no studies have been developed to demonstrate whether voluntary approaches are efficient. Some data illustrate administration and compliance costs. The Intel case suggests that Project XL may confer significant competitive advantages to Intel, but that the magnitude of the effect is impossible to measure.

Assessment data that have been developed suggest that the primary benefit of VAs may be intangible and, in any event, difficult to measure. Participants in 33/50 and Project XL all cite public opinion and/or regulatory goodwill as significant benefits. Improved goodwill may indirectly lower costs associated with permitting and reporting. Goodwill may allow influential firms to reduce regulatory adoption of more stringent regulation by agencies. Soft factors also may indirectly reduce administrative and abatement costs.

At a minimum, VAs such as Project XL have the potential to promote interaction among groups that act under status quo laws as adversaries. In the absence of explicit legal authority, industry, the EPA, and interested stakeholders must achieve consensus regarding project goals to minimize the potential for citizen lawsuits. In this regard, negotiated VAs in theory provide more opportunities for stakeholder participation than the status quo. In practice, though, implementation is hampered by the lack of clearly defined administrative, monitoring, and participatory procedures. Poor monitoring data have caused unilateral and negotiated approaches in particular to lack credibility among environmental groups and some industries. To promote greater trust and greater public participation, agencies must develop and use more robust monitoring and reporting measures to provide confidence that VA participants do deliver environmental results that are superior to those of conventional regulation.

NOTES

1 The U.S. Department of Energy administers about 20 voluntary climate change programs that have been examined elsewhere (Storey et al., 1996; Dowd and Boyd, 1998).

2 For an analysis of several prominent state voluntary programs, see Beardsley (1996).

3 Lévêque (1996) identifies three VA types: public voluntary, unilateral, and negotiated agreements. Public voluntary schemes refer to nonmandatory rules developed by a government body such as the EPA. Unilateral commitments refer to programs established by industry to encourage firms to achieve environmental improvements. Negotiated agreements refer to contracts between public authorities and industry. In contrast to public voluntary efforts, negotiated agreements contain specific targets and are legally binding.

4 The 19-percent figure is relative to releases and transfers only from participating firms, not total 1991 releases and transfers. The 19-percent difference in reductions by participating firms constitutes an 11-percent reduction relative to the *total* releases and transfers of 33/50 chemicals in the 1990 reference year. The total releases and transfers figure is a more appropriate baseline because it was the goal of the 33/50 program to reduce all discharges of 33/50 chemicals, not just those by a subset of firms.

5 I am indebted to Kathryn Harrison at the University of British Columbia for making this distinction.

REFERENCES

Arora, S., and T.N. Cason
  1995 *Why Do Firms Overcomply with Environmental Regulations? Understanding Participation in EPA's 33/50 Program.* Discussion paper 95-38. Washington, DC: Resources for the Future.

Boyd, J., A. Krupnick, and J. Mazurek
  1998 *Intel's XL Permit: A Framework for Evaluation.* Discussion paper 98-11. Washington, DC: Resources for the Future.

Citizen Fund
  1994 *Pollution Prevention or Public Relations?* Washington, DC: Citizen Fund.

Davies, J.C., and J. Mazurek
  1998 *Pollution Control in the United States: Evaluating the System.* Washington, DC: Resources for the Future/Johns Hopkins University Press.

Davies, J.C., J. Mazurek, N. Darnall, and K. McCarthy
  1996 *Industry Incentives for Environmental Improvement: Evaluation of U.S. Federal Initiatives.* Washington, DC: Global Environmental Management Initiative.

Dowd, J., and G. Boyd
  1998 *A Typology of Voluntary Agreements Used in Energy and Environmental Policy.* Office of Policy and International Affairs. Washington, DC: U.S. Department of Energy.

INFORM
  1995 *Toxics Watch 1995.* New York: INFORM.

Kappas, P.D.
  1997 The Politics, Practice and Performance of Chemical Industry Self-Regulation. Doctoral dissertation in political science. University of California, Los Angeles.

Lévêque, F.
  1996 *Environmental Policy in Europe - Industry, Competition and the Policy Process.* Cheltenham, Eng.: Edward Elgar.

National Academy of Public Administration (NAPA)
  1995 *Setting Priorities, Getting Results: A New Direction for EPA.* Washington, DC: NAPA.

1997 *Resolving the Paradox of Environmental Protection.* Washington, DC: NAPA.

Organization for Economic Co-operation and Development (OECD)

1999 *Voluntary Approaches for Environmental Policy: An Assessment.* Paris: OECD.

Storey, M., J. Dowd, G. Boyd, and T. van Dril

1996 Demand Side Efficiency: Voluntary Agreements with Industry. Working Paper 8, prepared for Annex I Expert Group on the United Nations Framework Convention on
Climate Change, December.

U.S. Environmental Protection Agency (EPA)

1996a *Partnerships in Preventing Pollution: A Catalogue of the Agency's Partnership Programs.* Washington, DC: EPA, Office of the Administrator.

1996b Project XL: Final Project Agreement for the Intel Corporation Ocotillo Site Project XL.
Office of Policy, Economics, and Innovation. [Online]. Available: http://www.epa.gov/
ProjectXL/implement.htm. [Accessed 12/01/01].

1996c *1994 Toxics Release Inventory.* Public Data Release. Office of Pollution Prevention and
Toxics. Washington, DC: EPA.

1997a *Pollution Prevention 1997: A National Progress Report.* Office of Pollution Prevention
and Toxics. Washington, DC: EPA.

1997b Risk *Reduction through Voluntary Programs.* Audit Report No. E1KAF6-05-0080-
100130, 3/19/97. Office of the Inspector General. Washington, DC: EPA.

1998 *Partners for the Environment: Collective Statement of Success.* Office of Reinvention.
Washington, DC: EPA.

2000 Project XL. Implementation and Evaluation. Office of Policy, Economics and Innovation. [Online]. Available: http://www.epa.gov/ProjectXL/implement.htm. [Accessed 12/
01/01].

2002 Partners for the Environment. Office of Policy, Economics, and Innovation. [Online].
Available: http//:www.epa.gov/partners. [Accessed 2/1/02].

U.S. General Accounting Office (GAO)

1994 *Toxic Substances: EPA Needs More Reliable Source Reduction Data and Progress
Measures.* GAO/RCED-94-93. Washington, DC: U.S. Government Printing Office.

1997a *Environmental Protection: Challenges Facing EPA's Efforts to Reinvent Environmental Regulation.* GAO/RCED-97-155. Washington, DC: U.S. Government Printing Office.

1997b *Global Warming: Information on the Results of Four of EPA's Voluntary Climate
Change Programs.* June 30. GAO/RCED-97-163. Washington, DC: U.S. Government
Printing Office.

APPENDIX 13-A

DESCRIPTION OF MAJOR VOLUNTARY AGREEMENTS

1. Green Lights Program

The EPA launched Green Lights in 1991. The program's goal is to prevent pollution by encouraging U.S. institutions to use energy-efficient lighting technologies. The program currently has more than 2,338 participants. Participants are required to survey their domestic facilities and upgrade lighting where it is profitable and improves or maintains lighting quality. According to the EPA, a profitable project is one that—on a facility aggregate basis—maximizes energy savings while providing an annualized internal rate of return (IRR) that is greater than 20 percent. This target is a "floor" rather than a ceiling; most lighting upgrades yield 20 to 40 percent IRR, according to the EPA. Participants must complete their lighting upgrades within 5 years of joining.

2. 33/50 Program

Established by the EPA in 1991, 33/50 is the first major public voluntary pollution reduction initiative in the United States. Concluded in 1995, the 33/50 Program encouraged companies to voluntarily reduce emissions of 17 target chemicals by 33 percent by 1992 and by 50 percent by 1995. Firms achieved the final 50 percent reduction goal in 1994—a year ahead of schedule. According to the EPA, the initiative helped to eliminate 700 million pounds of toxic waste. Approximately 1,300 companies participated in the initiative.

3. Project XL

President Clinton directed the EPA to create Project XL in 1995. The program is designed to give individual regulated sources (e.g., industrial facilities) relief from some regulatory requirements in exchange for environmental performance superior to that required by command-and-control regulation. The case-by-case projects are achieved through negotiation between firms and regulators, subject to stakeholder approval. Project XL is the only voluntary initiative in the United States that contains legally binding provisions. As of April 1998, seven XL projects were underway. Another nine were under negotiation. Four were in the proposal phase. Since 1995, 30 proposed projects have been rejected by the EPA or withdrawn.

14

Industry Codes of Practice: Emergence and Evolution

Jennifer Nash*

S ince the late 1980s, a number of trade associations in the United States have established codes of management practices with a twofold purpose: to improve members' environmental performance and to demonstrate this improvement to critical public audiences. Trade association codes call on firms to move beyond regulatory minimums and to continually improve their efforts in community involvement, pollution prevention, and product stewardship. Until recently, however, most trade associations had done little to monitor the extent to which members actually were putting codes into practice or to sanction those who failed to implement required practices.

Trade associations in the United States are voluntary associations of firms within a single industry (Bradley, 1965). Securing and maintaining members is an abiding preoccupation for trade associations, which depend on membership support to fund their budgets. Although individual members may want citizens and regulators to view the environmental conduct of their industry favorably, they may not believe that improving their own firm's environmental performance is in their self-interest (Olson, 1965). Members who feel pressure to improve their environmental performance may simply quit the trade association. To what extent, then, is it possible for trade associations to regulate the environmental conduct of their members?

---

*This chapter has been prepared with support from the U.S. Environmental Protection Agency, Emerging Strategies Division. The views expressed, as well as mistakes and omissions, are the authorís, not EPAís. Two students provided valuable research assistance: Anand Patel and Stephanie Okasaki. Thanks to Philip Byer and John Ehrenfeld for helpful comments.

This chapter is divided into three parts. The first part explores the question of why certain trade associations in the United States have developed environmental codes for their members. The second part considers the effectiveness of trade association codes in improving environmental performance. The third part offers conclusions about the direction in which trade association codes appear to be evolving and where they may be achieving results. In the past, trade association codes served primarily as defensive measures to improve public opinion and forestall public regulation. Now, however, trade associations are imposing codes on their suppliers and distributors as a condition for doing business. Trade associations are adding measures to observe the environmental practices of their business partners, and to sanction, with a decision to do business elsewhere, those who do not live up to code requirements.

CHARACTERISTICS OF TRADE ASSOCIATIONS THAT REGULATE THE ENVIRONMENTAL PERFORMANCE OF THEIR MEMBERS

Trade associations are nonprofit organizations of business competitors in a single industry (Bradley, 1965). In the United States, they have historically served two functions: enhancing the collective welfare of members through lobbying and legal action and providing direct service to members through educational programs, market information, and group discounts. Trade associations rely on membership support in order to operate. Membership is voluntary. Most trade associations raise their operating revenues from fees and dues assessed on members. Boards of directors, made up of executives from member firms, set policies for the groups.

Of the thousands of trade associations that operate at the national level in the United States only about seven have developed codes of environmental management practice, listed in Table 14-1.[1] In this chapter, discussion focuses primarily on codes of practice in the chemical industry, with references to other trade association codes to draw out similarities and differences. The efforts of the National Paint and Coatings Association and the National Association of Chemical Recyclers are not discussed, although they merit attention.[2]

By taking on the role of environmental regulator of its industry, a trade association runs the risk of alienating member firms. Firms can enjoy many of the collective benefits of membership, such as economic benefits that may result from trade association lobbying activities, without joining. Why, then, have some trade associations imposed environmental codes on their members? Codes have emerged in industries that citizens and government perceive lack self-control, cannot be trusted, or are inherently unsafe. Only in those industries have trade associations taken on the role of regulator of their members' environmental practices.

The public's negative perception of the chemical industry drove the American Chemistry Council (ACC, formerly known as the Chemical Manufacturers

TABLE 14-1   Codes of Environmental Management Practice Promulgated by U.S. Trade Associations

Trade Association	Code Name and Year Established
American Chemistry Council (ACC)— formerly Chemical Manufacturers Association (CMA)	Responsible Care, 1989
National Association of Chemical Distributors (NACD)	Responsible Distribution Process (RDP), 1991
National Association of Chemical Recyclers (NACR)	Responsible Recycling, 1993
National Paint and Coatings Association (NPCA)	Coatings Care, 1996
American Petroleum Institute (API)	Strategies for Today's Environmental Partnership (STEP), 1990
American Forest & Paper Association (AF&PA)	Sustainable Forestry Initiative (SFI), 1994 Environmental, Health and Safety Principles, 1995
American Textile Manufacturers Institute (ATMI)	Encouraging Environmental Excellence (E3), 1992 Quest for the Best, 1993

Source: Nash (1999).

Association) to develop the Responsible Care Program in 1989. Public opinion polling at that time showed that a large portion of the public believed the chemical industry had no self-control, did not listen to the public, and did not take responsibility for its operations (Rees, 1997). Before 1970 the chemical industry had been essentially free to manage its environmental impacts as it saw fit. By 1980, after congressional passage of the major environmental statutes, this freedom was gone; the perception among chemical industry managers was that the industry was run—not just regulated— by government environmental protection officials (Hoffman, 1995).

A defining event for the chemical industry's public image problem was the 1984 massive chemical release in Bhopal, India, that killed thousands of people. The huge oil spill from the *Exxon Valdez* oil tanker in March 1989 focused public attention on the hazards of the oil industry. Not only was the reputation of the Exxon company damaged, but the public perception of the entire industry fell significantly, prompting an editorial in an oil industry trade journal to urge firms to adopt a "group approach [toward building public trust] . . . mean[ing] more than companies' acting responsibly alone" (*Oil & Gas Journal,* 1990). In 1990 the American Petroleum Institute launched its environmental code, Strategies for Today's Environmental Partnerships, in response.

Public perception of the forest and paper industry parallels in many respects views about chemicals and petroleum. During the late 1980s and early 1990s, chief executive officers of the largest U.S. forest and paper companies commissioned extensive public opinion research to probe public attitudes. The results were dismaying. Many people, about 55 percent of those asked, believed the industry did not practice sustainable forestry. An even larger percentage found the industry was doing a "poor job" in its efforts to protect wildlife, conserve resources, protect air quality, and protect lakes and steams (American Forest & Paper Association [AF&PA], 1998). AF&PA board members, like their counterparts at ACC, decided that public relations alone would not dissipate these concerns. "Credibility can be enhanced only if we have clear behavioral changes and our message communicates this change," the board members noted (AF&PA, 1998:10).

Public opinion spurred environmental regulation. In June 1990 the U.S. Fish and Wildlife Service ruled to list the northern spotted owl as a threatened species. This decision eliminated timber harvesting from about 9 million acres of land in the Pacific Northwest, the owls' habitat (Bossong-Martines, 1999c). In addition to the Endangered Species Act, the Clean Air Act and Clean Water Act have had a substantial impact on forestry companies. Compliance with federal and state environmental regulations has required significant capital spending. Firms have been required to add secondary treatment plants, control plant emissions, reduce the use of elemental chlorine, and fulfill recycling commitments. Environmental spending has accounted for about 14 percent of capital outlays made by the U.S. forest and paper industry since the late 1980s, according to the U.S. Department of Commerce (Bossong-Martines, 1999c).

As this discussion suggests, codes have been developed by industries that citizens and government perceive are not capable of responsibly managing the unintended consequences of their practices on their own. The challenge to the textile industry's legitimacy has come from a different source: low-cost textile production in developing countries. The major focus of the American Textiles Manufacturing Institute (ATMI) has been to fight for import quotas, tariffs, and trade agreements favorable to the industry. It has enacted numerous campaigns to build public support for textiles and clothing manufactured in the United States—its "crafted with pride in the U.S.A." program, begun in 1983, is its longest sustained promotional effort (Morrissey, 1999). It launched Encouraging Environmental Excellence (E3) in 1992 to publicize the environmental accomplishments of members. The association hoped to use the program to distinguish members' products from imports that might be produced under less environmentally responsible conditions. Unlike the codes of the ACC and the AF&PA, however, adoption of E3 is not a requirement for membership in the ATMI. Members may choose whether or not to adopt this code, and about one-third of the trade association's members participate. The firms that take part tend to supply customers such as The Gap, Eddie Bauer, and Levi's, which have established codes of conduct of their own (Islam, 1999). E3 founders believed that demonstrating environmental responsibility to these customers might help, over time, to strengthen their business relationships with these customers.

The public's negative perception of the chemicals, petroleum, and forestry industries helps to explain the decision of their trade associations to develop environmental codes of conduct. These codes are specifically designed to improve the environmental performance of member firms and to demonstrate this improvement to critical public audiences. But why does the public hold these industries in such low regard, while accepting the risks of other similar manufacturers? The public's relatively high regard of the pharmaceutical industry is a case in point. Pharmaceuticals are nothing more than chemicals specifically designed for human and other animal intake. The pharmaceutical industry has experienced its share of widely publicized problems arising out of unintended consequences. A 1998 study found that more than 100,000 people die each year in the United States as a result of side effects of drug therapies (Lazarou, 1998). Yet the public's perception of the pharmaceutical industry has less of the negative quality that characterizes its view of the chemical industry, and much less controversy is associated with the introduction and maintenance of drugs than of chemicals.[3] The pharmaceutical industry has no plans to implement an industry code because it would not fulfill a perceived need of its members.

Members of the public experience the benefits offered by the pharmaceutical industry firsthand whenever their health improves after taking a prescribed medicine. Unlike the pharmaceutical industry, which markets its products directly to consumers, industries that have developed codes tend to be commodity manufacturers. They sell to other firms that process their product into something else. Chemical products, for example, nearly always require further processing before marketing to end-users. Most chemical products go through several manufacturing processes, often undertaken at different firms, before final sale (Rees, 1997). Similarly, many firms in the oil, wood pulp, and textile industries rarely market their products directly to consumers (Bossong-Martines, 1999b, 1999c). They rely on intermediaries to manufacture their products into forms that consumers buy. Public opinion polling by the chemical manufacturing industry has found that many Americans are aware of the risks, but not the benefits, associated with chemical manufacture, even though chemicals are used in the manufacture of hundreds of household products. Polling has shown that many Americans believe they "would be far better off without the chemical industry at all" (Deavenport, 1993:9). The same may hold true for other commodity manufacturing industries that have developed codes.

Firms in commodity industries tend to assume a collective identity in the public's mind. The problems of one company color public perception of the industry as a whole. Firms in commodity industries are therefore more likely to develop environmental codes, which are intended to improve the public image of the industry as a whole. This observation does not hold true, however, for the textile industry. Textile firms have adopted E3 not to improve the image of the entire industry, but to stand out from their competitors as environmentally excellent in order to appeal to customers for whom strong environmental performance is a business need.

The structure of the chemical, oil, and forestry industries offers a further explanation of the emergence of codes. These industries tend toward an oligopolistic structure, with a small number of very large firms dominating the industry (Bossong-Martines, 1999a, 1999b, 1999c). These large firms internalize a large portion of the collective reputation of the industry (Olson, 1965). Large firms are more visible and therefore held responsible for the behavior of the collective. Also, large firms have sufficient resources to cover the relatively high fixed costs of code development.

The chemicals, petroleum, and forestry industries have used codes as defensive strategies to protect themselves from external interference in the form of public regulation. These industries have faced particular problems interacting with the public because of the high environmental impacts of their operations, public distrust, and an inability to demonstrate the value of the products they manufacture. Firms in these industries have been painted with the same brush of environmental irresponsibility, no matter what their actual performance. They have used codes in an attempt to develop a new public identity based on the values of responsibility, caring, partnerships, excellence, and sustainability.

This discussion suggests several hypotheses concerning the conditions that lead trade associations to develop environmental codes. First, industries adopt voluntary codes only if pressed by public opinion or to meet customer demands for strong environmental performance. Second, commodity industries and industries dominated by a few large firms may be more likely to develop codes than industries that market their products directly to consumers or that are made up of small, heterogeneous organizations. Third, codes function mainly to deflect regulation rather than reduce environmental impact (Harrison, this volume, Chapter 16).

EFFECTIVENESS OF TRADE ASSOCIATION ENVIRONMENTAL CODES IN IMPROVING ENVIRONMENTAL PERFORMANCE

Can trade association codes actually lead to improvements in the environmental performance of members? Firms that belong to the trade associations that have developed codes have a common interest in fostering public approval and a favorable regulatory climate for their industry. They may have antagonistic interests when it comes to implementing environmental practices that impose costs on their operations. If rational, self-interested managers know that other members of their group are investing in environmental performance improvement, they may not make this investment themselves (Olson, 1965).

Do codes promote improvement or provide shields to hide poor performance? In this section, this question is explored through two approaches: by considering what codes, in theory, require firms to do, and by examining empirical evidence.

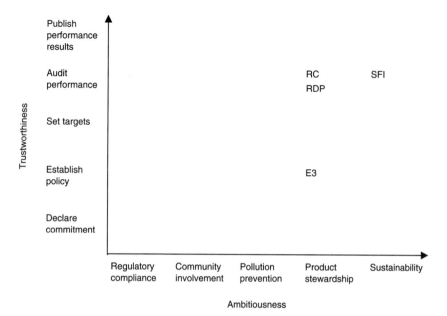

FIGURE 14-1   Trade association codes vary in the ambitiousness of the objectives they establish and their trustworthiness or reliability as guides for action. (E3=Encouraging Environmental Excellence; RC=Responsible Care; RDP=Responsible Distribution Process; SFI=Sustainable Forestry Initiative.)

Establishing Environmental Objectives for Managers

A key mechanism by which trade association codes of practice could change environmental performance is by changing the values of managers. Trade association codes could change values by establishing new environmental objectives for member firms. Firms that sign on to the Sustainable Forestry Initiative pledge to "promote habitat diversity" and "practice a land stewardship ethic" (AF&PA, 2002b). Firms that participate in Responsible Care must implement a pollution prevention program that achieves "ongoing reductions in wastes and releases, giving preference first to source reduction, second to recycle/reuse, and third to treatment" (American Chemistry Council, 2002b). These objectives, if taken seriously by managers, could change what they consider important and how they act.

The environmental objectives embodied in trade association codes can be visualized as a spectrum, as in Figure 14-1. These objectives range from compliance with regulation (a requirement for any organization, whether or not managers have signed on to a trade association code) to sustainable business practices. Regulatory compliance is a minimal level of ambitiousness, while sustainability represents the most ambitious environmental objective for firms. All of the trade association codes listed in Table 14-1 call on firms to practice product stewardship, an environmental objective near the ambitious end of the spectrum. Product stewardship guidelines call on firms to extend their responsibility for environmental protection beyond their fencelines and to oversee the environmental practices of their suppliers and customers.

How trustworthy are these trade association calls to action? Trustworthiness implies that there is consistency between espoused objectives and managers' actions. Codes vary in the degree to which they require specific practices geared toward achieving code objectives. At a minimum, trade associations simply require that managers declare their commitment to code objectives. The textile industry's Encouraging Environmental Excellence code requires only that members describe how they have "worked with suppliers and customers to address environmental concerns" (American Textile Manufacturers Institute, 2002). The chemical distributors' code, in contrast, specifies that members must "work with end-use customers to foster proper use, handling, and disposal of products commensurate with product risk" and "cease doing business with customers whose practices are inconsistent" with the code (National Association of Chemical Distributors, [NACD], 1997:8). The chemical distributors' code calls for actions that are consistent with stated objectives. It is therefore more trustworthy than the textile industry's code. The trustworthiness, or reliability, of codes to bring forth action consistent with stated objectives is depicted in Figure 14-1. Declaring commitment, establishing policies, setting targets, auditing performance, and publishing performance results correspond to higher levels of trustworthiness.

Trade associations provide discretion to members to meet code commitments in their own way, at their own pace. Importantly, with the exception of the requirement in some codes to achieve regulatory compliance, codes do not set performance standards. For example, ACC's distribution code requires that companies "implement...chemical distribution risk reduction measures that are appropriate to the risk level" (American Chemistry Council, 2002b). Companies use their own judgment about what constitutes an "appropriate" response.

Ensuring Performance Through Trade Association Oversight

The role of trade associations in monitoring members' code adoption, and sanctioning members that fall behind, has begun to take shape in recent years. This evolution is particularly apparent for codes in the chemical industry, Responsible Care and Responsible Distribution Process. The ACC board of directors voted to require Responsible Care adoption as a condition of membership in 1989. At first there was no deadline for implementation, and individual members' progress was known only to a consultant hired to tabulate results for the membership as a whole. In 1996 the ACC board set December 31, 1999, as the date when all members were expected to have fully implemented all management practices. The same year the board decided to disclose the names of firms whose Responsible Care programs were lagging to the board's Responsible Care committee. These firms were contacted by board and staff members and urged to do more. Reportedly, some firms resigned under pressure to improve Responsible Care performance. ACC's position is that it has not expelled any members. In 1998, the ACC began to require firms to establish at least one performance goal and to publicly report progress toward meeting it (American Chemistry Council, 2002c). In June 2000 the board decided to rank some aspects of members' code performance on a scale of 1 to 191 (the number of member companies), and distribute this ranking to its membership (Doyle, 2000).

In 1994, ACC introduced the option of management systems verification (MSV) to ensure that a firm has a system in place to meet code requirements, but not to assess the performance of these systems. For example, an MSV for Responsible Care would ensure that a company had a documented plan for responding to chemical transportation incidents. It would not evaluate the effectiveness of the plan. ACC has hired a private consultant, Verrico Associates, to conduct all MSVs for members. Verrico assembles a verification team made up of chemical industry managers and selected external stakeholders. ACC requires that the team include a community participant. The team interviews company personnel who have been assembled into panels that combine functional areas. For example, a panel of managers from risk assessment, distribution, and sales might be brought together and asked questions concerning the company's product stewardship activities. ACC's protocol for MSV lists the questions each panel is to be asked. The panel responsible for product stewardship activities, for example, is asked, "How does your company assess risk for existing products?" and "How do you track the performance of your customers and review it with them?" The verification team also walks around the plant, randomly interviewing employees, and talks with facility neighbors, suppliers, and distributors. Verrico Associates prepares a report of "findings and opportunities" identified through the verification. The report is owned by the company, and managers decide with whom they will share it (ACC, 2002a).

MSV is discretionary for ACC members. As of July 2000, approximately half of all members had had their Responsible Care programs reviewed. A recent development in the automobile industry may encourage more chemical companies to undergo verifications. In September 1999, Ford and General Motors (GM) announced that they will require all of their first-tier suppliers to be certified to ISO 14001, the international environmental management system developed by the International Organization for Standardization (ISO). ACC staff members have negotiated with automakers to convince them that Responsible Care is at least equivalent to ISO 14001. Ford remains skeptical of ACC's verification procedures. In early negotiations with ACC, Ford is insisting on having an independent third party conduct an ISO audit. As one possible solution, Verrico Associates may partner with ISO certification companies for future verifications (Schmitt, 2000).

In response to manufacturers' questions about the environmental practices of distributors, in 1991 the National Association of Chemical Distributors (NACD) launched a program of its own called Responsible Distribution Process (RDP). The system for monitoring and sanctioning established by NACD goes beyond Responsible Care in several respects. For NACD, management systems verification is mandatory, not discretionary. NACD uses third parties rather than industry peers to conduct the verifications. NACD has a history of suspending and terminating memberships for noncompliance, while ACC staff members emphasize that they work with lagging firms to improve their performance. Finally, NACD's verification system includes an option for review of environmental performance (NACD, 2002b). ACC's review only ensures that a management practice is in place.

Initially NACD required biannual self-assessments from members, the first of which was due on July 1, 1992. The NACD board of directors suspended the memberships of several companies for not meeting this deadline (Morris, 1993), although all of these companies later fulfilled NACD's requirements and rejoined the trade association (Morris, 1995). In October 1994, NACD began to require companies to mail their environmental policies to Underwriter Laboratories, a third-party verifier, to ensure compliance with RDP. The memberships of three companies were terminated in 1995 for refusing to participate (Morris, 1995). In May 1998, NACD voted to require members to submit to on-site, third-party audits of their management systems by Science Applications International Corporation (SAIC). This review went beyond mail-in policy verification by ensuring that code management practices were actually in place. Nine companies had memberships terminated for refusing to undergo this on-site review. Although NACD publicly states that it has terminated some firms' memberships, the association refuses to make public the names of these members.

The impetus for these requirements came from NACD's membership. Many NACD members were frustrated by the demands placed on them by supplying chemical manufacturers. Contracts signed with ACC members granted manufacturers free license to audit distributor facilities. Auditing distributors' environmental and safety practices is required by Responsible Care. NACD members were encouraged to adopt a unified, third-party auditing protocol to put an end to the logistic problems faced by distributors having to undergo different assessment protocols from each one of their suppliers. In addition, NACD chose to form its own auditing protocol, rather than adopt a protocol created by ACC, because many members had suppliers outside of ACC (Morris, 1997).

NACD members are required to undergo verification every 3 years, a cycle that began in January 1999. By April 2000, SAIC had conducted more than 120 verifications. Companies found by SAIC to have deficient management systems are given one year to correct identified problems and pay for an additional reverification. As of April 2000, SAIC had found that three companies required reverification. Verified companies are granted ownership of SAIC's report. This report usually is not made available to the public.

As already noted, management systems verifications ensure that a firm has management practices in place, but do not assess how well those practices are actually working. A group of ACC members has maintained that MSVs do not provide sufficient assurance. These manufacturers have negotiated with NACD to create an additional form of performance verification to be used for distributors that handle particularly hazardous chemicals. The chemical manufacturers that have participated in these negotiations with NACD are Dow Chemical, Eastman Chemical, ExxonMobil, FMC Corporation, Shell Chemical, Stepan, and Vulcan Chemical. Negotiations have resulted in a protocol called Site Class Verification (SCV) (NACD, 2002a).

NACD staff members explain that ACC members are under pressure to fulfill their product stewardship code, which requires that they ensure that distributors live up to the environmental protection practices of Responsible Care. The SCV process helps manufacturers decide whether a distributor is a suitable business partner. Although an MSV might indicate that a distributor had a documented procedure for unloading hazardous chemicals from trucks, for example, SCV would describe how trucks were actually unloaded at a distributor's facility. The costs of Site Class Verifications are paid by a group of 18 chemical manufacturers. Before establishing or renewing a business relationship, these manufacturers can obtain an SCV report on the distributor's environmental conduct. SCVs, unlike MSVs, are not required by NACD as a condition of membership because not all distributors do business with this group of chemical manufacturers.

The programs trade associations are using to monitor and sanction code performance are depicted in Figure 14-2. No U.S. trade association yet requires public disclosure of the results of verifications. Although the textile industry has not established a verification program, planning is underway to put such a program into place.

Empirical Evidence

Just as relatively few government-sponsored voluntary programs have been subject to careful evaluation (Mazurek, this volume, Chapter 13; Harrison, this volume, Chapter 16), only a handful of published studies have documented how firms respond to trade association codes. In 1995 a team of researchers explored Responsible Care adoption at 16 mid-sized firms (Howard et al., 2000). Authors found four general types of responses: drifters, promoters, adopters, and leaders. Drifters were companies that said Responsible Care had little impact on their activities. Changes were limited to documenting existing practices. Promoters, who used Responsible Care mainly to promote a strong environmental reputation to external stakeholders, saw Responsible Care as an adjunct to existing

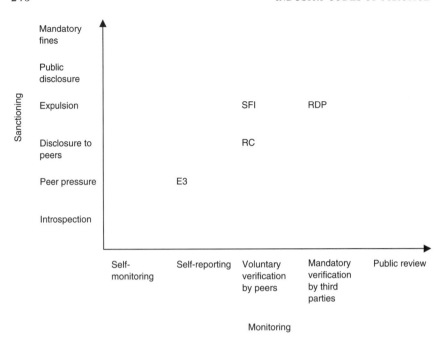

FIGURE 14-2   Trade associations use a range of approaches to monitor code adoption and to sanction laggards. (E3=Encouraging Environmental Excellence; RC=Responsible Care; RDP=Responsible Distribution Process; SFI=Sustainable Forestry Initiative.) Source: Lenox (1999).

environmental programs.   It reinforced what they were already doing, but did not cause them to rethink their activities.   People in this group spoke of Responsible Care as "formalizing" and "standardizing" what they already did.

Adopters were firms that saw Responsible Care as a valuable tool for improving their environmental practices.   Not only were environmental and communications staff handling Responsible Care activities; product managers, designers, and marketing staff also were involved.   Finally, leaders spoke about Responsible Care being a "whole new way of thinking."   They believed that their environment, health, and safety practices were strong prior to Responsible Care, but the initiative offered a way to go further.   In these firms, significant resources had been applied to Responsible Care implementation, and senior management took an active role in overseeing it.

While noting substantial variation in adoption practices, Howard et al., 2000, also found that a number of practices had been implemented by virtually all of the companies interviewed.   The most significant common practice was increased involvement by employees in local community relations.   Many interviewees expressed the view that interacting with the community was the whole purpose of Responsible Care.

A second common response was in the area of distribution practices. All of the participating companies said they now require much more of their distributors than they had before Responsible Care. All 16 companies had put in place an audit system to assess their carriers' safety and handling practices. They require distributors to provide them with documentation of their procedures, and in many cases chemical company employees inspected their distributors' facilities. Several of the firms had offered training programs to distributors, and a handful had ceased to use distributors that did not meet criteria under Responsible Care.

Responsible Care's impact on toxic emissions was studied by King and Lenox (2000). These researchers compared toxic releases reported to the U.S. Environmental Protection Agency's Toxics Release Inventory of Responsible Care firms and chemical firms that do not participate in Responsible Care during the period 1990 through 1996. The authors found that firms that participate in Responsible Care reduce their toxic releases no faster than comparable chemical firms that do not participate. They argue that the lack of mechanisms for observing and sanctioning individual firm performance has led to free-riding by low-performing firms. Although some ACC members are improving environmental performance faster than the norm, a large group is lagging behind, slowing progress for the group overall. The authors conclude that the "commons" being protected by Responsible Care is not the "physical commons" (King and Lenox, 2000:713) of a clean and healthful environment. Rather, Responsible Care is intended to protect a "reputational commons" (King and Lenox, 2000:713) that has been weakened by the industry's past environmental practices. Without the threat of sanctions by informed outsiders, opportunism has eroded Responsible Care's effectiveness.

It is important to note some of the limitations of the Howard et al. and King and Lenox studies. Both studies report results from the years prior to ACC's recent attempts to improve its oversight of members' Responsible Care progress. Management systems verifications, introduced in 1994, only recently have become common practice. King and Lenox's study only observes changes in toxic releases, while Responsible Care addresses many other aspects of environmental performance. King and Lenox do not attempt to assess those aspects of Responsible Care that Howard and colleagues identified as particularly robust—community participation and oversight of distributors. Yet the studies suggest that ACC board and staff members have more work to do to ensure that Responsible Care functions as a reliable system of industry self-regulation. Studies suggest that firms adopt Responsible Care in their own way, at their own pace, and that results in terms of environmental performance vary substantially.

This discussion of the effectiveness of trade association codes in improving environmental performance suggests two hypotheses. First, effectiveness depends on the ambitiousness of the objectives that trade associations set, and the degree to which the code is designed to foster actions that are consistent with these objectives. Second, effectiveness depends on the strength of trade association monitoring and sanctioning programs (Herb et al., this volume, Chapter 15).

CONCLUSION: WHERE CODES MAY BE ACHIEVING RESULTS

Trade associations have developed environmental codes to demonstrate to critical public audiences that members are voluntarily controlling their environmental behavior. Empirical studies of Responsible Care suggest that this code— the most highly developed of all U.S. trade association efforts in environmental self-regulation—has failed to reliably improve firms' internal management practices. When it comes to how environmental management is practiced within the plant, Responsible Care appears to reinforce existing norms rather than bring about higher standards. Adoption practices appear to vary substantially, depending on managers' preexisting commitments to environmental protection.

Trade associations are in a constant battle for membership and must walk a fine line between being inclusive and commanding minimal standards. The mechanisms they have developed for monitoring and sanctioning laggards have been limited. Trade associations are taking steps to strengthen these areas, but their ability to establish authority over members is uncertain. With the exception of Responsible Distribution Process, which requires external verification as a condition of membership, firms choose whether to have their management systems externally verified. About half of ACC's members have had their systems reviewed, and about 36 of AF&PA's membership have taken this step (AF&PA, 2002a). Those who choose management systems verification own the results and need not share them.

When managers of a firm know their environmental performance will not be observed, it may be in their rational self-interest to invest less heavily in environmental performance improvement than their competitors (Olson, 1965). Studies of Responsible Care adoption suggest that some members are using this code to deflect criticism and hide performance, although ACC's recent steps to establish performance goals and improve monitoring may, over time, change this result.

Although managers' responses to Responsible Care vary with respect to internal operations, this code has fostered a fairly uniform response in the ways managers interact with external constituencies. Responsible Care has had a strong impact on managers' oversight of their distributors. One manifestation of this impact is the decision by the NACD to develop an environmental code of its own, based on Responsible Care. This code includes programs to observe and sanction members that are considerably stronger than programs of other trade associations. Chemical distributors know that, to fulfill the requirements of Responsible Care, chemical manufacturers will need to review their distributors' environmental practices. To appeal to these suppliers, and establish some control over the review process, distributors have initiated their own program. The NACD example suggests that, as the motivation for developing and maintaining an environmental code shifts from defensive public relations to an appeal to customers, code requirements also may shift. As codes are used increasingly in business transactions, trade association programs for monitoring and sanctioning members' performance may become more common and effective.

This discussion suggests that some aspects of environmental performance may be more amenable to private regulation than others. An industry's management of the environmental practices of its suppliers, distributors, and customers may pose fewer conflicts than its management of its own members. Trade associations appear to be fostering a system of what could be called *lead industry regulation,* rather than industry self-regulation. Under a system of lead industry regulation, large firms that internalize a large portion of the collective reputation of the industry, such as members of ACC, establish environmental management practices for the industries and firms that do business with them, such as those represented by NACD. Ford and GM's decision to require their suppliers to become certified to ISO 14001 is a further example of lead industry environmental regulation.

A direct and effective sanction is simply to discontinue a business relationship. Through its distribution and product stewardship codes, ACC members have used this sanction effectively, and NACD has responded with a code of practice that complements Responsible Care. NACD's sanctioning authority over its own membership is substantially greater than ACC's. Its members have learned that environmental performance is a component of their business success, and they may believe they benefit from an environmental code that clarifies customer expectations and reduces transaction costs.

Although in the past trade association codes served primarily as defensive measures to improve public opinion and forestall public regulation, codes are now assuming a role in business. Trade associations are adding measures to observe the environmental practices of their business partners, and to sanction, with a decision to do business elsewhere, those who do not live up to their codes. This observation leads to a final hypothesis: Environmental codes may be most effective when large, publicly recognized businesses enforce them on their trading partners.

Will stronger monitoring and sanctioning programs lead to better environmental results? Lead industry regulation may repeat the problems of the public environmental regulatory system: inefficiency, rigidity, and limitations in scope. Empirical studies of the effectiveness of codes such as Responsible Distribution Process, which incorporate the elements of monitoring and sanctioning still missing from Responsible Care, are needed to assess the role of these efforts in environmental protection.

In undertaking this research, it will be important to compare the environmental performance of firms participating in trade association codes with similar firms that do not (King and Lenox, 2000). Through such comparisons it may be possible to understand the features of codes that do the most to foster performance improvement: ambitious and trustworthy objectives, or stringent sanctioning and monitoring. An additional research strategy would be to take advantage of the "natural experiment" (Coglianese and Nash, 2001) of customer mandates for code adoption. Customers of ACC members, for example, are expected to adopt management systems that achieve the objectives of Responsible Care. Researchers could ask whether firms that distribute the products of ACC members achieve higher levels of environmental protection than those that distribute the products of non-ACC firms. Such studies might be helpful to environmental regulators as they consider the role of trade association codes in public policy. Trade associations that require their trading partners to implement code practices are assuming the role of environmental regulators of their supply chain. The success—or failure—of their attempts to use codes to achieve higher levels of environmental protection could provide valuable lessons to public-sector regulators.

Trade association codes of environmental management practice are proliferating and growing stronger. Research has only begun to test the potential of these codes as tools in environmental protection. Understanding the role of trade association codes will become increasingly important as more organizations require code adoption as a condition of business.

NOTES

1 These environmental codes were identified in a 1999 survey of trade associations (Nash, 1999). Since that time, several trade associations, such as the National Association of Metal Finishers, American Furniture Manufacturers Association, and American Portland Cement Alliance, have shown interest in code development and have begun to launch programs. Other trade associations, such as the Steel Manufacturers Association, have established guiding environmental principles.

2 For a fuller discussion of the codes listed in Table 14-1, see Nash (1999).

3 Additional factors may explain the public's relatively positive view of the pharmaceutical industry despite the risks its products pose. For example, exposure to the risks of pharmaceuticals is voluntary (Slovic, 1987), while exposure to the byproducts of chemical manufacturing is rarely a matter of choice.

REFERENCES

American Chemistry Council
     2002a Management Systems Verification. [Online]. Available: http://www.americanchemistry.com/ [Accessed February 14, 2002].
     2002b Responsible Care General Information. [Online]. Available: http://www.americanchemistry.com [Accessed February 14, 2002].
     2002c Responsible Care Performance Goals. [Online]. [Available: http://www.americanchemistry.com/ rcperformance [Accessed February 14, 2002].

American Forest & Paper Association

    1998   Sustainable Forestry for Tomorrow's World: 1998 Progress Report on the American Forest & Paper Association's Sustainable Forestry Initiative (SFI) Program. Washington, DC.

    2002a AF&PA Member Companies and Licensees That Have Completed or Committed to SFI Third Party Certification. [Online]. Available: http://www.afandpa.org/forestry/sfi_frame.html [Accessed February 14, 2002].

    2002b SFI Program: A Bold Approach to Sustainable Forest Management. [Online]. Available: http://www.afandpa.org/forestry/sfi_frame.html [Accessed February 14, 2002].

American Textile Manufacturers Institute

    2002   E3: Encouraging Environmental Excellence. [Online]. Available: http://www.atmi.org/programs/e3.asp [Accessed February 14, 2002].

Bossong-Martines, E.M.

    1999a Chemicals: Basic. *Standard & Poor's Industry Surveys* 167(2).

    1999b Oil & gas: Production & marketing. *Standard & Poor's Industry Surveys* 167(10).

    1999c Paper & forest products. *Standard & Poor's Industry Surveys* 167(15).

Bradley, J.F.

    1965   *The Role of Trade Associations and Professional Business Societies.* University Park: Pennsylvania State University Press.

Coglianese, C., and J. Nash

    2001   Toward a management-based environmental policy? Pp. 222-234 in *Regulating from the Inside: Can Environmental Management Systems Achieve Policy Goals?* C. Coglianese and J. Nash, eds. Washington, DC: Resources for the Future Press.

Deavenport, E.W.

    1993   Taking the Fear Out of Chemicals. Speech to National Association of Chemical Distributors Annual Meeting, Laguna Niguel, CA, Dec. 1, 1993. Chemical Manufacturers Association, Arlington, VA.

Doyle, D.

    2000   Perspectives on Industry Environmental Codes. Presentation at U.S. Environmental Protections Agency workshop, Industry Environmental Codes of Conduct: What Do They Mean for Public Policy? Washington, DC, June 15.

Hoffman, A.J.

    1995   The Environmental Transformation of American Industry: An Institutional Account of Organizational Evolution in the Chemical and Petroleum Industries (1960-1993). Doctoral dissertation, Department of Civil and Environmental Engineering and Sloan School of Management, Massachusetts Institute of Technology, Cambridge, MA.

Howard, J., J. Nash, and J. Ehrenfeld

    2000   Standard or smokescreen? Implementation of a voluntary environmental code. *California Management Review* 42(2):63-82.

Islam, M.

    1999   Assessing the Role of Self-Governance in Promoting Environmental Responsibility: A Case Study of the Encouraging Environmental Excellence Program at the American Textile Manufacturers Institute. Master's thesis, Technology and Policy Program at the Massachusetts Institute of Technology, Cambridge, MA.

King, A., and M. Lenox

    2000   Prospects for industry self-regulation without sanctions: A study of responsible care in the chemical industry. *The Academy of Management Journal* 43(4):698-716.

Lazarou, J.

    1998   Incidence of adverse drug reactions in hospitalized patients. *Journal of the American Medical Association* 279(15):1200-1205.

Lenox, M.
   1999   Industry Self-Regulation for Environmental Performance. Presentation to 1999 Green-
           ing of Industry Conference, Self-Regulation Workshop, Chapel Hill, NC, November
           16.
Morris, G.D.L.
   1993   Responsible Care: Distributors get on board. *Chemical Week,* July 7/July 14, p. 45.
   1995   Deadlines keep members on their toes; pursuing external verification. *Chemical Week,*
           July 5/July 12, p. 60.
   1997   Associations confront verification realities. *Chemical Week,* January 1/January 8, p. 51.
Morrissey, J.A.
   1999   *Making It Happen: The First 50 Years of ATMI.* Washington, DC: American Textile
           Manufacturers Institute.
Nash, J.
   1999   *The Emergence of Trade Associations as Agents of Environmental Improvement.* Report
           prepared for U.S. Environmental Protection Agency Emerging Strategies Division. Cam-
           bridge, MA: Massachusetts Institute of Technology.
National Association of Chemical Distributors
   1997   *The National Association of Chemical Distributors: The Responsible Distribution Pro-
           cess.* Arlington, VA: National Association of Chemical Distributors.
   2002a   Comparison of National Association of Chemical Distributors' MSV and SCV. [On-
           line]. Available: http://www.nacd.com/rdp/msvscv.htm [Accessed February 19, 2002].
   2002b   Responsible Distribution Process. [Online]. Available: http://www.nacd.com/rdp/
           default.htm [Accessed February 19, 2002].
Oil & Gas Journal
   1990   Get off environmental sidelines. *Oil & Gas Journal* 88(1).
Olson, M.
   1965   *The Logic of Collective Action: Public Goods and the Theory of Groups.* Cambridge,
           MA: Harvard University Press.
Rees, J.
   1997   Development of communitarian regulation in the chemical industry. *Law and Policy*
           19(4):477-528.
Schmitt, B.
   2000   Fans root for MSVs, others stay on sidelines. *Chemical Week,* July 5/July 12, p. 49.
Slovic, P.
   1987   Perception of risk. *Science* 236(17):280-285.

15

Harnessing the "Power of Information": Environmental Right to Know as a Driver of Sound Environmental Policy

Jeanne Herb, Susan Helms, and Michael J. Jensen

A REVOLUTIONARY CONCEPT

T he Emergency Planning and Community Right-to-Know Act (EPCRA) of 1986 (under the Superfund Amendment and Reauthorization Act, Title III) delivered a new concept in federal environmental policy: the concept of government as a broker of information to which the public has the "right to know." Up until that time, environmental policy at the state and federal level was dominated by air-, water-, and hazardous waste-specific "command-and-control" regulatory approaches. Those conventional approaches oversaw industry operations from a fragmented perspective; overly prescribed expensive, technology-based solutions to industry rather than providing incentives for sound and lasting environmental performance; and did not recognize the role or value of public accountability. The initial reports from the Toxics Release Inventory (TRI) under EPCRA in 1987 were eye opening to most observers both within and outside the U.S. Environmental Protection Agency (EPA) (Wolf, 1996). The sheer size of allowable emissions reported under TRI, the cross-media transfers of pollutants, and the uncertainty of risk implications led the public, regulators, and industry managers to question the viability of traditional regulatory approaches.

TRI's results were so significant, in fact, that they contributed to a rethinking of regulatory mechanisms within and beyond the EPA (1997). Although legislation predating TRI, such as the 1974 Freedom of Information Act, may be said to have established the public right to access information, it was the Toxics Release Inventory that shortcut the bureaucratic information request process and made data truly accessible. It made popular the notion that "providing public access to environmental databanks is an innovative effort to reduce the role of big government bureaucracy" (Khanna et al., 1998:33). The concept of "the power of information" has become a cornerstone of EPA efforts to develop environmental policies that not only deliver better environmental results, but also do so more cost-effectively (U.S. Environmental Protection Agency, 1999). Some argue that the concept of public right to know has emerged as "a political idea that justifies mandatory and quasi-mandatory 'voluntary' reporting programs that the Agency could only imagine a few years ago" (Outen, 1999). Right-to-know laws since have expanded from environmental health (such as TRI, Occupational Safety and Health Administration (OSHA) material data safety legislation, and carcinogens under California's Proposition 65) to cover issues as diverse as discrimination (such as in the Home Mortgage Disclosure Act) and customer service statistics (airline delay and baggage-handling records) (Graham, 2000).

Triggered by the 1984 chemical accident in Bhopal, India, Congress enacted EPCRA based on the philosophic underpinnings that citizens who are informed about hazardous chemicals in their communities can make more educated decisions about their own protection. Yet putting chemical release data that are gathered via TRI into the public domain is intended to have an even broader effect than on citizens in local communities. Although the political rhetoric surrounding EPCRA was on empowering local citizens to make personal decisions, the sheer public nature of the data is intended to drive environmental performance at industrial facilities by affecting a host of sectors in society, as illustrated in Figure 15-1.

SUCCESSFUL RESULTS

TRI generally is seen as a tremendous success. The U.S. General Accounting Office estimated that "over half of all [TRI] reporting facilities made one or more operational changes as a consequence of the inventory program" (U.S. General Accounting Office, 1991:61). The EPA credits a 40-percent reduction in toxic chemical releases to TRI (Outen, 1999). This may be an exaggeration, because many state and local regulations and programs targeted chemicals that were listed under TRI. However, the fact that an effect of such magnitude could be attributed to this single law is astounding, particularly when compared with the small impacts of much more expensive and complicated legislation (Fung and O'Rourke, 2000).

TRI's strengths also are accompanied by weaknesses. The minimal oversight of the EPA in TRI reporting and interpretation is a significant (if initially unintentional) feature of the policy, which has other consequences besides taxpayer savings. The EPA sets no standards, gives only a limited technical review to the self-reported data, and has an extremely weak monitoring and enforcement system—inspections are limited to a small percentage of reporting firms per year (Fung and O'Rourke, 2000). Because the universe of chemicals regu-

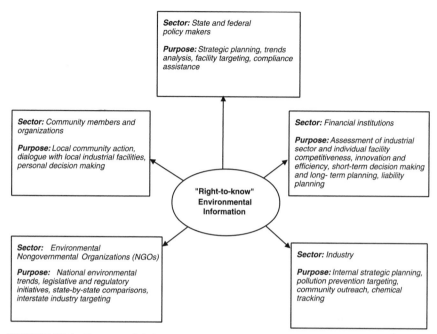

FIGURE 15-1   Impacts of right to know on societal sectors.

lated under TRI remains small compared to the universe of chemicals in commerce, companies can in some cases switch to similar, off-list chemical variants to avoid reporting (Dernbach, 1997). In addition, "phantom reductions" have been reported due to changes in reporting methodologies such as redefining on-site recycling as in-process recovery (Nathan and Miller, 1998). In what is perhaps TRI's greatest shortcoming, nearly a third of companies each year simply fail to report (Fung and O'Rourke, 2000).

The general hands-off approach to TRI data also has consequences on the user end. The data gathered under TRI sometimes are considered difficult for citizens to understand and fully utilize (Grant, 1997; Helms, 1997). Reporting firms contend that the data can be misrepresentative without context. Critics also rightly emphasize that chemical volumes are alone poor indicators of human health risks. These interpretive weaknesses increasingly are being addressed by both governmental and nongovernmental tools, as noted later in this chapter.

The deepest criticism leveled at TRI, and right to know generally, questions the underlying basis of disclosure—that available information will lead to more rational decision making. Often there will be some scientific or practical dispute regarding what interpretation of the data is rational. Information also can be used in negative ways, such as undermining proprietary business rights or for terrorist sabotage—an argument used to restrict EPA disclosure of the potential impact of chemical accidents (Tetreault, 2000). Do the new risks accompanying disclosure compare with the old risks hidden before disclosure? Does right-to-know information create another threat, or by helping to reduce toxics, eliminate the root of this threat?

Despite its widely recognized limitations, TRI continues to be viewed as an environmental success story. Its success is recognized by policymakers, industrial managers, and environmentalists (Hearne, 1996). Popular and academic research points to the tangible impact that TRI has had on the various societal sectors; TRI is now believed to be used regularly by citizens, community organizations, organized environmental groups, industry managers, state and federal agencies, lawyers, investment advisors, and the media, as will be described in the following paragraphs (Fung and O'Rourke, 2000).

Community Members and Organizations

One early survey indicated that TRI was being used in a variety of ways, including:

- Advocating for legislative and regulatory changes,
- Comparing a facility's emissions against permit records,
- Exerting public pressure on facilities,
- Planning for emergencies,
- Supporting direct negotiations between industry and citizens (Lynn and Kartez, 1997).

National environmental organizations have documented examples in which citizen organizations have directly used TRI data in organizing community efforts, negotiating with individual facilities, and advocating for new environmental programs (Working Group on Community Right to Know, 1991). They have also produced guidance for local community groups to use in applying TRI as a tool in community action efforts (Wise and Kenworthy, 1993).

Financial Institutions

Investment, insurance, and debt markets operate based on corporate disclosure of financial statistics and risk (for example, the information regulated by the Securities and Exchange Commission). Corporate environmental information, including TRI emissions, is used to a limited extent for purposes such as gauging liability, investment decision making on an ethical basis, or evaluating firm or industry standing in regard to strategic environmental opportunities and threats. Event studies indicate that environmental disclosures, including TRI release data, can affect stock prices in the short term (Hamilton, 1995). Studies also seem to indicate a small but positive correlation between firms' longer term stock prices and their environmental performance, which encourages the concept that financial markets may at some point provide incentives for firms to improve their environmental behavior (Konar and Cohen, 1997).[1]

Industry

Part of the philosophic underpinning of TRI is that the public disclosure of industry environmental performance will motivate industry to take actions to prevent itself from being viewed as a poor environmental performer. Analyses both inside and outside EPA point to such results, while acknowledging that it is impossible to draw a direct and conclusive connection between TRI alone and industry environmental management. One survey reported that 67 percent of firms reduced TRI released per unit of production and noted that protecting the health of facility employees and complying with state or federal regulations outranked all other factors as to why they reduced releases (Santos et al., 2000). Less research is available pertaining to the effect that TRI data have as an effective tool for internal industrial strategic planning. However, information that is available tends to note that TRI can serve as one of many tools for industry to internally identify priorities, engage in dialogue with the local community, and track toxics (Kerr et al., 2000). TRI also has played an integral role in voluntary pollution reduction efforts within industrial sectors, such as the American Chemistry Council's Responsible Care program.

National Environmental Organizations

TRI has become the cornerstone of several leading national environmental organizations' efforts to track industrial environmental performance trends, promote local citizen access to environmental information, undertake direct negotiations with industrial facilities (Natural Resources Defense Council, 1999), and promote changes to environmental policies. Organizations that are active users of the data include the Working Group on Community Right to Know, National Environmental Trust, Environmental Defense, Natural Resources Defense Council, the U.S. Public Interest Research Group, Good Neighbor Project, and the watchdog group, OMB Watch. In addition to applying data gathered under TRI, part of these national organizations' efforts involve making the data more accessible, contextualized, and useful to citizen organizations (see http://www.scorecard.org).

State and Federal Policy Makers

Clearly, TRI has been a tremendous resource for government agencies at the federal and state levels. It contributed to the design and development of several voluntary initiatives, including the EPA 33/50 and Performance Track programs. It is an important internal planning tool for various regulatory and nonregulatory initiatives at the EPA, and it provides an annual assessment of toxic chemical releases nationally. States have employed the data in a variety of ways, including developing pollution prevention programs, implementing compliance assistance efforts, formulating internal strategic plans, and tracking industrial pollution trends (Kerr, 2000; Aucott et al., 1996).

INFORMATION AS A DRIVER OF ENVIRONMENTAL POLICY

The general success of TRI leads us to consider how to best maximize the concept of using environmental information and its disclosure to cost-effectively drive improvements in environmental results. Several elements are critical to the successful implementation of environmental information. Although they are interrelated and mutually reinforcing, it is useful to disaggregate them and consider them separately, as one might examine each link in a chain. Broadly speaking, it is critical to have high-quality information and a means of distributing and sharing that information, then empowering people to act on that information. These general stages are listed in Figure 15-2.

Refining information **content,** for example, requires not only that regulatory agencies demand more emissions data from facilities, but that the data requested help answer important questions; that the data are clear with respect to units, time periods covered, and information that may be excluded (e.g., fugitive emissions); and finally, that the data are put into context with regard to a facility's contribution to a particular issue as a whole, changes over time, and potential health impacts.

Similarly, exploring information **uptake** more finely reveals that for information to be distributed effectively, it needs to be widely available to those with and without computers, and people need to know the information exists. In this arena, environmental information managers can borrow tools from the marketing industry to make their product as user-friendly and widely recognized as possible.

Information may satisfy the curiosity of some people, but it will do little as a lever for change unless users are empowered with the tools for **action.** Specifically, users need both a meaningful reason (motivation) to track down information and the skills to take advantage of it, whether it influences individual choices such as the products they purchase or where they live, or actions at a broader level such as dialogue with polluters, watchdog coalitions, letter campaigns, or legal actions. Information is power, and the goal of improving environmental information should be to empower all stakeholders to make fully informed decisions.

Finally, although the stairstep diagram appears to be a unidirectional model, it is important to revisit each step over time so that each step can be enhanced by what society learns. For example, if people are not motivated to retrieve available information, then providers should listen to those potential users and determine what the barriers might be. Then providers will be able to address the clarity and content or the availability and user-friendliness of the information.

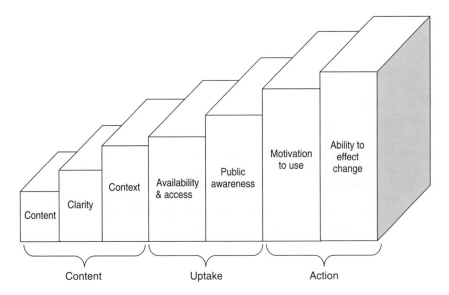

FIGURE 15-2   Information disclosure: Elements of success.

Thus, achieving success is not a matter of one trip up the steps, but a matter of making numerous revisions along the way.

FUTURE ENVIRONMENTAL POLICIES

Recognizing the contributions that have been made by TRI and the right-to-know concept overall, and considering the necessary elements to ensure the success of environmental information, it is timely to consider what new policies are needed to maximize the impact that environmental information can play as a driver of policy. Four recommendations follow:

1. Promote states' use of TRI and general right-to-know concepts. TRI is a federal program that is, appropriately, not delegated to states. In addition, states are not provided federal funds in implementing right-to-know programs (Grant, 1997). We are in an era of increasing devolution of responsibility for management of environmental regulatory programs from the federal to the state level. States need to be educated in the value of right to know and TRI, specifically. Incentives, including financial incentives, should be provided to states that reflect right-to-know concepts in implementation of federally delegable programs. Training and guidance should be provided to states in how to effectively use the information resources that are currently available in strategic planning and day-to-day management of regulatory programs.

2. Expand the concept of right to know beyond TRI. Clearly, the "power of information" has been evidenced in the TRI experience. The EPA, Congress, and the states should consider that experience when developing new initiatives by ensuring that information disclosure and public accountability components are included. This is particularly relevant in this time of regulatory reinvention at the state and federal levels. It is one thing for the EPA and the states to "talk the environmental information talk," but certainly another to walk the walk. New voluntary initiatives that are being rolled out need to contain robust public accountability—right to know— components to ensure that the regulatory incentives often provided by these programs are balanced by sound public tracking of industrial performance and program results.

3. Improve the data that are already collected. Although there is a wide array of activities at the state and national levels to improve information management, there is little being done to actually improve the quality of the data and answer the question: "What information do we really need?" Rather than continuing to propagate an environmental information infrastructure through investment of millions of dollars in systems to manage data with inherent limitations, we should be reconsidering not only *how* data are reported, but *what* data are reported (Helms, 1999). Our goal should be not just to ensure that reported data are efficiently stored and made retrievable, but we also should be enhancing data quality to ensure that data are accurate and can actually help us track compliance and overall environmental performance. A new reporting structure is needed that can have TRI and the right-to-know (RTK) concept as its basis and measurement of overall environmental performance and multimedia pollution prevention (Geiser, 1998). Such a structure can be created to fill in the gaps while, at the same time, eliminating redundancies and unnecessary data.

4. Provide context for data. Although efforts such as the Environmental Defense scorecard and RTK net are tremendous steps forward in providing the public with a context for the mass of information available, they both still face inherent limitations. TRI has been criticized for being misleading. Scorecard has been criticized for being unscientific. Nevertheless, these limitations have to be accepted in the light of no better alternative being available. A recommitment needs to be made to have policymakers, the environmental community, and industry come together to reexamine how to best provide the public with a full understanding of the nature of information to which the public has a right to know. Without a vision for how the public can interpret environmental information, systems cannot be put into place to improve the accessibility and interconnectiveness of TRI and other relevant data.

Within a decade, TRI has moved the concept of right to know from a vague advocacy term to a fundamental policy tool. The success of TRI offers important insights into how information disclosure, in general, can be enhanced to make right to know an even more effective tool in current efforts to reinvent environmental regulation. However, a clear recommitment from the EPA and

Congress to the concept of right to know, a plan for improving the content of environmental information as well as its collection, storage, and retrieval, and a vision for the role of the public and public accountability in environmental policy are needed to ensure that we improve on the success of TRI and maximize the concept of right to know in environmental policymaking.

NOTE

1 For a survey of literature on this issue, see Reed (1998).

REFERENCES

Aucott, M., D. Wachpress, and J. Herb
    1996   *Industrial Pollution Prevention Trends in New Jersey.* Trenton, NJ: Department of Environmental Protection.
Dernbach, J.C.
    1997   The unfocused regulation of toxics and hazardous pollutants. *Harvard Environmental Law Review* 21:1-82.
Fehr, S.
    1999   Worries about public disclosure, threat of terrorism. *The Washington Post,* October 10.
Fung, A., and D. O'Rourke
    2000   Reinventing environmental regulation from the grassroots up: Explaining and expanding the success of the Toxics Release Inventory. *Environmental Management* 25(2):115-127.
Geiser, K.
    1998   Can the pollution prevention revolution be restarted? *Pollution Prevention Review* 8(Summer):71-80.
Graham, M.
    2000   Regulation by shaming. *Atlantic Monthly* 285(4):36-40.
Grant, D.S.
    1997   Allowing citizen participating in environmental regulation: An empirical analysis of the effects of right-to-sue and right-to-know provisions on industry's toxic emissions. *Social Science Quarterly* 78(4):859-873.
Hamilton, J.T.
    1995   Pollution as news: Media and stock market reactions to the toxics release inventory data. *Journal of Environmental Economics and Management* 28:98-113.
Hearne, S.A.
    1996   Tracking toxics: Chemical use and the right to know. *Environment* 38:4-33.
Helms, S.
    1997   *Chemical Use Information Disclosure: Seeking Common Ground Among Business and Environmental Stakeholders.* Boston: Tellus Institute.
    1999   Report card. *Environmental Forum* 16(6):21-29.
Kerr, G.
    2000   Materials Accounting Project, Materials Accounting and P2: A Good Team? Unpublished paper prepared for the Anderson & April, Inc., National Pollution Prevention roundtable, August 15.
Khanna, M., W. Rose, H. Quimio, and D. Bojilova
    1998   Toxics release information: A policy tool of environmental protection. *Journal of Environmental Economics and Management* 36:243-266.

Konar, S., and M.A. Cohen
    1997   Information as regulation: The effect of community right to know laws on toxic emis-
            sions. *Journal of Environmental Economics and Management* 32:109-124.
Lynn, F.M., and J.D. Kartez
    1997   Environmental democracy in action: The Toxics Release Inventory. *Environmental
            Management* 18:511-521.
Nathan, T.E., Jr., and C.G. Miller
    1998   Are Toxic Release Inventory reductions real? *Policy Analysis* 32:368A-374A.
Natural Resources Defense Council
    1999   *Preventing Industrial Pollution at Its Source: The Final Report of the Michigan Source
            Reduction Initiative.* Dillon, CO: Meridian Institute.
Outen, R.B
    1999   *Designing Information Rules to Encourage Better Environmental Performance.* Arling-
            ton, VA: Jellinkk, Schwartz & Connolly, Inc.
Reed, D.J.
    1998   Green shareholder value: Hype or hit? In *Sustainable Enterprise Perspectives.* Wash-
            ington, DC: World Resources Institute.
Santos, S.L., V.T. Covello, and D.B. McCallum
    2000   Industry response to SARA Title III: Pollution prevention, risk reduction, and risk
            communication. *Risk Analysis* 16(1):57-65.
Tetreault, S.
    2000   Risk reports won't be on Net. *Las Vegas Review Journal* (September 4). [Online].
            Available: http://www.lvrj.com/lvrj_home/2000/Sep-04-Mon-2000/news/14203814.html
            [Accessed: March 21, 2002].
U.S. Environmental Protection Agency
    1997   *EPA Strategic Plan.* EPA/190-R-97-002, September.
    1999   *Aiming for Excellence. Actions to Encourage Stewardship and Accelerate Environmen-
            tal Progress.* EPA/100-R-99-006, July.
U.S. General Accounting Office
    1991   *Toxic Chemicals: EPA's Toxics Release Inventory Is Useful but Can Be Improved.*
            GAO/RCED-91-121. Washington, DC: U.S. General Accounting Office.
Wise, M., and L. Kenworthy
    1993   *Preventing Industrial Toxic Hazards: A Guide for Communities.* New York: INFORM.
Wolf, S.M.
    1996   Fear and loathing about the public right to know: The surprising success of the emer-
            gency planning and community right to know act. *Journal of Land Use and Environ-
            mental Law* 11(2):217-325.
Working Group on Community Right to Know
    1991   Working Notes on Community Right to Know. September-October. Washington, DC.

16

Challenges in Evaluating Voluntary Environmental Programs

Kathryn Harrison

ovel environmental policy instruments, including tradable permits, voluntary programs, and more flexible approaches to regulation, have been embraced by governments throughout the world in recent years. Rigorous evaluation of these diverse approaches is clearly essential if we are to draw any conclusions about which approaches offer the most effective basis for future policy. The focus of this chapter is on the particular challenges associated with evaluation of voluntary programs. Because it is not possible to provide a summary of the effectiveness of the wide variety of voluntary programs that have emerged in recent years within the scope of this brief chapter, I propose instead to focus on some of the challenges that arise in evaluation of voluntary programs, illustrated with three case studies. The discussion that follows focuses on conceptual issues in evaluation of voluntary programs, setting aside for the moment questions of political will and adequacy of administrative resources to conduct program evaluations (Gormley, 2000).

Policy evaluation is a difficult task, one that is too seldom undertaken for traditional regulatory as well as novel voluntary programs. However, rigorous evaluation of voluntary programs is arguably particularly important, because the historical failure of markets and voluntarism to address environmental problems and resource depletion creates a heavy burden of proof for those who advocate voluntary alternatives to regulation (Andrews, 1998; Leiss and Associates, 1996). Unfortunately, evaluation of voluntary programs is simultaneously especially difficult, because it is plagued by problems of data availability, credibility, and self-selection. I will argue that as a result of these factors, there has been a tendency to overstate the effectiveness of voluntary programs.

In the chapter that follows, I discuss three questions that arise in any evaluation: (1) What is being evaluated? (2) What criteria should be used in evaluation? (3) What baseline should be used for the evaluation? Then I illustrate issues that arise in that discussion with case studies of the U.S. 33/50 program, the Canadian Accelerated Reduction/Elimination of Toxics (ARET) program, and the Canadian National Pollutant Release Inventory.

WHAT IS BEING EVALUATED? A TYPOLOGY OF VOLUNTARY PROGRAMS

The Organization for Economic Co-operation and Development (OECD) (1999) distinguishes among three types of voluntary programs: negotiated voluntary agreements, public voluntary programs (also sometimes referred to as "voluntary challenge" programs), and unilateral (i.e., nongovernmental) programs. Among these, voluntary agreements are closest to regulation. Voluntary agreements are characterized by strong expectations on the part of government that nongovernmental parties to the negotiated agreements (typically industry) will comply, and such agreements often are prompted by an explicit or implied threat of regulation should the voluntary approach fail. Voluntary agreements typically are negotiated by government and industry, whether an individual firm or an entire sector, though other stakeholders also may be involved. Although most voluntary agreements are nonbinding, some take the form of legally binding contracts.[1]

In contrast to voluntary agreements, governmental efforts to persuade target groups to change their behavior via voluntary challenges or public voluntary programs typically involve less arm twisting. Although voluntary agreements have been prominent in Europe in recent years,[2] voluntary challenge programs are more common in North America. Examples include the U.S. Environmental Protection Agency's (EPA's) 33/50 program and various Partners for the Environment programs, and, in Canada, the ARET Challenge and the Voluntary Challenge and Registry for greenhouse gases.[3] The central difference between a voluntary agreement and a voluntary challenge is that the latter is an open-ended public challenge that applies widely, but that no particular actor is expected to accept. Thus, the EPA 33/50 program called on any and all firms releasing 17 specified toxic substances to reduce their discharges of those substances. In practice, requirements of participation in challenge programs tend to be very flexible. For instance, in both the 33/50 and ARET programs, participants were expected to commit to make some reductions of the listed substances, but they were not required in practice to meet any particular performance standard.

Information dissemination is often central to voluntary challenge programs, which typically offer participants an incentive of public recognition. A related approach, and one that has buttressed both the 33/50 and ARET programs, is the mandatory pollutant release or "community right-to-know" inventory (discussed also by Herb and colleagues, this volume, Chapter 15). Such inventories require that polluters publicly report their discharges, but do not require reductions of those discharges per se. Rather, the intent of the approach is to facilitate public pressure on firms to make voluntary reductions. Inventories can also document changes in releases reported to complementary voluntary challenge programs.

In addition to these government-sponsored voluntary programs, a wide variety of nongovernmental or unilateral programs exist (many of which are reviewed by Nash, this volume, Chapter 14). In many respects, nongovernmental programs parallel the state-based voluntary programs described earlier, but with someone other than government doing the coercing or encouraging. Participants in nongovernmental voluntary programs vary from trade associations to nonpartisan third parties such as the International Organization for Standardization (ISO) to environmental groups. A critical distinction among nongovernmental programs (to be discussed) is whether they focus directly on environmental performance objectives or rather on management systems as a tool to improve environmental performance.

No single unidimensional typology can capture all the relevant differences among voluntary programs. There is thus substantial variation within each of the categories listed.[4] However, accepting OECD's three-way typology as a useful starting point, it is apparent that the kind of voluntary program being evaluated has several implications for the evaluation. The first issue concerns the credibility of the evaluation, which can be especially problematic in the case of nongovernmental programs sponsored by firms or trade associations. In such cases, evaluations by the program's sponsors understandably may be viewed with the same skepticism that would greet students not only grading their own exam papers, but also writing the exam questions in the first place. Verification by independent third parties is thus essential to lend credibility to the evaluation.

That is not to suggest that program evaluations by government are immune to credibility challenges. Government evaluations of voluntary agreements negotiated bilaterally by industry and the state may be met with particular skepticism because as Ayres and Braithwaite (1992:55) note, "The very conditions that foster the evolution of cooperation are also the conditions that promote the evolution of capture and indeed corruption."[5] Openness of the evaluation process and, ideally, participation by environmentalists and other third parties in negotiating voluntary programs in the first place will be critical in establishing credibility (Gunningham and Grabosky, 1998).

Distinctions between voluntary programs also are relevant to program evaluation because conclusions about the effectiveness of one voluntary program may not apply to another very different program. In this regard, the typology offered earlier is at best a starting point, given the diversity of voluntary approaches within each category. For example, lessons learned from legally binding Dutch covenants negotiated to achieve environmental targets previously set by the state may not apply to nonbinding "gentlemen's agreements" in which the goals themselves may be subject to negotiation, let alone to voluntary challenges or nongovernmental programs.[6]

CRITERIA FOR EVALUATION

An obvious question for any program evaluation is what evaluative criteria should be used. Although it is clear that evaluation of any environmental policy instrument should consider its effectiveness in achieving environmental objectives, policymakers also routinely consider other societal objectives in devising and evaluating public policies. Evaluations of voluntary programs thus have tended to apply multiple criteria, including environmental effectiveness, cost-effectiveness for both government and business, equity, democratic accountability and public participation, promotion of innovation, and adaptability to future challenges (European Environmental Agency, 1997; Davies and Mazurek, 1996; Harrison, 1998). Many studies of voluntary approaches also have emphasized "soft effects," such as improvement of the business-government relationship, building trust among stakeholders, and promotion of a more environmentally sensitive business culture (European Environment Agency, 1997; European Commission, 1996; Nash and Ehrenfeld, 1997).

Among these criteria, it is useful to draw a distinction between ultimate objectives, such as equity, efficiency, and reduction of environmental impacts, and intermediate or instrumental objectives, such as fostering innovation and improving business-government relationships. It is striking that even ambitious attempts to evaluate voluntary programs often have been able to say little about attainment of objectives such as reduction of environmental impacts and cost-effectiveness, and thus have focused on instrumental goals (e.g., European Environment Agency, 1997). Indeed, some analysts of voluntary programs focus almost exclusively on these "soft" objectives.[7] This emphasis on intermediate objectives is not unique to voluntary programs. Gormley (2000) reports that in evaluating regulatory programs, government agencies also tend to focus on easier-to-measure "outputs" (such as inspections and permitting activity) to the exclusion of "outcomes" (such as emissions) or "impacts" on human health or the environment.

The focus on intermediate objectives arguably is more problematic with respect to voluntary than regulatory programs, however. The emphasis on soft effects often has been necessitated by a lack of data on environmental outcomes. This can be particularly problematic with respect to industry-sponsored programs, where if data on environmental performance do exist, they tend to be proprietary. However, it also can be an issue for government-sponsored voluntary programs which, unlike regulatory programs, typically do not rely on coercion to compel disclosure of monitoring nor threaten inspections to verify firms' reports. The extent of this problem was demonstrated by a study of 137 voluntary agreements by the European Commission in 1997, which found that 118 had no requirement for firms to report the results of their compliance monitoring, and 47 had no requirements for monitoring at all (European Commission Directorate, 1997). Based on a review of 88 industry-sponsored voluntary initiatives worldwide, Börkey and Lévêsque (1998) concluded that industry initiatives typically made no provisions for monitoring or public reporting. In other words, evaluations of voluntary codes of practice often have focused on intermediate objectives simply because they have no data with which to assess performance relative to environmental objectives. Although lack of environmental quality data is often problematic for evaluation of regulatory programs as well, at least the latter typically require monitoring and reporting by industry and also include provisions for independent verification by the state.

That might be acceptable if there was a sufficient basis for the assumption that "soft effects," such as business-government cooperation, will in fact deliver "hard" environmental benefits. As Mazurek speculates (this volume, Chapter 13), cooperation may not only foster success in voluntary programs, but may spill over and improve business-government relations in traditional regulatory programs. However, in the absence of evidence, it seems premature to draw that conclusion. I have argued elsewhere that business-government cooperation also can be detrimental to the environment (Harrison, 1995), while the fine line between cooperation and capture already has been noted (Ayres and Braithwaite, 1992:55).

Closely related to this distinction between ultimate and intermediate criteria is the distinction between voluntary programs that establish environmental performance objectives and "voluntary environmental management systems," which leave the setting of performance objectives up to individual participants and focus instead on management approaches to ensure that those goals are achieved. The latter approach is most common among industry-sponsored programs (Nash, this volume, Chapter 14). For example, the Canadian Responsible Care codes of practice commit participants to "be aware of all effluents and emissions to the environment, monitor those for which it is necessary, and implement plans for their control when necessary," but do not specify which emissions should be monitored nor what degree of control should be applied. Although the program provides for independent audits of firms' compliance, the audit focuses exclusively on whether the facility has all the elements of the environmental management system in place, not on its environmental performance per se.[8]

It might be argued that this emphasis on intermediate objectives is quite appropriate for nongovernmental codes of practice intended for use as internal management tools and that it is thus unfair to hold such programs to the broader societal standards appropriate for public policy. That is certainly true to the extent that environmental management systems are limited to use as internal tools. However, the complaint seems disingenuous in light of firms' efforts to publicize to customers and regulators their compliance with management systems such as ISO 14001 and Responsible Care. More importantly, to the extent that nongovernmental voluntary programs are being considered as a complement to or substitute for government policies, it is entirely appropriate for the state to hold them to the same standards, even if firms set less ambitious objectives for themselves.

BENCHMARKS FOR EVALUATION

A critical question in evaluating voluntary programs concerns the appropriate baseline against which to measure performance. One can, in fact, envision several different levels of analysis of environmental effectiveness (European Environment Agency, 1997). The first, and simplest to achieve, is a measure of change (e.g., in environmental performance, costs to government, costs to the private sector, and so on) relative to a reference year. A second and more rigorous standard is to ask to what degree the observed changes are attributable to the program in question, acknowledging that actors' behavior might have changed anyway in response to other factors, including market forces and other concurrent government policies. This standard necessitates the difficult task of developing a counterfactual "business-as-usual" baseline that would have prevailed in the absence of the voluntary program.

Finally, the third and most difficult standard of analysis is to compare the impact of the voluntary program to what would have happened if an alternative approach had been pursued, a level of analysis even more challenging than establishing the "business-as-usual" baseline. Although a full-blown comparison is obviously very difficult, greater attention to this issue is nonetheless warranted. Coercive regulatory programs often tend to be criticized against a yardstick of complete compliance and economic efficiency. In contrast, there is a tendency for proponents of voluntary programs to hail any accomplishments above the current baseline as a success.[9] This is appropriate if the voluntary program is the only realistic alternative to the status quo. However, to the extent that voluntary and regulatory programs are being considered as substitutes, the relevant question is how they compare to each other in practice.

In constructing counterfactuals (i.e., "what would have happened if" scenarios), a critical issue that typically arises concerns the motivation for voluntary action by business. Voluntary environmental improvements may be undertaken by a profit-oriented firm for three reasons. First, the firm may be motivated to change its behavior to save money, such as by reducing energy costs or by recovering valuable reactants, regardless of any associated environmental benefits. Second, a firm may be motivated by market pressures for environmental improvements from investors, lenders, insurers, customers, or workers.[10] Third, a firm might opt for voluntary measures beyond those prompted by cost savings and "green" market pressures in order to forestall or avoid mandatory regulation or legal liability. A critical question in that case is whether the incentive for voluntary action lies in reducing costs by accomplishing the same environmental objective with greater flexibility than allowed by traditional regulation, or rather in reducing costs by meeting less demanding environmental objectives.

Which of these three motives applies in a given case has implications both for the kind of voluntary program that is likely to be effective and the kinds of issues that will arise in program evaluation. When there are direct financial savings to be realized by reducing environmental losses, businesses have strong incentives to correct their behavior even in the absence of government intervention. However, government-sponsored information dissemination and technology transfer programs nonetheless may help firms recognize and take advantage of waste reduction opportunities and energy efficiency improvements (Storey, 1996). Similarly, firms facing green market pressures also have incentives to change their behavior regardless of the existence of governmental voluntary programs. Nonetheless, government verification of voluntary efforts, such as through certification and awards programs, may enhance those market forces by lending greater credibility to firms' claims with consumers. Similarly, government programs to disseminate information about *all* firms' behavior can help environmentally motivated consumers and investors identify leaders and laggards alike.

With respect to voluntary actions motivated either by environmental market pressures or direct cost savings, it is important to bear in mind that firms have incentives to change their behavior regardless of the existence of a voluntary program. Indeed, voluntary programs invite self-selection and thus prompt questions of whether participating firms are simply the ones that were already inclined to comply anyway. The challenge is thus to assess the marginal impact of the program on firms' behavior. A program evaluation that simply assesses the change in behavior relative to a reference year rather than a business-as-usual baseline will tend to overstate program effectiveness.

A similar issue of self-selection also arose in the context of energy conservation programs adopted in the 1970s. Some analysts evaluating those programs argued that failure to control for self-selection inevitably would result in overstatement of program benefits, because individuals or firms undertaking voluntary behavioral changes tend to be those facing the lowest costs of compliance (Weinstein et al., 1989). In contrast, Keating (1989) argued that self-selection could result in understatement of program benefits, because those who opt to participate tend to be those who perceive the greatest benefits from participation. In that case, comparison of any extra actions undertaken by participants relative to actions by nonparticipants would not provide an accurate assessment of the program's benefits because, in the absence of the program, participants actually would have done less than the "control" nonparticipants. However, as Keating noted, understatement of effectiveness seemed to be more problematic with respect to programs that offered subsidies for energy conservation, and indeed for those who had already invested in desired energy conservation measures would not be eligible. In contrast, overstatement was more problematic in the absence of such inducements, where those who had already assumed compliance costs not only would be allowed, but also would be most likely to participate. The latter is more consistent with contemporary voluntary challenge programs.

The situation is quite different when firms are motivated to undertake voluntary actions by a threat of regulation. In this case, the challenge for government is not merely to enhance existing market incentives, but to create those incentives in the first place in the form of a credible and sustained threat of regulation (Glachant, 1994:47; Gibson, 1999:244). There is reason to believe that government coercion is still a powerful influence on business environmental behavior. A survey by the European Commission found that the potential to forego or postpone regulation was cited as the most important benefit of voluntary environmental agreements by roughly two thirds of industry respondents (European Commission, 1997). Similarly, a Canadian business survey reported that 95 percent of firms cited compliance with regulations as one of the top five factors motivating their environmental improvements.[11] One implication is that in evaluation of voluntary agreements in particular, the applicable comparison is not necessarily the status quo or even the business-as-usual baseline, but rather what would have happened had regulations been adopted instead. The significance of coercion in motivating firms' behaviors also suggests a need to look at whether concurrent regulatory policies have contributed to any changes in behavior observed relative to the reference year.

The problem is that those conducting an evaluation of a voluntary program typically do not know what is motivating firms targeted by the program. Different firms are likely to have different motivations, and the same firm may be motivated to varying degrees by each of the three incentives noted earlier. Firms have incentives to understate their environmental releases and impacts whether they face voluntary or regulatory programs. However, it may be particularly difficult to unpack the degree to which progress is attributable to a voluntary program as compared to regulation. Once a regulatory program is established requiring a comparable level of compliance for all facilities affected, there is little reason for a firm to overstate what it "would have done anyway." In contrast, firms participating in a voluntary program in order to forestall regulation have an interest in convincing regulators not only of how much they have done but also how they have sufficient market incentives to maintain that behavior in the absence of regulation. As a result, although the need to establish a credible baseline requires that the evaluator understand what is going on in the heads of the target population, it is by no means clear that one can get a credible answer simply by asking.

CASE STUDIES

Three prominent voluntary programs—the U.S. 33/50 program, the Canadian Accelerated Reduction/Elimination of Toxics (ARET) program, and the Canadian National Pollutant Release Inventory (NPRI)—are reviewed in this section. No claim is offered that these three programs represent the universe of voluntary programs, though it is noteworthy that they are often advanced as particular success stories. Rather, the case studies are used to illustrate some of the issues already introduced, including the issue of credibility in self-reporting, the importance of establishing a common reference year, and the importance of questioning the degree to which progress relative to that reference year can be attributed to the program.

33/50[12]

In 1991, the EPA challenged the business community to voluntarily reduce its releases and transfers of 17 high-priority chemicals by 33 percent by the end of 1992 and by 50 percent by the end of 1995, dubbing the program "33/50." Consistent with the previous discussion of voluntary challenges, requirements for participation were very flexible. A firm needed only to write to the EPA pledging some degree of reduction of its discharges of any of the 33/50 chemicals. In turn, the EPA would provide a certificate of appreciation and recognize 33/50 participants in its annual report on the Toxics Release Inventory (TRI).

Although only 13 percent of firms contacted by the EPA agreed to participate, the goal of a 33-percent reduction relative to the reference year was achieved a year early, by the end of 1991, and the 50-percent reduction goal also was achieved early, by the end of 1994 (EPA, 1999). By the target year of 1995, reductions of 55 percent had been achieved. That has prompted many, including the EPA, to cite the 33/50 program favorably as an example of the potential of voluntary approaches. However, when one considers whether reductions were achieved relative to a "business-as-usual" baseline, the benefits of the 33/50 program are not as clear.

As noted in Table 16-1, the first concern is that when the program was launched in 1991, the EPA chose 1988 as the reference year, because it was the most recent year for which discharge data were available. As a result, more than one quarter of the reported reductions occurred before the program's inception.

TABLE 16-1   Analysis of Reductions of Discharges and Transfers of 17 Chemicals Reported to the 33/50 Program

	1988 Reference Year	1990 Reference Year
Goal	50%	NA
Total reduction by 1995 target year	55%	NA
Reductions after start of program	40%	47%
Reductions by participants	27%	32%
Excess reductions by participants relative to nonparticipants	9%	11%

Source: EPA (1999).

Excluding these reductions, the total reductions reported to the 33/50 program were 40 percent relative to the 1988 reference year or 47 percent relative to a more meaningful 1990 reference year.

The significant downward trend in releases even before the inception of the program leads one to ask what was motivating those reductions and to what degree those trends continued during the 33/50 years. In fact, even firms that elected not to participate in the program made substantial reductions in their discharges and transfers of 33/50 chemicals over the course of the program. Indeed, more than one quarter of the reductions reported by the 33/50 program were made by nonparticipating firms. Although it is conceivable that some nonparticipating firms were motivated by the 33/50 program to reduce their releases, perhaps because they interpreted the program as a subtle threat of regulation rather than an opportunity for credit, the fact that firms were making deep reductions even before the program was introduced would lead one to look first for other explanations. Excluding nonparticipants, reductions relative to a 1990 reference year fell by 32 percent, as shown in Table 16-1.

Circumstantial evidence does indicate that the 33/50 program encouraged firms to make reductions over and above what they would have made otherwise. Firms participating in the 33/50 program reduced their discharges of 33/50 chemicals by 55 percent from 1990 to 1995, compared to 36 percent for nonparticipating firms. This 19 percent "extra" reduction by participating firms translates to an 11 percent reduction relative to the total releases and transfers of 33/50 chemicals in the 1990 reference year.

The question of self-selection remains, however. The fact that participants in the 33/50 program made greater reductions than nonparticipants does not necessarily indicate that those reductions were prompted by the 33/50 program. Firms already inclined to make substantial reductions of 33/50 chemicals, whether in response to negative publicity associated with mandatory reporting of discharges to TRI, market forces, cost savings, or concurrent regulatory requirements, simply may have been the ones inclined to sign on for credit. This is supported by Arora and Cason's (1996) finding that the larger a firm's releases and transfers, the more likely it was to participate in 33/50 because these are the firms that would be expected to be subject to the greatest pressure in response to the release of TRI data, even in the absence of the 33/50 program. It is also problematic that none of the analyses of the 33/50 program conducted to date systematically have controlled for the effects of concurrent regulations. Yet the EPA (1999) reports that reductions of two Montreal protocol substances included in the 33/50 program alone account for 27 percent of reductions from 1990 to 1995.[13] Moreover, O'Toole et al. (1997) found that stringency of state regulations was one of the two most important factors in accounting for state-level reductions of 33/50 chemicals as reported to TRI. In summary, although it is not possible to conclude with confidence the precise benefits of the 33/50 program, it appears to have prompted discharge reductions of the specified chemicals of less than 11 percent, well below the 55 percent often attributed to the program.

ARET

The Canadian ARET Challenge, launched in 1994, is similar in many respects to the 33/50 program, though more ambitious. After the details of the challenge were negotiated by the government and some industry representatives, a broader challenge was issued to all firms to reduce discharges of some 30 chemicals considered to be toxic, persistent, and bioaccumulative by 90 percent by the year 2000, and of 87 others with one or more of these characteristics by 50 percent by the same year. Characteristic of a voluntary challenge program, there was no threat of penalties for failure to achieve those goals. Indeed, as in 33/50, firms that choose to participate were not required to commit to the full 90-percent and 50-percent reductions.

Preliminary assessments of the impact of ARET based on the first 4 years of participant reports are promising. By the end of 1998, action plans had been received from 316 industry and government facilities (ARET, 2000). Discharges of all ARET substances had been reduced 67 percent relative to base year levels. Reduction levels of the class of substances targeted for 50-percent reductions were surpassed 4 years ahead of schedule, with further reductions promised.

As with 33/50, however, the degree to which these reductions are attributable to the ARET program is unclear. The base year problem is exacerbated in the ARET case, because each participating facility can pick its own base year anytime after 1987. This option allows firms to claim credit toward the ARET program for discharge reductions they made as much as 6 years before the program's inception, and to strategically choose a year with particularly high discharges to maximize apparent reductions (Gallon, 1998). In fact, ARET program figures indicate that roughly half of the 67-percent reductions claimed by the end of 1998 had been achieved before the program was even launched.[14]

As with 33/50, there are also questions about whether the reductions attributed to ARET are in fact voluntary, and if so, whether they are attributable to the program. The ARET program notes that one quarter of base year emissions were substances regulated under the Canadian Environmental Protection Act (CEPA), and a further 11 percent were substances undergoing evaluation for potential regulation under CEPA.[15] Thus one might question whether the ARET reductions are an artifact of regulation. Harrison and Antweiler (2001) found that facilities reporting on-site releases of regulated substances to the Canadian National Pollutant Release Inventory reported greater reductions of those substances over time than of other substances. There is also anecdotal evidence that significant reductions by individual sources can be attributed to regulation.[16] Moreover, as with 33/50 and TRI, voluntary reductions may have been driven less by the positive publicity associated with the ARET challenge than the negative publicity associated with public reporting of discharges of the half of the ARET chemicals covered by the NPRI. Finally, the absence of any provisions for third-party verification of firms' own claims of discharge reductions is troubling, particularly for the half of ARET chemicals not covered by the regulatory NPRI program (Leiss and Associates, 1996).

NPRI

The discussion of the 33/50 and ARET programs notes the potentially confounding influence of toxic pollutant release inventories in both countries. Firms may have been motivated less by the positive incentives in the form of public recognition offered by the voluntary challenge programs as discussed earlier, and motivated more by negative incentives (shame) generated by the pollutant release inventories that underpinned them. Such inventories have increased in popularity since the EPA established its TRI in 1988. The Canadian government created its NPRI in 1993, and similar inventories have since been established in Australia and the European Union. It is noteworthy that such inventories are not exclusively voluntary programs in that all facilities—leaders and laggards alike—are required by law to report their toxic releases to the state. However, the raison d'etre for those regulations compelling disclosure is to promote voluntary action. Pollutant inventories are predicated on the assumption that, armed with more complete information about firms' environmental practices, consumers, workers, and investors will be empowered to use markets and social networks to pressure firms to voluntarily reduce their releases (Herb and colleagues, this volume, Chapter 15).

The impact of such inventories, measured in terms of the least challenging baseline of change relative to the base year, has been highly encouraging. Total releases reported to the TRI declined by 46 percent from 1988 to 1995 (Natan and Miller, 1998). Releases reported to the Canadian NPRI similarly declined by 36 percent in the first 3 years of the program, from 1993 to 1996. This success has led some commentators to conclude that discharge inventories that prompt voluntary action are more effective than regulation (Gunningham and Grabosky, 1998:64; Organization for Economic Cooperation and Development [OECD], 1996; Fung and O'Rourke, 2000). However, a more critical look at these inventories raises a number of questions for program evaluation.

First, it is important to acknowledge uncertainty about the extent to which the reductions relative to reference years are real (Hearne, 1996). Although accurate reporting is legally mandated, individual facilities prepare their own reports to both TRI and NPRI with minimal oversight by regulators. Indeed, facilities are not required to measure their own discharges; they can estimate them using techniques of varying reliability. Thus, an apparent decline in releases from a facility from one year to the next may simply represent adoption of alternative estimation methods. There is also concern that reported reductions often reflect a shell game, in which facilities merely recategorize waste streams so they no longer have to be reported (Natan and Miller, 1998). This could be especially problematic for the Canadian NPRI, which did not require reporting of recycling and reuse in the first few years of the program and thus may have invited facilities to reclassify their waste streams as "recycling" to evade reporting.

There has been a tendency to assume that the reductions revealed by TRI and NPRI have been voluntary responses to market mechanisms (e.g., Tietenberg, 1998:593; Fung and O'Rourke, 2000). However, the picture becomes less clear when one attempts to assess the degree to which the observed changes relative to base years are attributable to information dissemination. With respect to the U.S. TRI, few studies have considered the impact of regulatory programs on TRI discharges. Shapiro (1998, 1999) found state regulatory efforts to be among the most important determinants of reductions of TRI releases over time. Similarly, Khanna and Damon (1999) found that liability under the federal Superfund legislation and anticipation of new hazardous air pollutant regulations under the U.S. Clean Air Act were among the most significant factors in explaining firms' TRI reports for 33/50 program chemicals. Finally, Santos, et al. (1996) found that regulatory compliance was one of the two reasons most frequently cited by facilities (the other being employee health) for reduction of their TRI releases and transfers. These studies suggest a need for greater attention to regulation and liability in accounting for the rapid progress apparent in both discharge inventories and the voluntary challenge programs predicated on them.

With respect to Canada's NPRI, analyses performed by Harrison and Antweiler (2001) are summarized in Figure 16-1. The top line in Figure 16-1 represents the sum of reported releases of all NPRI substances by all facilities that reported to NPRI in one or more of the years between 1993 and 1998. Consistent with experience with the TRI, there was a dramatic 36-percent decrease in the first 3 years from 1993 to 1996, though that has since been counteracted somewhat by growth in releases, resulting in an overall reduction of 30 percent from 1993 to 1998.

These reductions of both on-site releases and off-site transfers may be understated as a result of a growing number of facilities reporting to NPRI. Although some of these are new facilities contributing real increases in waste production, others are older facilities that either belatedly learned of the requirement to report to NPRI or were affected by a minor adjustment to NPRI reporting requirements in 1995. The exclusion of these older facilities from the totals in previous years thus may understate reductions over time. One way to control for this is to focus on the subset of facilities that reported in both 1993 and 1998 (though this has the disadvantage of excluding genuinely new facilities). As indicated by the middle line in Figure 16-1, these facilities are doing somewhat better than the totals for all facilities.

Among this subset of continuous reporters, a relatively small number of facilities account for a substantial fraction of the total reduction in onsite releas-

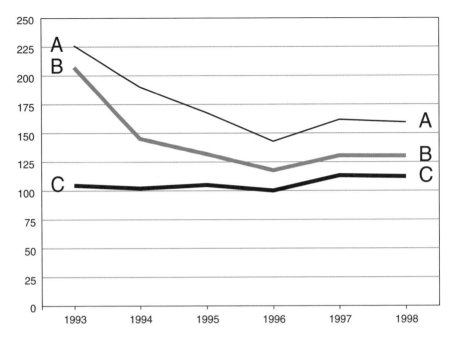

FIGURE 16-1   Trends in on-site releases (kilotons).
Note: A: total on-site releases from all facilities reporting to NPRI each year. B: total on-site releases from the "continuous reporter" subset of facilities, which reported to NPRI in each year 1993 through 1998. C: total on-site releases from "continuous reporters" excluding the pulp and paper industry and the Kronos facility in Varenne, Québec.
Sources: Harrison and Antwiler (2000). Reprinted with permission of Centre for Economic and Social Policy, University of British Columbia.

es (Harrison and Antweiler, 2001). Indeed, a single Quebec facility, Kronos Canada, which adopted process changes in response to regulatory enforcement actions undertaken by both the federal and provincial governments in the early 1990s (Picard, 1992; Hamilton, 1993), accounts for more than half of the total reductions. In addition to Kronos, many of the facilities that contributed the greatest reductions to NPRI were pulp and paper mills. It is noteworthy that the pulp and paper industry is the only industry that faced new discharge regulations at the national level in Canada during the past decade, and it also was subject to extensive reform of regulations and permits at the provincial level in the early 1990s (Harrison, 1996).

Because there are compelling reasons to conclude that reductions by these facilities were driven by regulation (not least among them these industries' resistance to behavioral change before regulatory enforcement), the final series in Figure 16-1 reports trends among the continuous reporters, with Kronos and the pulp and paper industry excluded. It is striking that once these most obvious impacts of national regulation are excluded, the impressive 37-percent reduction of on-site releases by continuous reporters evaporates, leaving a net *increase* of 7 percent. Although we cannot know whether an even greater increase in releases might have been observed in the absence of NPRI, this figure does suggest that the dramatic reductions of toxic discharges typically attributed to NPRI are largely the result of traditional regulation, rather than a voluntary response by firms to information disclosure.

A final question concerns the impact on human health and the environment of these emissions reductions. Pollutant release inventories publicly report the volume of releases, but do not provide an estimate of the risks such releases might pose to human health or the environment. Harrison and Antweiler (2001) take one step closer to such an evaluation by adjusting reported NPRI releases for toxicity using the EPA's Chronic Human Health Indicators.[17] After adjusting for toxicity, the 30-percent reduction in total on-site releases between 1993 and 1998 translates to a 9-percent *increase*. In other words, although the total weight of releases has declined, the toxicity of those releases apparently has increased, and has done so to a degree that this substitution effect may actually outweigh the benefits of declining releases. These findings, although far from constituting a complete risk assessment, reinforce the call for program evaluations to go beyond mere "outputs" and "outcomes" to consideration of "impacts."

CONCLUSIONS

The one clear area of consensus among students of voluntary approaches is that there has been too little attention to evaluation of either the economic or environmental benefits of voluntary programs (Storey, 1996; Davies and Mazurek, 1996; European Environment Agency, 1997; Beardsley, 1996; National Research Council, 1996; OECD, 1999; Mazurek, this volume, Chapter 13). In part, this reflects the novelty of voluntary approaches; it is simply too early to assess their effectiveness in many cases. However, it also reflects a pathology of unclear targets and inattention to the kinds of monitoring, verification, and public reporting needed to support program evaluation.

At a minimum, systematic evaluation of any program, whether regulatory, voluntary, or market based, demands a consistent reference year, clear expectations, monitoring and verification of environmental performance, and public reporting. Some of these characteristics present special challenges for voluntary programs, which all too often have had unclear expectations, inconsistent baselines, and few if any monitoring and reporting requirements. In contrast, regulatory programs typically have quite explicit and consistent standards, self-monitoring requirements, and provisions for independent compliance monitoring by the state. Nonetheless, these characteristics could in theory become standard practice for all government-sponsored voluntary programs as well.[18] Nongovernmental programs present a greater challenge, because private-sector sponsors may not be as inclined to conduct rigorous and public evaluations. Moreover, many nongovernmental programs are environmental management systems, which specify objectives in the form of management "inputs" rather than environmental performance "outputs." However, if participants in such programs seek either recognition or concessions from regulators, it is reasonable for the state to establish comparable expectations as for government-sponsored voluntary—and regulatory—programs.

Of course, program evaluation also requires political will and administrative resources. The greater effort made to evaluate the voluntary energy conservation programs in the 1980s may reflect the fact that many of those programs were predicated on government subsidies. With less public funding at stake (recognition is cheap after all), there simply may be less incentive to allocate resources to evaluation today, even though the opportunity costs of misdirecting scarce resources to a less effective policy instrument may be substantial. Ironically, the same impulse to reform regulation to lessen the burden on regulated interests that has given rise to voluntary programs simultaneously may have rendered evaluation of the effectiveness of those new programs more difficult. For instance, in an effort to minimize the burden on the private sector, the U.S. Paperwork Reduction Act of 1995 establishes a quite elaborate process of Federal Register notices and Office of Management and Budget approval for "information collection requests" should an agency want to contact 10 or more firms or individuals.

The issue of establishing a baseline for evaluation also remains a special challenge for voluntary programs. In the context of the literature on program evaluation, with voluntary programs we typically confront "quasi-experiments" involving nonequivalent groups (Cook and Campbell, 1979), in which the challenge is to separate the effects of "treatment" (i.e., the voluntary program) from differences resulting from lack of comparability between the "treated" and "untreated" groups. The tendency to assess environmental progress only relative to reference years almost certainly exaggerates program effectiveness, because some fraction of improvements typically would be attributable to market incentives and/or concurrent regulatory requirements in the absence of the voluntary program (to say nothing of the scenario in which the reference year is several years prior to the launch of the voluntary program). The problem is that development of a "business-as-usual" scenario in the absence of the voluntary program requires knowledge of what factors are motivating business behavior. Some recent studies have undertaken extensive interviews with program participants to address this issue.[19] Leverage also may be gained by considering trends prior to introduction of the program, looking for the impact of concurrent governmental programs, and using regression techniques to control for other factors that might explain self-selection.[20] Ideally, new programs could be introduced (perhaps initially at a pilot scale) in which a randomly assigned control group is ineligible for participation. Although none of these approaches offers a panacea, the risk of misallocating scarce resources to ineffective programs surely justifies greater effort.

NOTES

1 Although such contracts are legally binding, they can still be considered voluntary agreements because parties enter into the contract voluntarily.

2 For reviews of European experience, see European Environment Agency (1997), Börkey and Lévêsque (1998), and Rennings et al. (1997).

3 Mazurek's chapter in this volume (Chapter 13) and EPA (2001) both offer excellent overviews of U.S. challenge and information-based programs.

4 See Ehrenfeld, Response to Talking with the Donkey: Cooperative Approaches to Environmental Protection, and Harrison, A Response to Professor Ehrenfeld, *Journal of Industrial Ecology,* online letters to the editor, at http://www.yale.edu/jie.

5 Consistent with a close relationship between business and government, Dietz and Rycroft (1987) reported considerable career mobility between government and industry, but almost none between government and environmental nongovernmental organizations.

6 On the importance of the distinction between negotiation of goals versus means, see Glachant (1994).

7 For instance, Nash and Ehrenfeld (1997) evaluate environmental management systems against the criterion of whether they foster "cultural change" within a firm.

8 It is noteworthy that there is no requirement to make Responsible Care compliance audits public in the US program (Nash and Ehrenfeld, 1997). Indeed, the U.S. American Chemistry Council leaves it to individual member companies to define what constitutes "full implementation" for their own circumstances (Mazurek, 1998).

9 The EPA Partners for the Environment program's summary of benefits offers an example (see http://www.epa.gov/partners/partnerships.html). It is also noteworthy that the figures presented are based on self-reporting by individual "partners" and, in light of the discussion to follow, that no effort has been made to assess which of those benefits were prompted by the program or whether some fraction might have occurred even in the absence of the program in question.

10 Among these, second-order "green" pressures from investors, lenders, and insurers can be distinguished from first-order green pressures from customers, investors, and workers, in that second-order market pressures depend on the existence of first-order ones. However, second-order market pressure from investors, lenders, and insurers that is motivated by fear of civil liability or regulatory costs, rather than consumer or worker reactions, would fall in the third category of business motive, regulatory threat.

11 In contrast, factors such as cost savings, customer requirements, and public pressure were all cited by less than half the respondents (KPMG, 1994).

12 The analysis presented here was influenced significantly by the approach taken by Davies and Mazurek (1996).

13 The 1990 Clean Air Act Amendments also mandated regulation of discharges of volatile organic compounds (in order to achieve ground-level ozone objectives) and hazardous air pollutants, both of which could be expected to cover all 17 33/50 chemicals.

14 ARET reports total discharges of 39.4 tonnes in participants' various base years, 26.9 tonnes in 1993, and 13 tonnes in 1998. Thus, at least 12.5 of the 26.4 tonnes of reductions were achieved prior to the launch of the program in early 1994 (ARET, 2000).

15 The ARET program reports that 26 percent of base year emissions are CEPA "schedule 1" substances (ARET, 2000). The 11-percent figure was calculated using ARET base year data for the following substances on "priority substance list 2": chloroform, ethylene oxide, formaldehyde, acetaldehyde, 1,3-butadiene, phenol, and acrylonitrile.

16 Gallon (1998) notes that the 90 percent reduction in sulfur dioxide emissions claimed by INCO's Sudbury smelter under the ARET program were legally mandated.

17 Such an adjustment still does not take into account patterns of dilution or human exposure leading to chronic human health impacts, nor acute impacts on human health or environmental impacts.

18 Although such programs do not legally require participation, compliance with monitoring and reporting requirements can be made a condition for attaining program incentives (whether recognition or regulatory relief).

19 See, for example, information about the National Database on Environmental Management Systems (IDEMS) project (http://www.eli.org/isopilots.htm).

20 Obstacles to causal inference and statistical techniques that attempt to address those problems in the case of quasi-experiments involving nonequivalent groups are discussed in Chapters 3 and 4 of Cook and Campbell (1979).

REFERENCES

Andrews, R.N.L.
 1998   Environmental regulation and business 'self-regulation.' *Policy Sciences* 31:177-197.
Accelerated Reduction/Elimination of Toxics
 2000   *Environmental Leaders 3 Update.* Ottawa: ARET.
Arora, S., and T.N. Cason
 1996   Why do firms volunteer to exceed environmental regulations? Understanding participation in EPA's 33/50 Program. *Land Economics* 72(4):413-432.
Ayres, I., and J. Braithwaite
 1992   *Responsive Regulation: Transcending the Deregulation Debate.* New York: Oxford University Press.
Beardsley, D.
 1996   *Incentives for Environmental Improvement: An Assessment of Selected Innovative Programs in the States and Europe.* Washington, DC: Global Environmental Management Initiative.
Börkey, P., and F. Lévêsque
 1998   *Voluntary Approaches for Environmental Protection in the European Union.* Paris: Organization for Economic Co-operation and Development.
Cook, Thomas D., and D.T. Campbell
 1979   *Quasi-Experimentation: Design and Analysis Issues for Field Settings.* Boston: Houghton Mifflin.
Davies, J.C., and J. Mazurek
 1996   Industry Incentives for Environmental Improvement: Evaluation of U.S. Federal Initiatives. Unpublished manuscript prepared for the Global Environmental Initiative, Washington, DC.
Dietz, T., and R.W. Rycroft
 1987   *The Risk Professionals.* New York: Russell Sage Foundation
European Commission
 1996   Communication from the Commission to the Council and the European Parliament on Environmental Agreements. COM(96) 561.
European Commission Directorate
 1997   Study on Voluntary Agreements Concluded Between Industry and Public Authorities in the Field of the Environment. General III.01 – Industry. Unpublished final report, Brussels, Belg., January.

European Environment Agency
 1997 *Environmental Agreements: Environmental Effectiveness.* Copenhagen, Den.: European Environment Agency.

Fung, A., and D. O'Rourke
 2000 Reinventing environmental regulation from the grassroots up: Explaining and expanding the success of the toxics release inventory. *Environmental Management* 25:115-127.

Gallon, G.
 1998 Accuracy is optional in reporting voluntary success. *Alternatives* 24:12.

Gibson, R.B.
 1999 Conclusion. In *Voluntary Initiatives: The New Politics of Corporate Greening,* R. Gibson, ed. Peterborough, Ontario, Can.: Broadview.

Glachant, M.
 1994 The setting of voluntary agreements between industry and government: Bargaining and efficiency. *Business Strategy and the Environment* 3(2):43-49.

Gormley, W.T.
 2000 Environmental Performance Measures in a Federal System. Unpublished paper presented at the Annual Meeting of the Association for Public Policy and Management.

Gunningham, N., and P. Grabosky
 1998 *Smart Regulation: Designing Environmental Policy.* New York: Oxford.

Hamilton, G.
 1993 Kronos, two executives charged with polluting St. Lawrence. *The Gazette*, March 9, p. A4.

Harrison, K.
 1995 Is cooperation the answer? Canadian environmental enforcement in comparative context. *Journal of Policy Analysis and Management.* 14:221-244.
 1996 The regulator's dilemma: Regulation of pulp mill effluents in the Canadian federal state. *Canadian Journal of Political Science* 29: 469-496.
 1998 Talking with the donkey: Cooperative approaches to environmental protection. *Journal of Industrial Ecology* 2(3):51-71.

Harrison, K., and W. Antweiler
 2001 *Information Dissemination vs. Environmental Regulation: The View from Canada's National Pollutant Release Inventory.* Working Paper. Vancouver: Centre for Research on Economic and Social Policy, University of British Columbia.

Hearne, S.A.
 1996 Tracking toxics: Chemical use and the public's 'right-to-know.' *Environment* 38:5-9, 28-34.

Keating, K.M.
 1989 Self-selection: Are we beating a dead horse? *Evaluation and Program Planning* 12:137-142.

Khanna, M., and L.A. Damon
 1999 EPA's voluntary 33/50 program: Impact on toxic releases and economic performance of firms. *Journal of Environmental Economics and Management* 37:1-25.

KPMG
 1994 *Canadian Governmental Management Survey.* Toronto, Can.: KPMG.

Leiss W. and Associates
 1996 *Lessons Learned from ARET: A Qualitative Survey of Perceptions of Stakeholders. Final Report.* Working Paper Series 96-4. Ontario, Can.: Environmental Policy Unit, School of Policy Studies, Queen's University.

Mazurek, J.
 1998 *The Use of Unilateral Agreements in the United States: The Responsible Care Initiative.* ENV/EPOC/GEEI(98)25/FINAL. Paris: Organization for Economic Co-operation and Development.

Nash, J., and J. Ehrenfeld
    1997   Codes of environmental management practice: Assessing their potential as a tool for change. *Annual Review of Energy and Environment* 22:487-535.
Natan, T., Jr., and C.G. Miller
    1998   Are toxic release inventory reductions real? *Environmental Science and Technology* 3(8):368A-374A.
National Research Council
    1996   *Fostering Industry-Initiated Environmental Protection Efforts.* Washington, DC: National Academy Press.
Organization for Economic Co-operation and Development
    1996   *Pollutant Release and Transfer Registers (PRTRS): A Tool for Environmental Policy and Sustainable Development. Guidance Manual for Governments. OCDE*/GD(96)32. Paris: Organization for Economic Co-operation and Development.
    1999   *Voluntary Approaches for Environmental Policy: An Assessment.* Paris: Organization for Economic Co-operation and Development.
O'Toole, L.J., Jr., C. Yu, J. Cooley, and G. Cowie
    1997   Reducing toxic chemical releases and transfers: Explaining outcomes for a voluntary program. *Policy Studies Journal* 25(1):11-26.
Picard, A.
    1992   Plant to shut down for ignoring clean-up orders. *Globe and Mail,* May 30, p.A2.
Rennings, K., L. Brockmann, and H. Bergmann
    1997   Voluntary agreements in environmental protection: Experiences in Germany and future perspectives. *Business Strategy and the Environment* 6:245-263.
Santos, S.L., V.T. Covello, and D.B. McCallum
    1996   Industry response to SARA Title III: Pollution prevention, risk reduction, and risk communication. *Risk Analysis* 16:57-66.
Shapiro, M.
    1998   The Power of Information: Community Exposure and the Efficacy of State Information Regulation Policy. Unpublished paper presented at the Annual Meeting of the American Political Science Association, New York, October.
    1999   Equity and Information: Risks from Toxic Chemicals, Environmental Justice, and Information Regulation Policy. Unpublished paper presented at the Annual Meeting of the Association for Public Policy Analysis and Management, Washington, DC, November.
Storey, M.
    1996   Demand Side Efficiency: Voluntary Agreements with Industry. Annex I Expert Group on the UN FCCC, supported by the Organization for Economic Co-operation and Development and IEA, December.
Tietenberg, T.
    1998   Disclosure strategies for pollution control. *Environmental and Resource Economics* 11:587-602.
US Environmental Protection Agency
    1999   *33/50 Program: The Final Record.* EPA-745-R-99-004. Washington, DC: U.S. Environmental Protection Agency.
    2001   *The United States Experience with Economic Incentives for Protecting the Environment.* EPA-240-R-01-001. Washington, DC: U.S. Environmental Protection Agency.
Weinstein, R., R. Scott, and C. Jones
    1989   Measurement of 'free-riders' in energy conservation programs. *Evaluation and Program Planning* 12:121-130.

17

Assessing the Credibility of Voluntary Codes: A Theoretical Framework

Franco Furger

Over the past 10 years, there has been a steady increase in the number of voluntary environmental initiatives among private sector organizations.[1] In addition to the now well-known case of Responsible Care, several initiatives have been launched at the domestic and international levels. Examples include Coatings Care by the National Paint and Coatings Association, the Coalition for an Environmentally Responsible Economy (CERES) Principles, the development of the ISO 14000 standards of environmental management by the International Organization for Standardization, and the principles and criteria of sustainable forest management developed by the Forest Stewardship Council, a business-nongovernmental organization (NGO) alliance.

The proliferation of voluntary initiatives raises several difficult questions: Can these codes be trusted? How can their credibility be assessed? Should regulators rely on private initiatives to meet public policy goals, and if so what principles and criteria should inform agreements between the public and the private sector? In this chapter, I focus exclusively on the second question—on how the credibility of voluntary codes may be assessed. I explore what criteria may be used by regulators and the public to determine whether a voluntary program should be considered credible.

The discussion is conducted in fairly abstract terms and is not centered on any specific voluntary initiative, although I refer occasionally to specific cases to illustrate a point. Discussions of individual cases may be found in Prakash (2000), Howard et al. (2000), Furger (1997), and Rees (1994, 1997). The goal of such a general discussion is to develop a concept of credibility that is sufficiently robust to be applicable to a wide range of voluntary initiatives. It should enable decision makers in the public sector to assess the credibility of voluntary initiatives in a cost-effective way, even when environmental performance data are scarce or unavailable. Finally, it should prove helpful to decision makers involved in the design of voluntary agreements.

CLASSIC PERSPECTIVES ON "CREDIBILITY"

Public administration scholars usually conceive of credibility (or lack thereof), in formal and procedural terms, rather than substantive terms. For example, voluntary programs have been criticized for not including formal mechanisms of monitoring and compliance assurance for rarely incorporating sanctions short of expelling a member. Furthermore, in many cases voluntary codes are perceived as too generic to be enforceable. Finally, critics have faulted these initiatives for not including measures of environmental performance (see Nash, this volume, Chapter 14, for an in-depth discussion of these issues). Not surprisingly, environmental groups have labeled some of these initiatives as "greenwash" (Greer and Bruno, 1996).

I submit that a legalistic definition of credibility is largely responsible for obfuscating our understanding of voluntary codes and has hindered a fruitful debate over their role in public policy. Consider the claim that the codes are generic, prone to conflicting interpretations, and therefore impervious to vigorous enforcement. This characterization indeed does apply to many codes. However, there is no reason to assume that generic codes cannot be trusted. In fact, the generic nature of voluntary codes may increase rather than undermine their credibility. For example, an association representative recently pointed out that the generic nature of the code published by her trade association—the American Textile Institute—allows her to customize the program's requirements to her members' specific operational and technological circumstances.

From the trade association's point of view, customization has the potential to reconcile what public administration scholars have long thought to be the unavoidable tradeoff between effectiveness and accountability (Light, 1995; Osborne, 1988). The code provides a general, common framework for all member firms. It may be characterized as an architecture of environmental management. It enables member firms to systematically manage their environmental impacts. On the other end, the generic nature of the code makes it possible for the trade association manager to tailor the program's requirements to her members' specific operational, organizational, and technological conditions. Yet the code manager remains accountable to her members: Customization is based on a set of specific guidelines developed by the trade association and its members. Thus, generic codes may reflect the need to carefully balance effectiveness and accountability, rather than suggest deceptive intentions by trade associations. (In this paper, the term "effectiveness" indicates the ability of a trade association manager to customize generic code requirements to the specific operational conditions of individual member firms.)

A second important criticism has focused on inadequate mechanisms of monitoring and compliance assurance. Skeptics often have pointed out that the absence of formal enforcement mechanisms invariably will translate into "free-riding" behavior by member companies. To be sure, free-riding is an all too real possibility. But there are good reasons to believe that free-riding behavior is not nearly as widespread as this legalistic perspective suggests. The U.S. Environmental Protection Agency (EPA) reckons that compliance levels in the private sector average approximately 86 percent (Cohen, 1999). The rather limited efforts of this agency to monitor and enforce environmental laws and regulations cannot explain the high level of compliance with environmental regulations displayed by U.S. firms. There are reasons to believe the EPA estimates may be too high (U.S. General Accounting Office, 1990), but these differences don't invalidate the conclusion that these levels of compliance are inexplicably high.

If vigorous law enforcement is a poor explanation of law-abiding behavior, what may account for this outcome? As many legal scholars have shown, law-abiding behavior is not simply a matter of assessing the costs and benefits of complying with the law. Often, law-abiding behavior reflects internalized social norms (Etzioni, 2000; Tyler and Darley, 2000; Cooter, 1996; Tyler, 1990). Sanctions matter of course, but often it is extralegal sanctions rather than state policing that ensure "compliance" (Posner, 1996, 2000; Bernstein, 1992): Trade association managers, environmental professionals, top executives, community leaders, environmentalists, bankers, insurers, and large customers all may rely on a variety of extralegal sanctioning mechanisms to ensure the proper implementation of a voluntary code. What kind of extralegal sanctions may be available to industry insiders and to the industry "constituencies" are important questions that I discuss later in this chapter.

But what evidence is there that voluntary codes may be effective? In-depth evaluations of these initiatives are few and far between and often inconclusive. Indirect support for the effectiveness of voluntary codes is provided by the U.S. experience with voluntary standard-setting organizations.[2] There are many similarities between standard-developing organizations (SDOs) and trade associations. Both operate as bridging institutions between business interests and the public (Furger, 2001). Voluntary codes, just like private standards, are developed by private interests. Like voluntary codes, many of these standards are designed to provide a public good. And as for most voluntary codes, no formal enforcing mechanisms exist to ensure compliance with privately developed standards. Finally, most of these standards are not self-enforcing and therefore are exposed to free-riding.

Few attempts have been made to evaluate the credibility of standards developed by SDOs. The most thorough analysis of this question has been offered by Krislov (1997). This author deals extensively with the issues just introduced—the credibility of private standards, SDOs' enforcing efforts, and free-riding behavior. Based on the (limited) evidence available, he comes to the conclusion that "voluntary regulation is less than draconian, but by no means irrelevant or ineffectual" (Krislov, 1997:62).[3]

The preceding discussion suggests that it may be misleading to conceive of credibility in formal terms. The obvious alternative, to evaluate the impact of voluntary codes on the environmental performance of participating firms, is rarely a practical option. Compiling reliable and consistent data is fraught with difficulties. In addition, no consensus has emerged on how to define environmental performance.[4] Finally, these efforts rarely reach unambiguous conclusions and thus are of little value to decision makers in the public sector.[5]

The lack of good environmental performance data is only partially responsible for our rather limited understanding of voluntary codes. Without a robust theoretical framework, it is impossible to explain why some voluntary codes do contribute to improving the environmental record of participating firms while other comparable codes don't. Explaining these differences requires an in-depth understanding of the motives, beliefs, values, and incentives that inform the implementation of a voluntary code. In other words, to be of any practical use the concept of credibility must be defined in substantive rather than formal terms.

The chapter is organized as follows: In the next section, I discuss the implementation of voluntary codes as a case of private provision of a public good. The subsequent section forms the core of this chapter. In that section, I show that under certain circumstances, even large and geographically dispersed groups of firms may be able to provide a public good. I then discuss the limits of a purely economic approach to evaluating the credibility of voluntary codes and demonstrate the importance of negotiations and of processes of mutual learning to achieve credibility. I then summarize key findings and explore briefly how the concept of credibility developed in this chapter may inform the design of voluntary agreements.

THE PROVISION OF PUBLIC GOODS IN LARGE GROUPS

The adoption of a voluntary code by a trade association may be thought of as a collective action problem: Adopting the code requires all industrial firms to take costly measures, such as investing in pollution prevention technologies and implementing environmental management systems. Although the costs of adopting the code and thus of protecting the environment are borne by all industry members, the benefits of providing this public good are enjoyed by a much larger population.[6] Under these circumstances, standard economic theory predicts that this public good is unlikely to be provided.

This is a well-known story, told numerous times since Olson's (1965) classic, "The Logic of Collective Action." Applied to the current situation, the theory of collective action predicts that industry members will adopt a voluntary code only if two conditions are met: Every industry member is capable of monitoring other member firms, and the industry as a whole is able to sanction "free-riders," that is, member firms that enjoy the reputational and economic benefits of adopting the code, but avoid the implementation costs. These two requirements often have been taken to suggest that only small groups will succeed in providing a public good.

The literature on self-governance supports the view that only small groups succeed in managing problems of collective action. Accordingly, this literature usually focuses on small communities. Ostrom and her students have accumulated a wealth of empirical evidence to support the view that many communities indeed are able to successfully manage common-pool resources (a special case of a public good) without external intervention (Ostrom et al., 2002; Ostrom et al., 1999; Ostrom et al., 1994; Blomquist, 1992; Ostrom, 1990). Some legal scholars have repeatedly demonstrated that tightly knit communities are quite successful at providing public goods to their members. Ellickson, for example, has studied whaling (Ellickson, 1989) and farming communities (Ellickson, 1991).

The assumption that large groups are unable to provide a public good is rooted in a problematic definition of group size: If a group succeeds in providing a public good, it must be small and spatially concentrated. Conversely, failure to provide a public good indicates a large and dispersed population. There is some wisdom to this definition of group size. But just as Newtonian mechanics provides a satisfactory explanation for many, but not all, natural phenomena, the theory of collective action seems unable to explain why some large groups do manage to provide a public good, while smaller ones don't. Empirical evidence suggests that large groups are not as ill equipped to provide a public good as the economic theory of collective action predicts. Under certain circumstances, information can circulate quite efficiently even within large groups. And as I will show, the effectiveness of informal sanctioning mechanisms is not confined to small groups.

Consider the following two examples, the international marine industry and the international diamond trade. The marine industry provides an illustration of the claim that information can circulate quite efficiently even within large groups. The diamond traders for their part demonstrate that informal sanctioning mechanisms may be very effective at sustaining common norms, such as honest behavior, even in large groups. In an interview with this author, a representative of the Salvage association—an organization on which the marine insurance industry depends to conduct accident investigations—claimed that an instance of collusion between a shipowner and a shipyard in Singapore would be common knowledge among marine insurers in London within a week. In the international marine industry, information gathering has become the focus of specialized activities. For example, insurance brokers traditionally have played a very important role in providing information to marine insurers and shipowners. However, insurers don't rely exclusively on brokers. They depend on "agents" in many ports of the world to obtain firsthand information about prospective clients. They also leverage shipowners' knowledge about their competitors to overcome problems of asymmetrical information.[7] In sum, in an industrial sector as large as the international marine industry, one may indeed seem to be operating in a "small world."

The diamond traders of New York City provide an illustration of how a sizable group may be able to sustain common rules and norms without resorting to external legal action. The industry is organized into so-called trading clubs, also known as "bourses." The New York Diamond Dealers Club (DDC) is the largest in the United States and has approximately 2,000 members, most of whom are also members of the New York Jewish community. The most striking feature of the DDC is the almost exclusive reliance on extralegal rules and norms to conduct its business. Every aspect of the bourse is governed by the bourse bylaws, and business disputes are resolved through arbitration. The arbitrator's decisions are enforced by the business community, and external enforcement actions usually are not necessary (Bernstein, 1992; Coleman, 1988). Generally speaking, legal considerations and legal enforcement actions play only a minor role in this trade.

Neither the international marine industry nor the international diamond trade may be considered small groups, yet they display a surprising ability to address problems of collective action. They demonstrate that group size is a poor predictor of a group's ability to provide a public good. In the next section, I explore the relationship between group size and collective action in more depth. I show that the ability of a group to overcome problems of collective action can be explained by certain features of the social network in which the group members are embedded rather than by sheer group size.

ASSESSING THE CREDIBILITY OF VOLUNTARY CODES: A STRUCTURAL APPROACH

The Small World Phenomenon

The small-world (SW) phenomenon undermines the traditional distinction between small and large groups. This term refers to a familiar situation in which two strangers discover to their surprise that they share a common acquaintance. The term was coined in 1967 by Milgram. He showed that within the U.S. population, the average distance between any two individuals, measured by the average number of links separating them, is approximately six—hence the now famous "six degrees of separation" hypothesis (Milgram, 1967).

SW networks intuitively can be described as large groups that retain key characteristics of small groups. Members of SW networks experience their social environment as a small group. As for any other small community, the more cohesive this environment is, the more likely it is that an individual's conduct becomes public information. In addition, in cohesive groups, individuals care about their reputation and avoid actions that may damage it. Finally, more cohesive groups are more likely to sanction a violation of a group norm. The degree of cohesion of the social environment of an individual in the network is measured by the so-called clustering coefficient.[8]

It is not unreasonable to assume that the more cohesive these clusters are, the more likely they will be disconnected. Individuals belonging to different clusters are very unlikely to know each other, that is, to be connected through a link. This means that the average distance between two individuals each belonging to a different cluster may be very large.[9] Thus, one can expect that large social groups are characterized by a high clustering coefficient and a long characteristic path length, or by a small clustering coefficient and a short characteristic path length, but not by a high degree of clustering <u>and</u> a short characteristic path length. Watts has shown that this intuitive result is inaccurate. There exists a large class of networks with both high degree of clustering <u>and</u> a short characteristic path length. He has labeled these networks small-world networks (Watts, 1998, 1999a, 1999b). This counterintuitive result has a fairly straightforward explanation: High clustering and short characteristic path length are properties of highly clustered networks connected with one another through a small number of random ties. It is precisely these ties that account for the small-world phenomenon (Watts, 1999b).

The SW phenomenon is relevant to the current discussion for one main reason: Trade association members embedded in SW networks are much more likely to adopt and credibly implement a voluntary code. In addition, these trade groups will display a superior ability to solve problems of collective action—independently of group size and geographical dispersion. Let's discuss briefly why this may be the case.

Earlier I argued that efficient information exchange among industry members is a key requirement for a trade group to credibly implement a voluntary code. If the personal and professional ties among the members of a trade association constitute an SW network, information about individual members is more likely to be common knowledge among all members even if the membership is large and spatially dispersed. This is so because the shorter the characteristic path length, the easier it is for a bit of information to circulate outside the limited realm of a small cluster of member firms. I have also argued that a high degree of clustering increases the likelihood that another firm in the same cluster will apply some form of extralegal sanctions to the "defector." Thus, the higher the degree of clustering, the more likely it is that another industry member will sanction improper implementation practices.

Little currently is known about the network structure of specific industry groups.[10] Generally speaking, one can expect the structure of these networks to vary considerably from one trade association to the next. Nevertheless, one can predict whether the members of a trade group are embedded in an SW network by examining a few structural variables. For example, mature industries characterized by a stable industrial membership are more likely to be embedded in a well-developed SW network: Some degree of stability is needed for an SW network to emerge. Conversely, young industries characterized by fast technological change and with significant turnover in their membership may not be conducive to the formation of SW networks. Another important factor affecting the formation of SW networks is the labor market. A closed labor market, that is, a labor market that has significant entry and exit barriers but displays significant internal turnover, greatly facilitates the emergence of an SW network. On the other hand, an open labor market is likely to impede the formation of SW networks. The reliance on consultants and other professionals by the member firms also may favor the emergence of SW networks.[11]

The preceding discussion was based on several simplifying assumptions. I didn't distinguish between firms and individuals operating within these firms, and I haven't examined the role played by the trade association in the implementation process. In addition, I have not examined how information about an industry member may circulate, and what kind of extralegal sanctions may be available to various industry constituencies. In the next section, I begin to discuss these issues.

Information Gathering in Large Groups

The credibility of a voluntary code depends to a significant extent on the ability of a trade association to identify "recalcitrant" members—those industrial members with a poor record of code implementation. As noted earlier, identifying "defectors" is a difficult task because often codes do not include any monitoring mechanism.[12] Under these circumstances, a code manager depends on his or her ties to industry insiders to assess the credibility of self-audits, or to obtain information about specific industry members.

If a code manager operates in a large trade group, it is unlikely that she will have direct ties to all member firms. However, this is not necessarily a serious impediment to obtaining relevant information about the member firms. The discussion of SW networks suggests that direct ties to all members firms are not necessary. If the code manager is embedded in an SW network, she is likely to be separated from the relevant source of information by a small number of links. In plain English, even though she may not have a colleague working for the member firm she is seeking information about, chances are she will know somebody who does. If the second-order acquaintance is unable to provide the information needed, he may be able to help by tapping into his own network of colleagues and acquaintances. Interestingly, a direct tie with the firm under scrutiny may not be the most effective way to obtain sensitive information about that firm. A company official may find it problematic to provide sensitive information to a trade association representative, especially if that information can embarrass his company.

Code managers don't depend only on their professional networks to obtain relevant information. Competitive dynamics among member firms may produce unexpected results. Interviews with representatives of the National Paint and Coatings Association (NPCA) have shown that information about free-riding behavior may be reported to the trade association by firms that have been put at a competitive disadvantage by free-riders: Participating firms may have a strong incentive to ensure that their competitors won't gain an unfair competitive advantage by poorly implementing the code. How rival firms learn about their competitors' standards of implementation is not always clear, and the modality varies from industry to industry. For example, it appears that in the pharmaceutical industry, senior vice presidents for environment, health, and safety entertain cordial relations with each other and exchange relevant information. Representatives of large corporations have made similar statements on occasion.

Skillful code managers may renew and expand their SW networks in several ways: by organizing training events focused on implementation of the code; by convening meetings dedicated to regulatory, technical, and other issues of common interest; and by participating in annual industry meetings, national conferences, professional meetings, and social gatherings. Whether in reality code managers consciously pursue this strategy is an important question that has not yet received much attention.

But who is in a position to provide good, reliable information about a firm's efforts to implement a code? The most likely source of reliable information is neither the top executive nor the line personnel, but the mid-level manager responsible for the program's implementation. Individuals in top management positions are unlikely to share embarrassing or damaging information with trade association representatives. On the other extreme, line personnel are in a much better position to share relevant information with outsiders. However, they are not likely to have strong ties to trade association managers or to other industry practitioners. By contrast, mid-level managers are embedded in networks of professional ties that may include several other firms outside their local community. Unlike top management, their position may allow them to share information about their firm's deficiencies with their professional peers and with trade association representatives.[13]

Efficient information exchange is obviously an indispensable condition for maintaining the credibility of an industry code. But information alone doesn't guarantee that the members of a trade group will properly implement the code. In the absence of some form of sanctioning, the code is unlikely to become an effective tool of environmental management. In the next section, I examine possible sources of extralegal sanctioning and discuss their relevance.

Sources of Extralegal Sanctioning

The absence of explicit sanctions from most voluntary codes shouldn't be interpreted as a demonstration of untrustworthiness. As this section shows, the

credibility of voluntary codes is not predicated upon the trade association imposing formal, highly visible sanctions. Several other organizations operating in a
firm's institutional environment may play an important role in ensuring the
code's credibility.[14]

Extralegal sanctions may come in at least three forms. The improper implementation of the code's requirements may trigger guilt or shame, may damage a
firm's reputation, and may have negative economic consequences. Guilt and
shame speak to the moral dimension of individual choices, to the internalization
of moral values and norms. The economic literature tends to dismiss this element
of social regulation, but there is little doubt that over the past 20 years, the
business community as a whole has dramatically changed its attitudes towards
the environment. Of more immediate relevance to this discussion are sanctions
that may damage a firm's or an individual's reputation. A large body of literature
suggests that reputational considerations are a key aspect of many social interactions, including business dealings.[15] A damaged reputation can affect an individual or an organization in a variety of ways. It can translate into a loss of sales
and revenues, a lower stock valuation, higher insurance premiums, the loss of
friends and colleagues, and in extreme cases, a ruined professional career. Finally,
contractual and business relations also may become tools of extralegal sanctions,
such as when consumer boycotts occur or when the inadequate implementation
of the code translates into higher insurance premiums and more costly access to
financing.

Several organizational actors in the institutional environment of a firm may
play an active role in ensuring the proper implementation of a code. In addition
to the trade association, the parent company, other industry members, large customers (Walton et al., 1998),[16] financial institutions (insurance companies and
banks),[17] local communities, accounting firms and professional organizations,
national and international NGOs (Bendell, 2000), and international organizations all may play a significant role in promoting a credible implementation of
the code (Miles and Covin, 2000). Empirical studies that assess the role of these
constituencies in ensuring a credible code implementation are still rare, but tend
to support the view that their aggregate impact can be significant. For example,
in developing countries, local communities may play an important role in improving environmental protection in the absence of an effective regulatory system. Empirical studies have shown that industrial plants in communities with
active local politics are associated with lower environmental impacts (Arora and
Cason, 1999; Pargal and Wheeler, 1995).

Finally, how does a trade association deal with noncompliance? It is well
known that trade associations usually have a limited sanctioning capacity. A
trade association's ability to sanction noncompliance essentially reflects its bargaining power, that is, the costs and benefits associated with staying versus
leaving the association. A trade association's sanctioning capacity is strengthened if the membership provides highly valuable and exclusive services to the

member firms. On the other end, a firm's decision to leave a trade association may be associated with significant costs. Leaving the trade group may result in being perceived as a liability by potential business partners; may be interpreted by the public, environmental groups, and regulators as an indication of a poor environmental record; and may increase the costs of insurance and financing. National regulators are in an excellent position to improve the attractiveness of membership. For example, the recently launched EPA Performance Track program provides recognition and some degree of regulatory flexibility to a selected number of trade associations (Steve Sides, National Paint and Coating Association, personal communication, June, 2001).

Absent such enticements, trade association officials must resort to different strategies. For example, these officials may make a strategic use of their position as "information brokers" in the industry network. An information broker can wield considerable indirect sanctioning power by making sensitive and potentially damaging information about a recalcitrant member available to its direct competitors and/or to other firms. In essence, their position as information brokers confers on them the power to trigger what may be called an extralegal "sanctioning cascade." A sanctioning cascade is not without problems. Depending on the structure of the industry network, it can generate a considerable amount of gossip and outrage. Intense gossip can lead to information distortion and unfairly damage a firm's reputation.[18]

From Strategic Behavior to Learning Processes

The credibility of voluntary codes cannot be discussed exclusively in terms of information diffusion and extralegal sanctioning mechanisms. The incomplete nature of many voluntary codes reduces considerably the relevance of informal sanctions: For sanctions to be fairly imposed, there must be a consensus on what constitutes a "violation" or an "infraction." Only after the industry members have agreed on what constitutes a proper implementation of the code can the role of informal sanctions be explored. In this section, I focus on the role played by environmental managers in shaping the expectations related to implementation of the code. "Implementation" in this context takes on a specific meaning: It identifies the process by which environmental managers convert incomplete and open-ended code requirements into actual implementation practices. It consists of extensive and repeated discussions among environmental professionals, combined with regular information exchange. It is, in essence, a process of mutual learning that leads to a common understanding of what constitutes a proper implementation of the code. It is also a process that facilitates the establishment of new ties among environmental professionals ("horizontal" ties) and reinforces the relationship between these professionals and the trade association ("vertical" ties), thus contributing to the formation of what some scholars have called a "community of practice" (Gherardi et al., 1998; Brown and Duguid, 1991).

The term "common understanding" doesn't imply that the code requirements will ultimately translate into uniform implementation practices. "Common
understanding" in this context refers to acceptable ways of interpreting generic
code requirements in different technical, regulatory and business contexts.
Developing a common understanding of what constitutes acceptable interpretations of generic code requirements is key to reconciling conflicting demands for
accountability and flexibility, it enables customization of a consensus about differences rather than uniformity—a demonstration of the code's ability to reconcile effectiveness and accountability.[19]

Two opposing forces shape this negotiation process: on one end, the interests of the firms represented in these negotiations, and on the other end, the
allegiances to criteria of professionalism, technical competence, and impartiality
typical of a professional community. The tension between professionalism and
the firms' interest may lead to a temporary situation in which the environmental
professionals develop a common understanding of what constitutes the proper
implementation of the code that differs from the position taken by some of the
participating companies. These conflicts may be resolved in several ways.

One possible way to resolve these disputes is for the professional community to simply ignore this disagreement. This is not likely to be a durable solution
because as it creates a dangerous precedent. The conflict is also likely to resur-
face at a later time in a more virulent form. A second option is to lower expectations to avoid possible conflicts. This possibility often is identified as one of the
main reasons for distrusting private standards (Cheit, 1990). However, the evidence in support of this claim is not overwhelming. The third and final option is
to reallocate the responsibility for resolving this matter from the community of
environmental managers to the trade association (to be discussed later).

An agreement on what constitutes a proper implementation of a voluntary
code does allow environmental and code managers to determine whether a firm
is "complying" with the code's requirements, but it doesn't necessarily make the
use of extralegal sanctions more effective. A firm may have implemented a code
poorly for several reasons. For example, a firm may lack scientific and technical
expertise. A recent joint effort by the EPA and the American Chemistry Council
to determine the root causes of regulatory noncompliance concluded that "human error" and "procedures" (operating procedures not followed) are two of the
main causes of noncompliance (U.S. Environmental Protection Agency, 1999).
Implementing the code also may create complex organizational problems (Fryxell and Vryza, 1999). Finally, financial resources to properly implement the code
may be in short supply. Thus, one should distinguish between voluntary and
involuntary noncompliance.

Involuntary noncompliance demonstrates the limits of assessing the credibility of voluntary codes from a narrowly defined economic perspective. If involuntary noncompliance is a significant aspect of a poorly implemented code—
and there is significant evidence that implementing these codes can be

overwhelming for many small and medium-size firms—applying sanctions to noncompliant firms may be wholly ineffectual. If noncompliance can be attributed to limited technical and scientific expertise or to organizational barriers, the effective implementation of a voluntary code may depend on the availability of training and technical assistance and on processes of information sharing and mutual learning, rather than on sanctions.

A collaborative approach to the code implementation has a dramatic impact on the code manager's ability to gather accurate information about member firms. It also enables environmental managers to freely exchange with each other even sensitive bits of information about implementation problems. Conversely, legalistic and confrontational relations are likely to undermine the free flow of knowledge and information among environmental professionals, and between these professionals and their trade group.[20]

The preceding discussion shouldn't be interpreted as evidence that collaborative and confrontational approaches are mutually exclusive. Rather, it suggests that sanctioning noncompliance is itself a choice that must be based on carefully assessing its costs and benefits. Those who favor the imposition of strict sanctions all too often are oblivious to the possibility that these sanctions—legal or otherwise—may be associated with considerable costs. By sanctioning noncompliant behavior in an inappropriate or excessive way, a trade group may compromise its future ability to shape the implementation process. This means that the role of code managers is subtler than usually assumed. Code managers are neither "private cops" nor mere "cheerleaders." Rather, they may be characterized as mediators. Their central position in the industry network confers on them the power to address disputes among member firms over implementation issues, between industry interests and the public sector, and between environmental groups and member firms. In short, they become what some scholars have dubbed "crosscutting ties."[21]

CONCLUSION

In this chapter I have shown that the two dominant approaches to evaluating voluntary codes—determining the impact of a code on environmental performance and assessing the code requirements against formal criteria—may not provide satisfactory results. Accurate, public sources of information on environmental performance are rarely available. Furthermore, there is no consensus among scholars and practitioners on how to define measures of environmental performance that would allow meaningful interfirm comparisons. On the other end, evaluating voluntary codes on formal criteria alone may be misleading. As this chapter has shown, firms may undertake considerable efforts to implement a voluntary code, even though the code itself is not fully specified and/or doesn't include detailed reporting requirements and specific mechanisms of compliance assurance. This chapter suggests that a better way to assess voluntary codes is to focus on processes and structures that may facilitate or impede a credible code implementation. This includes industry networks, information diffusion and information exchange, extralegal sanctioning mechanisms, and interorganizational learning, among others.

The concept of credibility advanced in this chapter provides several lessons for the design of voluntary agreements. First, a contractual relationship between a regulatory agency and a trade association is likely to transform the relationship of the trade association to its members from collaborative to confrontational. In other words, it may crowd out collaborative dispositions and undermine information exchange and mutual learning. In addition, it may impair the trade association's ability to mediate among opposing interests. Whether regulators could avoid these outcomes by providing significant benefits to participating firms remains to be seen.

Second, most industrial codes may be characterized as an architecture of environmental management. Current voluntary codes are not designed to directly address highly specific environmental concerns. The incomplete, open-ended nature of many voluntary codes is typical of social obligations that are impervious to formalization. As a result, they are ill suited to become part and parcel of voluntary agreements.

Third, developing an in-depth knowledge of the implementation processes should enable decision makers in the public sector to focus on improving the self-regulatory and learning capacity of a trade association and its members, rather than concentrating on specific elements of the code. In less abstract terms, voluntary agreements should be designed to improve the problem solving capacity of the parties involved (Lindblom and Cohen, 1979).

NOTES

1 The term "voluntary initiative" is not particularly accurate, as some trade groups have made their codes a condition of membership. However, for the sake of clarity I will use this term throughout the chapter. Voluntary initiatives must be distinguished from "voluntary agreements." The latter usually refers to a formalized contractual arrangement between a private party, often a trade association, and a regulatory agency, while the former is defined as an institutional arrangement developed by a trade association for the benefit of its members.

2 In the United States, Standard Developing Organizations have been involved in setting health and safety standards in many industries and for many consumer products (Yilmaz, 1998; Krislov, 1997; Cheit, 1990; Hamilton, 1978; Hemenway, 1978).

3 One may attribute this result to fears of product liabilities. However, these fears may be compensated by concerns about antitrust laws.

4 See, for example, National Academy of Engineering (1999), Natan and Miller (1998), Ranganathan (1998), and Ditz and Ranganathan (1997).

5 Consider, for example, Responsible Care. Even though this code has existed for well over a decade, only one comprehensive evaluation has been conducted so far (King and Lenox, 2000). Their statistical sophistication and thoroughness notwithstanding, the authors of this study were unable to draw firm conclusions about the effectiveness of Responsible Care. Other efforts at evalu-

ating voluntary codes include Coglianese and Nash (2001), Cowton and Thompson (2000), Prakash (2000), and Nash and Ehrenfeld (1997).

6   This is so because by definition, nobody can be excluded from enjoying the benefits of a public good.

7   The term "asymmetrical information" identifies a situation in which access to critical information is uneven. Consider, for example, the market for used cars. Prospective buyers are usually unable to determine whether they are getting a "lemon," only the salesperson has that information. This is a classic case of asymmetrical information (Akerlof, 1970).

8   The clustering coefficient is a local property: If a node has n immediate neighbors, then this neighborhood defines a subgraph that has at most $n(n-1)/2$ edges, if the neighborhood is fully connected. The clustering coefficient of this subgraph is the fraction of this maximum that is actually realized. The clustering coefficient of the entire graph is the average of these fractions calculated for every node (Watts, 1999b:498).

9   The characteristic path length is defined as the "average number of edges that must be traversed in the shortest path between any two pairs of vertices in the graph" (Watts, 1999b:498). It is a *global* property of the network.

10   In this context, the literature on interlocking directorates is of limited help. In this chapter, I am concerned with networks of environmental professionals and top executives, and not with relationships among boards of directors.

11   Consultants and other providers of professional services may be described as "weak ties" (Granovetter, 1973). As I will show, individuals and organizations with weak ties to otherwise disjointed networks are in an excellent position to mediate between conflicting interests and to influence the outcome of these conflicts. An indirect illustration of this argument is provided by Dietz and Rycroft (1987).

12   This is not surprising because acceptance of these codes by member firms would be seriously damaged by any attempt to incorporate such mechanisms from the outset. Accordingly, trade association representatives carefully avoid any talk of "enforcement action."

13   See Canan and Reichman (1993) for a discussion of the role of mid-level managers and engineers in promoting the implementation of the Montreal protocol. Also see the discussion in Brown and Duguid (1991).

14   It is fashionable to discuss the role of organizations located in the institutional environment of a firm in terms of first-, second-, third-, and fourth-party inspections and certification. Although this terminology occasionally may be useful, it tends to obscure important differences within each category. For example, a well-reputed industry insider and an international consulting firm both can be described as "third parties." However, a small firm may find it wholly unacceptable to be surveyed and certified by an international consulting firm, but most likely would agree to be audited by an independent industry insider. An NGO would take the opposite view.

15   The details of this argument are more involved. The groundwork was laid by several economists in the 1980s (Mailath and Samuelson, 1998; Kreps et al., 1982; Kreps and Wilson, 1982). Economic historians have provided vivid illustrations of the role of reputational incentives in sustaining trade (Greif, 1989, 1991, 1993; Milgrom et al., 1990). During the 1990s, many legal scholars recognized the relevance of reputational incentives in sustaining social norms (Schwarcz, in press; Coffee, 2001; Posner, 1996, 2000; Choi, 1998; Bernstein, 1992; Ellickson, 1989, 1991; Charny, 1990). Political scientists have studied the role of reputation, guilt, and shame as a means of informal policing (Braithwaite, 1989, 1993). Surprisingly, sociologists—with few exceptions (Etzioni, 2000; Raub and Weesie, 1990)—have been slow to recognize the importance of this topic.

16   An increasing number of large industrial firms are requiring suppliers to achieve some degree of sustainability by adopting environmental management systems such as ISO 14000. The automobile industry is a case in point.

17   For example, the American Chemistry Council (ACC, formerly the Chemical Manufacturers Association) has negotiated better insurance rates for Responsible Care companies. Unfortunately,

this approach removes selective incentives for ACC firms to improve their environmental performance.

18 In the case of the diamond trade, this has occurred, as documented in Bernstein (1992:121). A diamond trader whose reputation had been damaged by baseless gossip was able to restore his good name by posting a rebuttal on the DDC's bulletin board.

19 Consider, for example, the following requirement, which is part of the NPCA Coatings Care program: "I.III.1 Occupational Safety and Health: 1.2 Plan and carry out periodic targeted inspections for conformity with site policies and practices." Obviously, a small firm will implement this requirement rather differently than a large industrial company. These differences, however, don't necessarily constitute evidence of inadequate code implementation.

20 A growing body of literature in economics demonstrates that confrontational interactions among the members of a group hamper the group's ability to cooperate. See, for example, Fehr and Gächter (2001), Falk et al. (2002), and Frey (1997). This literature also illustrates the relevance of processes of face-to-face communication in overcoming problems of collective action.

21 The relevance of crosscutting ties hardly can be overemphasized. For example, Lipset (1959) has shown that a key requirement of a functioning democracy is to have crosscutting ties among the various social and political groups. In his now classic *The Strength of Weak Ties,* sociologist Granovetter suggests that missing crosscutting ties among ethnic Italian families in a Boston neighborhood account for the inability of this neighborhood to organize against "urban renewal" (Granovetter, 1973:1373-1376; see also Granovetter, 2000:1079). Locke notes that missing crosscutting ties between business and labor in Turin, Italy, explain the very confrontational and ultimately self-defeating path of industrial restructuring at FIAT, the largest Italian carmaker. By contrast, in the case of Alfa Romeo, the existence of numerous crosscutting ties among the parties involved led to a cooperative restructuring process, with positive economic fallouts for the entire Milan area (Locke, 1995).

REFERENCES

Akerlof, G.A.
  1970   The market for 'lemons': Quality uncertainity and the market mechanism. *Quarterly Journal of Economics* 84(3):488-500.
Arora, S., and T.N. Cason
  1999   Do community characteristics influence environmental outcomes? Evidence from the Toxics Release Inventory. *Southern Economic Journal* 65(4):691-716.
Bendell, J., ed.
  2000   *Terms for Endearment: Business, NGOs and Sustainable Development.* Sheffield, Eng.: Greenleaf Publishing.
Bernstein, L.
  1992   Opting out of the legal system: Extralegal contractual relations in the diamond industry. *Journal of Legal Studies* 21(21):115-157.
Blomquist, W.A.
  1992   *Dividing the Waters: Governing Groundwater in Southern California.* San Francisco: Institute of Contemporary Studies Press.
Braithwaite, J.
  1989   *Crime, Shame and Reintegration.* Cambridge, Eng.: Cambridge University Press.
  1993   Shame and modernity. *British Journal of Criminology* 33(1):1-18.
Brown, J.S., and P. Duguid
  1991   Organizational learning and communities of practice: Toward a unified view of working, learning and innovation. *Organization Science* 2(1):58-82.
Canan, P., and N. Reichman
  1993   Ozone partnerships, the construction of regulatory communities, and the future of global regulatory power. *Law & Policy* 15(1):61-74.

Charny, D.
1990 Nonlegal sanctions in commercial relationships. *Harvard Law Review* 104:375-467.

Cheit, R.E.
1990 *Setting Safety Standards: Regulation in the Public and Private Sectors.* Berkeley: University of California Press.

Choi, S.
1998 Market lessons for gatekeepers. *Northwestern University Law Review* 92:916-966.

Coffee, J.C.J.
2001 The Acquiescent Gatekeeper: Reputational Intermediaries, Auditor Independence and the Governance of Accounting. Unpublished manuscript, Columbia Law School, Center for Law and Economic Studies, New York.

Coglianese, C., and J. Nash
2001 Bolstering Private-Sector Environmental Management. Issues in Science and Technology. [Online Edition]. Available: http://www.nap.edu/issues/17.3/ [Accessed September 28, 2001].

Cohen, M.A.
1999 Monitoring and enforcement of environmental policy. Pp. 44-106 in *International Yearbook of Environmental and Resource Economics,* Vol. III, T. Tietenberg and JH. Folmer, eds. Northampton, MA: Edward Elgar.

Coleman, J.S.
1988 Social capital in the creation of human capital. *American Journal of Sociology* 94:S95-S120.

Cooter, R.D.
1996 Decentralized law for a complex economy: The structural approach to adjudicating the new law merchant. *University of Pennsylvania Law Review* 144(5):1643-1696.

Cowton, C.J., and P. Thompson
2000 Do codes make a difference? The case of bank lending and the environment. *Journal of Business Ethics* 24(2, Part 2):165-178 I.

Dietz, T.M., and R.W. Rycroft
1987 *The Risk Professionals.* New York: Russell Sage Foundation.

Ditz, D.W., and J. Ranganathan
1997 *Measuring Up : Toward a Common Framework for Tracking Corporate Environmental Performance.* Washington, DC: World Resources Institute.

Ellickson, R.C.
1989 A hypothesis of wealth maximizing norms: Evidence from the whaling industry. *Journal of Law, Economics and Organization* 5(83):83-97.
1991 *Order Without Law: How Neighbors Settle Disputes.* Cambridge, MA: Harvard University Press.

Etzioni, A.
2000 Social norms: Internalization, persuasion, and history. *Law and Society Review* 34(1):157-178.

Falk, A., E. Fehr, and U. Fischbacher
2002 Appropriating the commons: A theoretical explanation. Pp. 157-191 in National Research Council, *The Drama of the Commons.* Committee on the Human Dimensions of Global Change, E. Ostrom, T. Dietz, N. Dolšak, P.C. Stern, S. Stonich, and E. Weber, eds. Washington, DC: National Academy Press.

Fehr, E., and S. Gächter
2001 Do Incentive Contracts Crowd Out Voluntary Cooperation? University of Southern California Center for Law, Economics & Organization Research Paper No. C01-3. [Online]. Available: http://papers.ssrn.com/abstract_id=229047 [Accessed February 15, 2001].

Frey, B.S.

    1997   *Not Just for the Money: An Economic Theory of Personal Motivation.* Cheltenham, Eng.: Edward Elgar.

Fryxell, G.E., and M. Vryza

    1999   Managing environmental issues across multiple functions: An empirical study of corporate environmental departments and functional co-ordination. *Journal of Environmental Management* 55(1):39-56.

Furger, F.L.

    1997   Accountability and self-governance systems: The case of the maritime industry. *Law and Policy* 19(4):445-476.

    2001   Global markets, new games, new rules: The challenge of international private governance. Pp. 201-245 in *Ruler and Networks: The Legal Culture of Global Business Transactions,* R. Applebaum, W.F. Felstiner, and V. Gessner, eds. Oxford, Eng.: Hart Publishing.

Gherardi, S., D. Nicolini, and F. Odella

    1998   Toward a social understanding of how people learn in organizations. *Management Learning* 29(3):273-297.

Granovetter, M.S.

    1973   The strength of weak ties. *American Journal of Sociology* 78(6):1360-1380.

    2000   A Theoretical Agenda for Economic Sociology. Unpublished manuscript. Palo Alto, CA: Stanford University, Department of Sociology.

Greer, J., and K. Bruno

    1996   *Greenwash: The Reality Behind Corporate Environmentalism.* Penang, Malay.: Third World Network.

Greif, A.

    1989   Reputation and coalitions in medieval trade: Evidence on the Maghribi traders. *Journal of Economic History* 49(4):857-882.

    1991   The organization of long-distance trade: Reputation and coalitions in the Geniza documents and Genoa during the eleventh and twelfth centuries. *American Economic Review* 83(3):525-548.

    1993   Contract enforceability and economic institutions in early trade: the Maghribi Traders' Coalition. *American Economic Review* (83)3:525-548.

Hemenway, D.

    1978   *Industrywide Voluntary Product Standards.* Cambridge, MA: Ballinger.

Howard, J., J. Nash, and J. Ehrenfeld

    2000   Standard or smokescreen? Implementation of a voluntary environmental code. *California Management Review* 42(2):63-82

King, A.A., and M.J. Lenox

    2000   Industry self-regulation without sanctions: The Chemical Industries Responsible Care Program. *Academy of Management Journal* 43(4):698-716.

Kreps, D.M., P. Milgrom, J. Roberts, and R. Wilson

    1982   Rational cooperation in the finitely repeated prisoners' dilemma. *Journal of Economic Theory* 27(2):245-252.

Kreps, D.M., and R. Wilson

    1982   Reputation and imperfect information. *Journal of Economic Theory* 27:253-279.

Krislov, S.

    1997   *How Nations Choose Product Standards and Standards Change Nations.* Pittsburgh: University of Pittsburgh Press.

Light, P.C.

    1995   *Thickening Government: Federal Hierarchy and the Diffusion of Accountability.* Washington, DC: Brookings Institution.

Lindblom, C.E., and D.K. Cohen
  1979 *Usable Knowledge: Social Science and Social Problem Solving.* New Haven, CT: Yale University Press.
Lipset, S.M.
  1959 Some social requisites of democracy: Economic development and political legitimacy. *American Political Science Review* 53(1):69-105.
Locke, R.M.
  1995 *Remaking the Italian Economy.* Ithaca, NY: Cornell University Press.
Mailath, G.J., and L. Samuelson.
  1998 Your Reputation Is Who You're Not, Not Who You'd Like to Be. CARESS Unpublished working paper. Philadephia: University of Pennsylvania, Department of Economics.
Miles, M.P., and J.G. Covin
  2000 Environmental marketing: A source of reputational, competitive, and financial advantage. *Journal of Business Ethics* 23(3, Part 1):299-311.
Milgram, S.
  1967 The small world problem. *Psychology Today* 1(2):60-67.
Milgrom, P.R., D.C. North, and B.R. Weingast
  1990 The role of institutions in the revival of trade: The law merchant, private judges, and the champagne fairs. *Economics and Politics* 2(1):1-23.
Nash, J., and J. Ehrenfeld
  1997 Codes of environmental management practice: Assessing their potential as a tool for change. *Annual Review of Energy and the Environment* 22:487-535.
Natan, T.E.J., and C.G. Miller
  1998 Are toxics release inventory reductions real? *Environmental Science and Technology* 32(15):368A-374A.
National Academy of Engineering
  1999 *Industrial Environmental Performance Metrics.* Washington, DC: National Academy Press.
National Research Council
  2002 *The Drama of the Commons.* Committee on the Human Dimensions of Global Change, Ostrom, E., T. Dietz, N. Dolšak, P.C. Stern, S. Stonich, and E.U. Weber, eds. Division of Behaviorial and Social Sciences and Education, National Research Council. Washington, DC: National Academy Press.
Olson, M.
  1965 *The Logic of Collective Action.* Cambridge, MA: Harvard University Press.
Osborne, D.E.
  1988 *Laboratories of Democracy.* Boston, MA: Harvard Business School Press.
Ostrom, E.
  1990 *Governing the Commons: The Evolution of Institutions for Collective Action.* Cambridge, Eng.: Cambridge University Press.
Ostrom, E., J. Burger, C.B. Field, R.B. Norgaard, and D. Policansky
  1999 Revisiting the commons: local lessons, global challenges. *Science* 284(5412):278-282.
Ostrom, E., R. Gardner, and J. Walker
  1994 *Rules, Games, and Common-Pool Resources.* Ann Arbor: University of Michigan Press.
Pargal, S., and D. Wheeler
  1995 *Informal Regulation of Industrial Pollution in Developing Countries: Evidence from Indonesia.* Washington DC: Policy Research Department, Environment, Infrastructure, and Agriculture Division, The World Bank.
Posner, E.A.
  1996 The regulation of groups: The influence of legal and nonlegal sanctions. *University of Chicago Law Review* 63(133-197).

    2000   *Law and Social Norms*. Cambridge, MA: Harvard University Press.
Prakash, A.
    2000   Responsible Care: An Assessment. *Business and Society* 39(2):183-209.
Ranganathan, J.
    1998   *Sustainability Rulers: Measuring Corporate Environmental and Social Performance*. Washington, DC: World Resources Insititute.
Raub, W., and J. Weesie
    1990   Reputation and efficiency in social interactions: An example of network effects. *American Journal of Sociology* 96(3): 626-654.
Rees, J.V.
    1994   *Hostages of Each Other: The Transformation of Nuclear Safety Since Three Mile Island*. Chicago: University of Chicago Press.
    1997   Development of communitarian regulation in the chemical industry. *Law and Policy* 19(4):477-528.
Schwarcz, S.L.
    in   Private Ordering of Public Markets: The Rating Agency Paradox. *University of Illinois*
    press   *Law Review*.
Tyler, T.R.
    1990   *Why People Obey the Law*. New Haven, CT: Yale University Press.
Tyler, T.R., and J.M. Darley
    2000   Building a law-abiding society: Taking public views about morality and the legitimacy of legal authorities into account when formulating substantive law. *Hofstra Law Review* 28(3):707-739.
U.S. Environmental Protection Agency
    1999   *EPA/CMA Root Cause Analysis Pilot Project: An Industry Survey*. EPA-305-R-99-001. Washington, DC: Office of Enforcement and Compliance Assurance.
U.S. General Accounting Office
    1990   *Improvements Needed in Detecting and Preventing Violations*. Washington, DC: U.S. General Accounting Office.
Walton, S.V., R.B. Handfield, and S.A. Melnyk
    1998   The green supply chain: Integrating suppliers into environmental management processes. *International Journal of Purchasing and Materials Management* 34(2):2-11.
Watts, D.J.
    1998   Collective dynamics of 'small-world' networks. *Nature* 393(6684):440-442.
    1999a   *The Dynamic of Networks Between Order and Randomness*. Princeton, NJ: Princeton University Press.
    1999b   Networks, dynamics, and the small-world phenomenon. *The American Journal of Sociology* 105(2):493-527.
Yilmaz, Y.
    1998   *Private Regulation: A Real Alternative for Regulatory Reform*. Cato Policy Analysis 303. Washington, DC: Cato Institute.

18

Factors in Firms and Industries Affecting the Outcomes of Voluntary Measures

*Aseem Prakash**

Voluntary programs pertain to policies that firms adopt even though they are not required to do so by law. In the past two decades, such programs have gained prominence in many Organization for Economic Co-operation and Development (OECD) member countries (Gibson, 1999; Haufler, 2001). In the United States, the Environmental Protection Agency (EPA) has sponsored more than 40 voluntary programs. In the European Union (EU), more than 300 such agreements can be identified, and in Japan 30,000 local-level programs (especially agreements negotiated between a firm and a municipality) have been reported (Borkey et al., 1998).

Voluntary measures have been designed, monitored, and enforced by government agencies (the EPA's 33/50 and Project XL programs), nongovernment groups (World Wildlife Fund's Forest Stewardship Council program), industry associations (American Chemistry Council's Responsible Care program), and individual firms (corporate environmental reporting).

They vary in their scope, targeting firms at the global level (ISO 14000), regional level (the EU's Environmental Management and Audit System [EMAS] standards), national level (Britain's BS 7750 standards), or subnational level (Ontario's industry-specific pollution prevention projects). Within these levels of aggregation, they could be specific to a firm (the compact between the International Federation of Building and Wood Workers and IKEA), or an industry (the mining charter of the International Council for Metals and the Environ-

---

*The author would like to express gratitude to Tom Dietz, Paul Stern, Dan Kane, and the two anonymous reviewers for comments on the previous draft.

ment), or cut across industries (the Coalition for Environmentally Responsible Economics initiative).

Voluntary measures could be market promoting (such as technical standards that reduce transaction and production costs) or market restricting (such as ratings by the Motion Picture Association of America regarding appropriateness of films for children). Two key issues in studying them as a category of policy instruments include under what conditions they arise (who demands and supplies them) and how effective they are in improving firms' environmental performance.

EMERGENCE OF VOLUNTARY CODES

Demand for Voluntary Measures

The popularity of voluntary measures can be traced to the changing preferences and strategies of a variety of actors on how to deal with negative environmental externalities. Voluntary measures enable regulators facing declining budgets to implement their mandates at lower costs (see Randall, this volume, Chapter 19). Unlike command-and-control policies that these measures are supposed to replace or supplement, regulators conceivably can achieve their policy objectives without the accompanying acrimony.[1] Citizens also enjoy an increased supply of environmental amenities without increased taxes. Voluntary measures seem to be in consonance with the political climate that calls for "reinventing government" (Osborne and Gaebler, 1992) and encourages self-regulation by the regulatees. The same political climate often leads many to view governments as suffering from competency deficits, thereby making them laggards in understanding complexities of modern technologies, let alone in regulating them.

Firms demand voluntary measures because, compared to command-and-control regulations, they get greater operational flexibility in designing and implementing their policies (National Academy of Public Administration, 1995; Majumdar and Marcus, 2001).[2] Voluntary measures that encourage firms to adopt stringent pollution standards also may increase profits inasmuch as pollution represents resource waste (Hart, 1995; Porter and van der Linde, 1995; for a critique, see Walley and Whitehead, 1994). For example, in the Green Lights program (now merged into the Energy Star program), the EPA provides technical information about efficient lighting practices to participating firms. Many firms have experienced substantial reductions in energy bills, often equivalent to a 50-percent return on investment (Borkey et al., 1998).

Voluntary measures may have marketing payoffs as well if they enable firms to compete on environmental quality (Arora and Cason, 1996; Charter and Polonsky, 1999)—especially important for firms that seek to sell "green products" (see Thøgersen, this volume, Chapter 5). Strategically, firms could attempt to preempt and/or shape environmental regulations if they themselves craft voluntary policies (Nehrt, 1998). If higher standards incorporated in voluntary measures lead to stringent regulations, technologically advanced firms could raise the cost of entry for their rivals (Salop and Scheffman, 1983; Barrett, 1991). Furthermore, if firms may require their suppliers to adhere to voluntary programs (for example, Ford requiring its suppliers to become ISO 14001 certified), joining a voluntary program may become a business necessity (see Rejeski and Salzman, this volume, Chapter 2, on this subject).[3]

In demanding voluntary measures, firms are driven by other critical motives as well. Most importantly, firms often adopt them to seek legitimacy from external stakeholders. Reputational benefits of being a good corporate citizen serve their long-term profit and nonprofit objectives (Hoffman, 1997). Of course, it is difficult to quantify reputational benefits. Consequently, top management's commitment—as reflected in their values, beliefs, and attitudes—is important because many voluntary measures cannot be justified on traditional economic criteria that require estimating rates of return and comparing them with a company's cost of capital (Prakash, 2000a; Nakamura et al., 2001).

The need for legitimacy varies across industries. Arguably, firms in pollution-intensive industries or industries with bad reputations of complying with environmental laws are more likely to demand voluntary measures (see Nash's discussion on Responsible Care, this volume, Chapter 14). In some cases, voluntary policies may be adopted simply because top managers consider them the "right things to do." Thus, economic explanations are often underspecified in explaining firms' responses to voluntary programs.

Voluntary measures also have their critics. Many environmental groups are suspicious of voluntary codes, viewing them to be outside public scrutiny. U.S.-based groups are accustomed to a policy environment shaped by the 1946 Administrative Procedures Act (APA), which provides for public involvement in the regulatory process. These groups fear that the processes of establishing voluntary codes are not adequately inclusive and transparent; this is a major concern if such measures replace or dilute traditional policy instruments whose development followed APA procedural guidelines. Voluntary measures also may lack teeth: To monitor compliance, they often involve self-audits by regulatees, not external audits by credible third or fourth parties.[4] Furthermore, by making laws and policy processes less adversarial, voluntary policies may lessen the recourse to the judicial setting, the arena of choice of environmental groups who view it as relatively "liberal" and certainly not "captured" by the industry (Vogel, 1995).

Supply of Voluntary Programs

Voluntary measures can be supplied—designed, established, and promoted—by governments (see Mazurek, this volume, Chapter 13), industry groups, nongovernment groups, or individual firms. An important factor influencing firms' incentives to adopt voluntary measures is whether their reputational benefits manifest as public, impure public, or private goods.[5] Based on the twin attributes of consumption, excludability and rivalry, products can be classified in four stylized categories: private goods (rival, excludable), public goods (nonrival, nonexcludable), common-pool resources (rival, nonexcludable), and impure public goods (nonrival, excludable) (Ostrom and Ostrom, 1977; National Reseach Council, 2002). Rivalry implies that it is difficult for two or more consumers to simultaneously consume (or enjoy the benefits of) a given quantity of a product. In contrast, multiple users can use nonrivalrous products such as roads, movie theatres, and public parks at the same time. Excludability implies that Consumer A, who has paid for the product, can prevent other consumers (who have not paid for it) from enjoying a product's benefits. If excludability were not possible, Consumer A would not be able to prevent "free riding" by others. As a result, the consumer would have few incentives to pay for the product in the first place. Thus, for the market mechanism to function, it is necessary that goods be excludable (Olson, 1965; Hardin, 1968).

Voluntary codes can be viewed as a category of impure public goods of two kinds: toll and club (Prakash, 2000b; for an extended discussion on impure public goods, see Cornes and Sandler, 1996). Toll goods such as movie theaters can be unitized; that is, consumers can reveal their preferences by paying for every additional unit. They are provisioned by levying a user toll (such as a movie ticket). In contrast to toll goods, the discrete consumption units of club goods cannot be priced (because it is difficult to estimate their marginal costs), and membership fees (that are based on average costs) finance their collective provision. Many voluntary measures can be conceptualized as club goods whose reputational benefits are nonrival and potentially excludable, and it is difficult to price their discrete units.

From a firm's perspectives, voluntary policies create excludable benefits (reduce resource waste, shape regulations and so on) as well as nonexcludable benefits (improve environmental quality), but impose private costs on them. Reputational benefits, often the key motivation to subscribe to a voluntary measure, are often nonexcludable, spilling over to other firms. Therefore, making reputational benefits excludable becomes a key issue in the institutional design. This can be accomplished by establishing boundary features (such as participation logo or certificate) that enable stakeholders to differentiate adopting firms from nonadopters, thereby transforming reputational benefits into excludable club goods.

However, akin to Olson's (1965) "privileged groups," leading firms unilaterally may supply voluntary clubs that create reputational benefits for the whole industry. This is because the gains accruing to them are significant enough that they are willing to tolerate free riding by others. For example, the Responsible Care program has been designed, adopted, and promoted predominantly by large chemical firms because they perceive themselves to be receiving most of the goodwill benefits that the program generates for the chemical industry (Nash, this volume, Chapter 14; Prakash, 2000b).

As examined by Mazurek (this volume, Chapter 13), some voluntary measures (such as Project XL) that are supplied by government agencies grant regulatory flexibility to firms that join them. In return, firms often agree to adhere to more stringent regulatory standards than those required by the statute. Thus, firms reap excludable benefits (regulatory flexibility, better relationships with stakeholders), but bear higher excludable costs. The problem arises as some stakeholder groups, being skeptical of such business-government relationships, oppose these policies. This reduces goodwill benefits accruing to firms, thereby making these voluntary clubs less attractive to them. Consider the tepid response of U.S. firms (as compared to European and Asian firms) to ISO 14000 standards. As of March 2001, 6,261 Japanese, 2,400 German, 2,010 British, 1,441 Spanish, 1,420 U.S., 1,370 Swedish, and 881 Taiwanese facilities were ISO 14000 certified (ISO World, 2001). If controlled for the size of the economy, the low levels of American acceptance become even starker. This voluntary code is sponsored by a nongovernment organization, the International Organization of Standardization, which seeks to promote uniform environmental standards across countries. To get the ISO 14001 seal (the membership card to this club), firms are required to have a third-party certification (at a sizable cost) of their environmental management systems. One would expect that the requirement of third-party audits would make ISO 14001 legitimate to stakeholder groups. Firms, of course, want these audits to be protected by attorney-client privileges, lest these audits uncover incriminating information. Many environmental groups oppose granting attorney-client privilege to these audits because firms may abuse this privilege. The EPA, therefore has, not granted this privilege, and as a result, U.S. firms have been lukewarm toward ISO 14000 standards (Kollman and Prakash, 2001).

DO VOLUNTARY CODES MATTER?

As with any policy instrument, voluntary codes should be examined in terms of their efficacy. This can be done at two levels: first, their adoption rates, given that firms are not obliged by law to join them, and second, how they influence firms' environmental performance.[6] The first dimension was discussed earlier in terms of demand and supply aspects and how these affect adoption rates. In operationalizing the second dimension, it is important to assess codes' impact in relation to alternatives—that is, to compare environmental performance cross-sectionally (adopters versus nonadopters) and longitudinally (within adopters, preadoption versus postadoption).

To provide a concrete example, take the case of the EPA's 33/50 program. Under Section 313 of the 1986 Emergency Planning and Community-Right-to-Know-Act (EPCRA), firms with manufacturing facilities in the United States are required to submit annual reports on their releases and transfers of specified chemicals. Because EPCRA also required the EPA to make these data available to the public, the EPA developed a computerized database known as the Toxics

Release Inventory (TRI). To encourage firms to reduce the releases of 17 specific TRI chemicals,[7] in February 1991, the EPA proposed a voluntary program called 33/50. Under 33/50, firms voluntarily committed to reducing the releases/transfers of these 17 chemicals by 33 percent by 1992 and by 50 percent by 1995, with 1988 as the baseline. The EPA contacted nearly 10,000 facilities, of which 1,300 (13 percent) agreed to participate in this program. Thus, the adoption rates were not high. However, the program did affect environmental performance: 33/50 chemical releases/transfers reduced by 15.4 percent during 1988-90, but by 46.9 percent during 1990-95 (1991 is the baseline because the program was launched in 1991). Furthermore, the releases/transfers of 33/50 chemicals dropped by 46.9 percent during 1990-95 as opposed to a reduction of 25.3 percent for TRI chemicals outside the purview of the program. Again, joining this voluntary program seemed to have a positive input on environmental performance (EPA, 1999).

CONCLUSION

Voluntary measures are an exciting development in the environmental policy landscape. They are demanded and supplied by a whole gamut of actors. Their acceptance among firms and stakeholders depends on how they cohere with extant institutions, and more importantly, how managers perceive the costs of adopting them in relation to their excludable reputational benefits. Thus, an important factor in influencing their adoption rates is whether their institutional design transforms reputational benefits from a public good to a club good. Future research must carefully examine their efficacy on various dimensions in relation to competing policy instruments.

NOTES

1 For reference, 70 percent of the EPA's decisions that reflect the command-and-control mode are challenged in courts (Reilly, 1999).

2 Voluntary programs also create positive externalities such as generating awareness and disseminating information about best environmental practices. Such externalities may not sufficiently persuade firms to invest resources in adopting them. However, from a policy perspective, the presence of such externalities can serve as a powerful incentive for the regulators to promote them.

3 The corollary then is that sponsors of voluntary programs should focus on recruiting the "big fish" that have extensive forward and backward linkages. Taking advantage of market power enjoyed by such actors, their market power exercised through the value chain networks (that are increasingly becoming the defining features of cross-border economic flows) can create significant environmental multipliers. Also see Furger (this volume, Chapter 17).

4 Gereffi et al. (2001) identify four kinds of audits signifying increasing levels of credibility with external stakeholders: first-party or internal audits, second-party audits done by managers from the industry association, third-party audits by external auditors paid for by the firm, and fourth-party audits by external auditors not paid for by the firm.

5 As pointed out earlier, monetary incentives—reduced costs or increased sales/profits—also

may serve as motivators. However, many believe that because firms have exhausted opportunities for reducing costs, voluntary measures need to be justified in terms of nonquantifiable benefits.

6 Though not discussed here, the costs of rulemaking, monitoring, enforcing, and sanctioning (the so-called transaction costs) and the dynamic impacts on innovation and productivity also are important in comparing the efficacy of various instrument types.

7 The rationales for targeting these chemicals were that: (1) they have significant adverse impacts on health and the environment; (2) they were used in large quantities by facilities; (3) their releases relative to their total usage were high; and (4) their usage as well as releases could be reduced by employing pollution-prevention technologies and practices.

REFERENCES

Arora, S., and T.N. Cason
    1996   Why do firms volunteer to exceed environmental regulations? Understanding participation in EPA's 33/50 program. *Land Economics* 72:413-432.
Barrett, S.
    1991   Environmental regulations for competitive advantage. *Business Strategy Review* 2:1-15.
Borkey, P., M. Glachant, and F. Leveque
    1998   *Voluntary Approaches for Environmental Policy in OECD Countries,* Paris: Centre d'economic Industrielle. [Online]. Available: http://www.ensmp.fr/Fr/CERNA/CERNA [Accessed July 14, 2001].
Charter, M., and M.J. Polonsky, eds.
    1999   *Greener Marketing.* 2nd ed. Sheffield, Eng.: Greenleaf Publishing.
Cornes R., and T. Sandler
    1996   *The Theory of Externalities, Public Goods, and Club Goods.* Cambridge, Eng.: Cambridge University Press.
Gereffi, G., R.G. Johnson, and E. Sasser
    2001   NGO-industrial complex. *Foreign Policy* (July-August):5-65.
Gibson, R.B., ed.
    1999   *Voluntary Initiative: The New Politics of Corporate Greening.* Toronto: Broadview Press.
Hamilton, R.W.
    1978   The role of non-governmental standards in the development of mandatory federal standards affecting safety or health. *Texas Law Review* 56:1329-1484.
Hardin, G.
    1968   The tragedy of the commons. *Science* 162:1243-1248.
Hart, S.J.
    1995   A natural resource-based view of the firm. *Academy of Management Review* 20: 986-1014.
Haufler, V.
    2001   *A Public Role for the Private Sector.* Washington, DC: Carnegie Endowment for International Peace.
Hoffman, A.J.
    1997   *From Heresy to Dogma.* San Francisco: New Lexington Press.
ISO World
    2001   The number of ISO14001/EMAS registration of the world. [Online]. Available: http://www.ecology.or.jp/isoworld/english/analy14k.htm [Accessed March 27, 2001].
Kollman K., and A. Prakash
    2001   Green by choice? Cross-national variations in firms' responses to EMS-based environmental regimes. *World Politics* 53:399-430.

Majumdar, S.K., and A. Marcus
    2001   Rules versus discretion: The productivity consequences of flexible regulation. *Academy of Management Journal* 44:170-179.
Nakamura, M., T. Takhashi, and I. Vertinsky
    2001   Why Japanese firms choose to certify: A study of managerial responses to environmental issues. *Journal of Environmental Economics and Management* 42:23-52.
National Academy of Public Administration
    1995   *Setting Priorities, Getting Results.* Washington, DC: National Academy of Public Administration.
National Research Council
    2002   *The Drama of the Commons.* Committee on the Human Dimensions of Global Change. E. Ostrom, T. Dietz, N. Dolšak, P.C. Stern, S. Stonich, and E.U. Weber, eds. Washington, DC: National Academy Press.
Nehrt, C.
    1998   Maintainability of first-mover advantages when environmental regulations differ between countries. *Academy of Management Review* 23:77-97.
Olson, M.
    1965   *The Logic of Collective Action: Public Goods and the Theory of Groups.* Cambridge, MA: Harvard University Press.
Osborne, D., and T. Gaebler
    1992   *Reinventing Government: How the Entrepreneurial Spirit Is Transforming the Public Sector.* Boston: Addison-Wesley.
Ostrom, V., and E. Ostrom
    1977   Public goods and public choice. Pp. 7-49 in *Alternatives for Delivering Public Services,* E.S. Savas, ed. Boulder, CO: Westview
Porter, M., and C. van der Linde
    1995   Toward a new conception of the environment-competitiveness relationship. *Journal of Economic Perspectives* 9:97-118.
Prakash, A.
    2000a *Greening the Firm: The Politics of Corporate Environmentalism.* Cambridge, Eng.: Cambridge University Press.
    2000b Responsible care: An assessment. *Business & Society* 39:183-209.
Reilly, W.K.
    1999   Foreword. Pp. xi-xv in *Better Environmental Decisions.* K. Sexton, A.A. Marcus, K. Easter, and T.D. Burkhardt, eds. Washington, DC: Island Press.
Rugman, A.R., and A. Verbeke
    1998   Corporate strategies and environmental regulations: An organizing framework. *Strategic Management Journal* 19:363-375.
Salop, S.C., and D.T. Scheffman
    1983   Raising rivals' costs. *American Economic Review* 73:267-271.
U.S. Environmental Protection Agency
    1999   The 33/50 Program: Final Record. [Online]. Available: http://www.epa.gov/opptintr/3350/3350-fnl.pdf [Accessed October 6, 2001].
Vogel, D.
    1995   *Kindred Strangers.* Princeton, NJ: Princeton University Press.
Walley, N., and B. Whitehead
    1994   It's not easy being green. *Harvard Business Review* (May-June):46-51.

19

The Policy Context for Flexible, Negotiated, and Voluntary Measures

Alan Randall

In recent times, the rhetoric of environmental regulation has shifted quite dramatically toward the use of flexible incentives, voluntary approaches, and the devolution of authority. These approaches cover a broad range of possibilities, which vary along several dimensions: the assignment of responsibility among different levels of government and between the public and private sectors, the choice of policy instruments, and the degree of rigor with which accountability is assigned and policies are enforced. It is important at this stage to ask which aspects of regulatory control should be flexible, voluntary, or devolved to other authorities, and what implications they pose for environmental quality.

REDEFINING THE ROLE OF GOVERNMENT

In the standard model of environmental regulation, a central government agency adopted environmental quality standards to protect a wide range of public interests, including the health and safety of its citizenry. The standard tools for achieving environmental policy targets included strict regulation of point sources of pollution and rigorous monitoring and enforcement. "Polluter pays" was a core principle. In practice, typically, particular control technologies were prescribed. Although nonpoint sources were, in aggregate, major contributors of effluents, they got something of a free pass. Addressing them within the "command-and-control" framework seemed too difficult—technically difficult because their diffuse nature impeded monitoring nonpoint effluents from particular sources and enforcing controls, and politically difficult because the long-standing public propensity to subsidize rather than regulate farming militates against regulatory controls and "polluter pays" for agriculture, a major source of nonpoint pollution.

After roughly two decades of this approach to pollution control, the seeds of change were evident. Command-and-control regulation of point sources had accomplished significant reductions in water pollutant loadings nationwide, but increasingly had come to be regarded as inflexible, inefficient, and stifling to innovation, all of which tend to increase pollution control costs. In addition, changes in the economy—the service sector grew faster than manufacturing, and more dispersed modes of manufacturing gained at the expense of traditional, concentrated forms—necessitated changes in the standard command-and-control modus operandi (Rejeski and Salzman, this volume, Chapter 2). The successes in reducing loads from point sources left the nonpoint sources increasingly conspicuous, as their proportional contribution to the remaining total loads grew. Although implementation even in the point source case seldom matched the rigor of the rhetoric, the language of "command and control" and "polluter pays" started to give way to a new and more gentle language emphasizing flexible incentives, voluntary approaches, negotiated solutions, and devolution of authority. Policy approaches within this very broad and ill-defined set that serve to soften the sharp edges of regulatory control also tend to deemphasize the distinction between point and nonpoint sources.

With these newer approaches in the mix, the policy tool kit offers a broad range of possibilities, and a consensus typology has yet to emerge (Andrews, 1998). The choice of policy tools is linked to the devolution question—not just the question of federal, state, or local jurisdiction, but also the division of authority and responsibility among governments, industry groups, and individual firms. The following incomplete list serves to illustrate the breadth of considerations that are relevant for thinking about the effectiveness of flexible and voluntary approaches.

- *Regulatory status.* Possibilities include regulations already in place, self-policing at the individual or group level with threat of regulation if the problem persists, and neither of the above.
- *Control objective.* The objective may be to control environmental performance, or to impose prescribed control technology. The details (the required level of performance, or the control technologies designated as acceptable) may be set by the regulator or negotiated among industry groups, firms, and government.
- *Incentives/penalties.* Possibilities include flexible incentives (taxes, markets in pollution reduction credits) aimed at inducing desired environmental performance, prescribed penalties for failure either to achieve the required level of performance or to implement the acceptable control technologies, negotiated penalties (such as warnings, time extensions), shaming, and none.

- *Monitoring.* Options include inspection by the regulator (with greater or lesser frequency and effectiveness), self-inspection with self-reports to the regulator, self-inspection with audit confidentiality and immunity, and self-monitoring at the individual or group level.
- *Enforcement.* Possibilities include enforcement by the regulator, self-enforcement at the individual or group level (and enforcement may vary from strict to lax), public pressure, and no enforcement.

Given the considerable diversity of arrangements that are placed in the category of flexible, negotiated, and voluntary measures, we cannot begin to discuss their effectiveness without paying close attention to particulars: The devil truly is in the details.

SOME THINGS WE HAVE LEARNED FROM ECONOMICS AND EXPERIENCE

Traditional economic theory has emphasized that rational individuals behave cooperatively only when the expected penalties for noncooperative behavior exceed the private benefits. This line of reasoning supports a generally skeptical attitude to self-monitoring, voluntary agreements, and the like. It is, however, supportive of flexible incentives and "smart" monitoring and enforcement strategies.

Flexible Incentives

Flexibility in the assignment of responsibility for pollution control and choice of control technology may be introduced via pollution taxes, trading in pollution reduction credits, and similar market-oriented policy instruments. This approach seeks to enforce a given level of environmental performance, while freeing managers to choose the means of compliance. By permitting firms to choose optimal technologies and factor combinations, this approach has been shown to reduce compliance costs and increase effectiveness in a broad range of situations. Perhaps more importantly in the long run, human creativity is unleashed in the search for even more cost-effective pollution control technologies. Economic theory predicts as much, and experience with pollution permit markets tends to confirm their powerful incentives for cost reducing innovations in control technology (Burtraw, 1996).

"Smart" Monitoring and Enforcement Strategies

Often the regulator is chronically underfunded for monitoring and enforcement, which threatens regulatory effectiveness. Nevertheless, "smart" monitoring and enforcement strategies have been developed that are promising in such situations. Here I summarize three results that support this conjecture. First, focusing monitoring pressure on prior offenders would reduce regulatory costs of compliance. Hentschel and Randall (2000a) support this conjecture with theoretical results and simulations based on real data on pollution control costs in the paper industry. Second, self-monitoring and self-reporting requirements would reduce regulatory costs of compliance. Note that this approach is a kind of devolution as firms self-monitor and self-report, but devolution is limited to the extent that the authorities require reports and maintain monitoring capacity sufficient to sustain a credible threat of inspection. Hentschel and Randall (2000a) provide theoretical results and simulations. Third, audit confidentiality and immunity laws tend to encourage firms to self-audit and clean up environmental problems discovered. But there is a caveat: Legislation typically permits firms to earn confidentiality or immunity by making a diligent effort at cleanup and, if relatively ineffective cleanup efforts count as "diligent effort," such provisions may undermine the advantages of these laws. Hentschel and Randall (2000b) provide theoretical results.

Richer People Demand More Environmental Quality

People function both as workers and consumers, and they value consumer goods and environmental quality. From these modest premises, we can draw several interesting conclusions. First, the relatively high income elasticity of demand for environmental quality (as incomes rise, demand for environmental quality rises even faster) limits the "race to the bottom" in a particular society. A state or local jurisdiction seeking economic advantage by weakening its environmental controls will find that, other things being equal, skilled workers demand higher wages to compensate for the poorer environment, thus undermining any gains from such a strategy (Blomquist et al., 1988). Empirical cross-state comparisons in the United States have demonstrated that lax environmental policies are not associated with any consistent advantage in terms of growth and prosperity (Levinson, 1996).

Second, richer societies may nevertheless engage in a race to export pollution to poorer, less sensitive places. It is not clear that this is reprehensible. Although pollution can be life threatening, so can abject poverty with its associated poor health care, malnutrition, and abysmal workplace safety. Postwar history in East Asia reveals several success stories (Japan, Taiwan, South Korea), in which societies jump started economic development by serving as pollution sinks, but quickly became more environmentally sensitive as incomes rose. The key point is that citizen demand for environmental quality serves as a powerful restraint on environmentally lax policy (Andreoni and Levinson, 2001).

Third, the same citizen demand for environmental quality will motivate the behavior of private firms, which will rationally seek good environmental reputations and—if the best way to gain a good environmental reputation is to earn it—

will do some environmental good in the process (Thøgerson, this volume, Chapter 5; Teisl et al., in press). Industry codes, such as ISO 14000 and Energy Star, apply the same principle at the industry level. Compliance provides a means of signaling "greenness" to consumers, and such codes have achieved some environmental successes (Nash, this volume, Chapter 14).

The Sustainability and Effectiveness of Collective Action

Economists now realize that their traditional "tragedy of the commons" skepticism about the sustainability and effectiveness of collective action was overdone. Empirical evidence tends to reject the hypothesis of pervasive collapse of the public goods and common property sectors. These sectors tend to be inefficient yet to avoid total collapse (Ostrom, 1990), and game theory provides some powerful reasons why (Axelrod, 1984). This line of research shows the potential for creating stable market incentives that encourage group cooperation, as opposed to individual competition, and provides optimism for overcoming many kinds of "market failures." As we better understand the role of repetition in games, reputation effects, etc., we become more optimistic about the prospects for voluntary agreements, industry codes, and performance regulation in cases where individual performance is not readily observable. For example, Randall and Taylor (2000) suggest the feasibility of performance-based instruments, with monitoring of collective performance at the subcatchment level, for nonpoint-source pollution control. If such a scheme proved workable, farmers would gain the considerable savings in control costs that come with the switch to policy instruments that reward pollution-control performance rather than subsidize specified technologies.

Negotiated Solutions and Voluntary Agreements

Most people discussing flexible environmental policy have in mind something much more flexible than the flexible incentives described earlier, which—while introducing flexibility by assigning the lion's share of the abatement task to low-cost abaters and allowing flexibility in choice of control technology—adhere rigidly to the specified environmental performance standard. Negotiated and voluntary agreements, which are very much a part of the European scene and figured prominently in the 2000 U.S. presidential election campaign, focus on flexibility in setting and enforcing performance expectations as well as in choosing methods to achieve improved performance. What does it mean to have environmental regulation where standards are flexible, negotiated, or voluntary?

Although there is some U.S. experience with voluntary approaches (Mazurek, this volume, Chapter 13), some European countries have made considerably greater commitments to voluntary agreements. The European experience with these more flexible policies is instructive. For example, Bruyninckx (2001) has evaluated the voluntary agreements that are widely used by the Flemish government to specify what is expected of Flemish municipalities in the way of environmental performance. At the outset, it is important to recognize that these agreements are perhaps better characterized as negotiated agreements: The provincial government, which had long neglected environmental policies aimed at municipalities, made it clear in 1992 that municipalities henceforth would be expected to commit to environmental improvement objectives. After about 8 years of operating experience with this policy, Bruyninckx concluded, 38 percent of municipalities had met all of their commitments. He considers this a significant accomplishment given that the policy is relatively new, the baseline level of environmental performance was low, and many of the municipalities in the noncompliant 62 percent had met at least some of their commitments. He observes that these kinds of agreements are more likely to be effective if:

- policy objectives can be accomplished even if there remains significant nonparticipation of potential actors (to put it another way, approaches stronger than "voluntary" agreements are necessary if success requires that everyone comply);
- the goals are clearly defined in terms of performance indicators;
- the agreements have a motivational or norm-setting aspect that can move other actors in the same direction;
- the obligations of all parties are clearly defined; and
- a serious enforcement mechanism is in place.

Conclusion

Traditional economic theory supports flexible incentives, and "smart" monitoring and enforcement strategies, but is skeptical toward many of the policy innovations discussed earlier. However, recent conceptual advances and a growing body of empirical experience support cautious optimism about voluntary or negotiated agreements, industry codes, green marketing, and performance-based rather than technology-based instruments for nonpoint pollution control. Nevertheless, all of these rather optimistic results depend on the presence of an effective regulator, at least in the background. The only firm economic-theoretic result I know of, concerning voluntary agreements, has a similar flavor: Voluntary agreements can work if there is a credible threat of regulation should self-policing fail (Segerson and Miceli, 1998).

WHAT SHOULD GOVERNMENT DO?

These economic-theoretic results and empirical observations allow us to draw some fairly strong tentative conclusions about the appropriate stance of government toward environmental control instruments.

1. Some essential functions of government.
- Maintain a capacity for meaningful monitoring (even effective self-monitoring requires a credible threat of regulatory inspection).
- Maintain a credible threat of regulation should self-policing fail.
- Maintain a regulatory and/or legal liability system ready to throw the book at blatant polluters and repeat offenders.

These minimal functions of government in the environmental sphere remain essential because it would be naively incautious to rely entirely on good intentions, green consumerism, and social pressures to achieve consistent environmental performance that is costly to polluting firms, consumers, and public agencies.

2. Some good things government can do.
- Regulate ambient performance standards for those environmental attributes that really matter for protection of human and ecosystem health.
- Provide broad freedom for firms and individuals to find cost-effective methods of compliance. Policy is responsive to costs, benefits, and economic impacts, so innovations that lower the costs of compliance enhance the political viability of strong environmental policies. Furthermore, a strong economy encourages consumers to demand higher environmental quality and motivates firms and industry groups to seek "green" reputations.
- Encourage experiments with a broad range of flexible policy instruments, while reserving the right to terminate experiments and reject policy instruments that fail.
- Cultivate a legal and political environment that encourages polluting industries, firms, and public agencies to accept the obligations of environmental good citizenship, and an active, informed citizenry to reward and punish firms and industries on the basis of environmental performance.

The approach recommended here is hierarchical: Serious environmental regulation to protect human and ecosystem health is the foundation, and flexible incentives are the favored instruments where feasible. That much is standard economics. Yet, because we economists are no longer quite so sure we have all the answers, the policy framework should be open to experiments with a variety of flexible policy instruments, and good environmental citizenship should be encouraged for industries, firms, public agencies, and individuals.

3. Might voluntary agreements, industry codes, green marketing, and so on, be sufficient by themselves?
  Innovations such as voluntary (or negotiated) agreements, industry

codes, and green marketing should be viewed as promising additions to the environmental tool kit, but they should supplement, not supplant, the regulatory framework. They make a nice frosting on the regulatory cake. But the cake itself must be there.

REFERENCES

Andreoni, J., and A. Levinson
    2001   The simple analytics of the environmental Kuznets curve. *Journal of Public Economics* 80(2):269-286.
Andrews, R.N.L.
    1998   Environmental regulation and business "self-regulation." *Policy Sciences* 31:177-197.
Axelrod, R.
    1984   *The Evolution of Cooperation.* New York: Basic Books.
Blomquist, G.C., M.C. Berger, and J.P. Hoehn
    1988   New estimates of quality of life in urban areas. *American Economic Review* 78(1):89-107.
Bruyninckx, H.
    2001   Voluntary Environmental Agreements between the Flemish Community and Municipalities. Unpublished paper presented at the Environmental Policy Conference, Ohio State University Environmental Policy Initiative, Columbus, OH, April 16-18.
Burtraw, D.
    1996   The SO_2 emissions trading program: Cost savings without allowance trades. *Contemporary Economic Policy* 14:79-94.
Hentschel, E., and A. Randall
    2000a An integrated strategy to reduce monitoring and enforcement costs. *Environmental and Resource Economics* 15:57-74.
    2000b Audit Confidentiality and Immunity Laws: The Problem of Diligent Effort. Presented at annual conference, European Association of Environmental and Resource Economists, Rethymnon, Greece, June 28-30.
Levinson, A.
    1996   Environmental regulations and manufacturers' location choices: Evidence from the census of manufacturers. *Journal of Public Economics* 62(1-2):5-29.
Ostrom, E
    1990   *Governing the Commons: The Evolution of Institutions for Collective Action.* Cambridge, Eng.: Cambridge University Press.
Randall, A., and M.A. Taylor
    2000   Incentive-based solutions to agricultural environmental problems: Recent developments in theory and practice. *Journal of Agricultural and Applied Economics* 32:221-234.
Segerson, K., and T.J. Miceli
    1998   Voluntary environmental agreements: Good or bad news for environmental protection? *Journal of Environmental Economics and Management* 36(2):109-130.
Teisl, M., B.E. Roe, and R. Hicks
    in   Can eco-labels tune a market? Evidence from dolphin-safe labeling of
    press canned tuna. *Journal of Environmental Economics and Management.*

20

Understanding Voluntary Measures

Thomas Dietz

S everal important lessons about voluntary arrangements involving the private sector emerge from the previous chapters. First, it is clear that far more attention must be paid to evaluation of voluntary efforts. This task will be facilitated by the longstanding tradition of evaluation research in the social sciences. Second, voluntary agreements present fundamental challenges to social theory, and a better understanding of them will require more integrative concepts of organizational behavior than are currently available. In particular, understanding voluntary agreements will require tentative solutions to the aggregation problem of linking the behavior of individuals to that of organizations. Third, voluntary agreements may provide important opportunities for meeting new environmental policy challenges, but for the most part, the voluntary agreements in place today are dealing with problems of the sort that traditionally have been handled by command-and-control regulation. Some creative institutional design will be required to engage voluntary agreements as tools for dealing with as yet uncontrolled forms of pollution, such as that from nonpoint sources. Finally, more attention must be paid to the comparative analysis of voluntary agreements. Although it can be difficult to generalize across political economies and cultures, other industrial nations are engaged in interesting experiments with voluntary agreements, and much could be learned from this experience.[1] Furthermore, some of the most important voluntary agreements, such as the International Standards Organization (ISO) environmental protocols, are international in character. In this chapter, I will try to address each of these lessons, suggesting where research must move in the future if we are to have the knowledge base necessary to support innovative environmental policy.

POLICY EXPERIMENTS AND EVALUATION RESEARCH

As Mazurek (this volume, Chapter 13), Nash (this volume, Chapter 14), and especially Harrison (this volume, Chapter 16) note, the initial claims of success for some innovative voluntary agreements should be viewed with skepticism. When care is taken to learn what changes in environmental impact can legitimately be attributed to a voluntary program rather than to other factors, the estimated effects may be less impressive than those contained in first reports that are less rigorous in methodology. Thus in developing and assessing the merits of voluntary agreements in the future, it would be wise to think carefully about methodological and conceptual issues of evaluation.

Environmental policy evaluation is still in its infancy. The amount of research in the peer-reviewed literature is still small, though some evaluations also are found in the "gray" literature of internal agency documents and consulting reports. A major effort is underway to assess the effectiveness of international environmental treaties (Young, 2001). But if systematic evaluation is a new issue for voluntary agreements and other environmental programs, evaluation has been routine practice for many social programs.[2] A great deal can be learned from this experience that will enhance our understanding of all environmental programs, especially programs involving the private sector.

Avoiding Hubris

More than thirty years ago, Campbell (1969) argued that nearly all policies and programs should be considered experiments. Clearly environmental policies are social experiments. We never have enough information about the particular circumstances of implementation nor a complete enough set of models and theory to predict with great accuracy the outcomes of an innovation. As a result, we would be wise to treat changes in policy as opportunities to learn rather than assuming we know the outcomes in advance. We need informed humility, not hubris, in approaching new policies and programs. Informed humility requires that we build into the design of a policy evaluation research that can inform us about what has happened and why. In that way we will build a cumulative knowledge base that will allow better designed reforms in the future. This same general point has been made in the context of adaptive ecosystem management and social learning (Gunderson et al., 1995). Our knowledge is imperfect, things change, and the only wise way to proceed is one that is reflexive, allowing for trial, analysis, and revision. Our reflexivity can be improved greatly if attention is paid to evaluation while a program is being designed rather than attempting to construct an evaluation after crucial design decisions have been made and opportunities to collect baseline data have been lost.

Estimating Program Effects Is Difficult

Only rarely in social or environmental programs can we conduct true experiments with random assignment of subjects (individuals, communities, or firms) to treatment and control groups. Such assignment typically would be cumbersome and perhaps illegal and/or unethical. In the absence of such an experimental design, it can be difficult or impossible to attribute causal influence to a program because many other factors also are influencing the behavior of interest.

A substantial part of the literature on evaluation research deals with the problem of assessing causality and estimating true program effects in the absence of an experimental design. Researchers refer to "quasi-experiments" in which there is some variation in who participated and did not participate in programs, but there is not random assignment to experimental and control groups of the sort that facilitates strong causal inference.[3] As Harrison notes (this volume, Chapter 16), selection bias can be a particular problem—firms who join voluntary programs may be very different from those who do not. Appropriate statistical models can help in dealing with selection bias, but such bias is still a major issue in most evaluations of voluntary agreements (Heckman, 1976, 1979). Although it may not always be possible to implement the ideal design, or even have sufficient data for the best possible statistical analysis, careful attention to the factors that interfere with estimating true program effects can help in avoiding gross errors and allow for proper caveats in reporting the estimates that are possible.

Nor is the design question a purely statistical matter. Policy and normative considerations enter into determining the proper basis for making a comparison. The design of evaluation research must address the issue of what constitutes a "fair" comparison to use as a basis for judging the success or failure of a voluntary program. The overall success of a voluntary agreement in reducing environmental impact needs to consider growth in the amount of production and consumption activity that can be attributed to an industry. A seemingly successful voluntary agreement could reduce impact per unit consumption or production even as overall impact increases as a result of growing population, affluence, and consumption per unit affluence.

Evaluations should consider the improvement in environmental performance that is likely to have occurred over time without any intervention. Evaluations should also consider the reasonable upper bound of improvement set by factors such as the rate of replacement of capital equipment. They also should attend to possible tradeoffs in impacts—for example, a reduction in emissions of a targeted pollutant, at the cost of an increase in emissions of one not targeted. Integrated measures of environmental impact, such as the "ecological footprint," may be useful in assessing changes in overall impact (Rees and Wackernagel, 1994; Rees, 1992; Wackernagel and Rees, 1996; Wackernagel et al., 1999). Early evaluations that did not consider these factors were perhaps too generous in their assessments. In the future we can anticipate an important debate about what comparison standards are appropriate. Some of that debate is foreshadowed in the chapters of this section, where the issue of what to measure in evaluating voluntary programs is discussed.

Thinking About Program Goals

The most obvious goal of voluntary agreements is to reduce environmental impact. But in evaluating voluntary programs, the obvious goal may not be obtained quickly or directly. As Herb and colleagues (this volume, Chapter 15), Harrison (this volume, Chapter 16), Furger (this volume, Chapter 17), and Prakash (this volume, Chapter 18) emphasize (see also Dietz and Stern, this volume, Chapter 1), a benefit of voluntary agreements may be a general change in organizational culture among participating firms that in turn leads to environmental protection receiving a much higher priority than otherwise would be the case. Changes in culture that reduce environmental impact may act more slowly than the implementation of new command-and-control regulations. But such cultural changes have large and broad effects in the long run. The implication for program evaluation is that there may be program outcomes that are worth monitoring in addition to short-term reductions in pollutant generation or resources used.

Of course, measuring and evaluating such alternative or interim goals should not be a substitute for assessing success or failure with regard to the primary goal—reduction in environmental impact. Indeed, one of the standard criticisms of voluntary agreements is that they produce good feelings, good publicity, increased awareness, and little if any change in environmental impact. In some circumstances it may be appropriate to design programs whose goal is to change organizational culture in order to provide the substrate for changes that reduce impact. If that is the goal, the program should be clear about what will and will not be achieved. It also may be useful to assess interim goals such as changes in organizational culture, establishment of environmental accounting systems, and so forth because they may be good indicators that a voluntary agreement is on the way to producing the desired environmental benefits. Interim indicators that are known to be good predictors of environmental outcomes may be useful as evaluation tools and as secondary program goals.[4]

Of course, even when a voluntary agreement program does not reduce environmental impact over what might be achieved by command-and-control regulation, it may be able to achieve the same change in impact with lower costs. Such savings can prevent increases in consumer prices that might be associated with environmental protection, provide both more funds for further investment in environmental quality, and reduce future resistance to addressing environmental problems. From the viewpoint of society as a whole, these spinoffs of cost-effective environmental regulation are beneficial. So one criterion for evaluating voluntary agreements is not just their overall environmental impact, but the cost per unit of pollution reduction.

The possible importance of secondary and interim goals such as changed organizational culture and practices leads us to the distinction between summative and formative evaluation. Summative evaluation looks at the overall success or failure of a program with regard to its most important goals. It provides information intended to influence decisions about expanding or contracting or even eliminating the program based on what has been achieved. In contrast, formative evaluation is intended to provide guidance to program administrators and participants while the program is in process. The goal is to suggest mid-course corrections. Formative evaluation usually focuses more on program process and interim indicators of progress than on ultimate outcomes. It intends to provide advice rather than a judgment. Developing successful voluntary agreements will require both formative and summative evaluation. Participants and managers can benefit from research that helps them do better, but policymakers and society as a whole must be able to assess how well a voluntary agreement approach is working in a given domain.

To the extent that voluntary agreements are intended to produce changes in organizational culture and routines, it may be necessary to rethink not only the variables that should be considered as outcome criteria, but also the statistics used to summarize the behavior of participants in a program or agreement. The mean and its analogs (conventional regression analysis, analysis of variance, and other versions of the general linear model) are appropriate if the intent of an evaluation is to focus on the sum of benefits to society. This follows because the mean times the size of the population is the sum of a variable across all members of the population. Thus analyses focused on the mean give a sense of the total benefits from a program. But in some circumstances, such as when a program is considered a demonstration to point the way toward best practices, it may make more sense to focus on the tails of the distribution of organizational performance. Those firms that have done exceptional jobs in reducing environmental impact and those that have made the least progress may provide more information than firms whose performance under the program is "typical" (at the center of the distribution). The methods available for statistical analysis of "outliers" are not as well developed as that for understanding the center of a distribution, but increases in computational power are making more tools available for such analyses.[5] A particularly helpful evaluation design might include a summative evaluation using procedures based on the mean performance to assess the overall effectiveness of a program with an analysis of the outliers. The outliers then could become the subject of case studies to determine why those organizations have done so well or so poorly compared with typical performance—a formative evaluation goal.

WHY PARTICIPATE?

The problem of altruism versus self-interest has been one of the most vital themes in the social sciences over the last quarter of the twentieth century and is central to any analysis of voluntary agreements.[6] As Harrison (this volume, Chapter 16), Furger (this volume, Chapter 17), Prakash (this volume, Chapter 18), and Randall (this volume, Chapter 19) detail, the interplay between altruism and self-interest is also central to the motivations for participating in voluntary agreements. Given a voluntary agreement, profit-maximizing firms have substantial incentive to "free-ride" by signing on to the agreement, but not taking any costly steps to reduce environmental impact. Such free-riders receive most of the benefits of the agreement—good public image, freedom from intrusive regulation—without bearing the costs. If many firms follow this narrowly rational strategy little reduction in environmental impact will occur. As Olson (1965) noted more than 35 years ago, there will be an undersupply of public goods, in this case environmental protection, compared to what would be optimal.

Although the free-rider model is compelling in its logic, decades of research also have shown that it is not always a realistic model of human behavior. Codes of silence are observed by prisoners, commons are managed sustainably, and people make sacrifices for the collective good.[7] In the case at hand, firms sometimes change their behavior as a result of participation in voluntary agreements even when they could free-ride. To design better voluntary agreement programs, we need a better understanding of why firms act so as to reduce environmental impacts. Achieving that understanding will require careful theoretical and methodological investigations of the behavior of organizations. Although research on participation in voluntary agreements is in its early stages, several key issues already have emerged.

Voluntary Participation Is Not Wholly Voluntary

As Randall (this volume, Chapter 19) and Nash (this volume, Chapter 14) emphasize, when a voluntary agreement is developed for a serious environmental problem, the threat of command-and-control regulation often is on the horizon. Participation in the voluntary agreement may be the result of comparing the costs and benefits of the agreement requirements with those that would flow from a regulation that can be anticipated if the voluntary agreement is not successfully implemented. Even when the threat of government regulation is not present, other firms who are customers may require products that meet environmental requirements. As Rejeski and Salzman note (this volume, Chapter 2), some corporate buyers dwarf their suppliers and can bring about substantial changes in products and production processes.

Nor are corporate customers the only source of pressure. Even the largest retail firms may fear boycotts motivated by environmental concern, while, as Nash notes, environmentally friendly products may allow for niche markets that are advantageous to those who produce them. Finally, it is increasingly common for environmental standards, such as those incorporated in ISO 14001, to influence access to export markets.[8] Firms that hope to penetrate markets where such standards have substantial sway will act to meet the requirements and may alter products and production processes beyond what is required in their other markets simply because such uniformity is simpler. So voluntary action may be "coerced" by concern with future government regulation, or by concern with present or anticipated actions by either corporate customers or consumers. A number of studies document such effects on firms' environmental policies (Henriques and Sadorsky, 1996; Nakamura et al., 2001).

Voluntary action may also be induced by regulatory or other environmental policy tools. As Herb and colleagues note (this volume, Chapter 15), one outcome claimed for the U.S. Toxics Release Inventory, a regulatory program requiring only the production of information, is voluntary efforts by firms to reduce their output of listed pollutants. "Green" labeling programs (see Thøgersen, this volume, Chapter 5), whether developed by government or nongovernment groups, can provide a similar stimulus for voluntary action. Such possibilities for synergism between voluntary measures and other policy instruments are important considerations for both program design and evaluation.

Networks Matter

At least since C. Wright Mills's *The Power Elite* it has been clear that major corporations typically do not make decisions in isolation from one another, even though direct collusion is prohibited by U.S. law.[9] Trade or industry associations may be especially important links in these networks, as Furger (this volume, Chapter 17) and Nash (this volume, Chapter 14) note. To some extent, their influence comes through acting as the representatives of a business sector in policy debates.[10] In that role, they may convey information about future government regulations that may encourage preemptive voluntary action. But they may also play a role of sharing information across firms, mediating and encouraging voluntary action.

Linkages across firms that do business with each other also can be very consequential. Such transactions, even when governed by contracts, require trust, a point noted by Furger and Nash. Trust may be damaged not only by actions that directly affect the interests of a firm, but also by observing that a trading partner exhibits public behavior that is out of line with industry norms. Like any other social actor, firms often compare their performance to that of their peers and those with whom they have frequent contact, and modify their behavior to conform to the norm, or even to be perceived as a leader and an innovator. Thus effective participations in voluntary agreements may be encouraged by other members of a firm's network directly and may be rewarded by a good reputation within the network. Conversely, free-riding may bring informal sanctions from trade associations and other firms.

Individuals Matter

Decisions within firms are made by individuals who bring their personal beliefs, values, and norms to bear as well as the corporate interests. At least since Galbraith (1967), there has been debate about the degree to which managers have autonomy from stockholders in steering the corporation. As noted in this volume, Chapter 1, many senior corporate managers are members of cohorts that have been exposed to environmental arguments since college or before. Nakamura et al. (2001) and Henriques and Sadorsky (1996) have found that the environmental concerns of corporate managers influence at least some aspects of firms' decisions.

Protecting the Environment May Protect the Bottom Line

Analyses from industrial ecology often argue that much environmental impact generated by the production process is a result of poor design and represents wasted resources (Socolow et al., 1994). Better engineering could both reduce environmental impact and save money.[11] If these views are internalized by a firm, then voluntary agreements become a mechanism for receiving public approval for implementing reduced impact, finding flexibility within government regulations to innovate, and sharing information with peer firms, while the core motivation remains cost reduction. Andrews (1994) suggests that adoption of industrial ecology strategies by firms can be greatly facilitated by voluntary agreements. We agree, but suggest that the reverse also may be true—that the insights from industrial ecological analysis may be a major motivation to participate in voluntary agreements.

Organizational Structure Matters

Standard models from microeconomics begin with the assumption that the firm is a unitary entity that makes decisions to maximize profits or some other measure of economic performance. If this is the case, then explaining why firms would participate in voluntary agreements reduces to explaining how voluntary agreements might enhance economic performance. This is the essence of the rational actor model that underpins the analysis in several of the preceding chapters. But at least since Simon (1947), the assumption that organizations, including private-sector firms, can be seen as purely rational utility maximizers has been dubious.

Two problems can be raised about the rational utility maximizer view of the firm. One is the assumption that organizations act to maximize profits. Other decision rules may be more plausible. The second, related problem is viewing the organization as a unified decision making entity, whatever the decision rule applied might be.[12] Who makes decisions is, of course, related to the problem of what criteria are used in making decisions. If Galbraith is correct in suggesting that in modern corporations managers have considerable independence from stockholders, then executives may be free to pursue goals they consider desirable, perhaps for ethical or career reasons, without strict adherence to profit maximization. Nakamura et al. (2001) contrast models of profit maximization with those of utility maximization that take into account managers' preferences and concerns, and find that both are relevant to the environmental protection decisions of Japanese firms. DeCanio & Watkins (1998) found that patterns of stock ownership had some influence on decisions to participate in the U.S. Environmental Protection Agency's (EPA's) Green Lights energy conservation program.

NEW TOOLS FOR OLD PROBLEMS

As the literature reviewed in previous chapters makes clear, voluntary agreements are being used primarily to deal with environmental problems that previously have been addressed with either command-and-control regulation or with market-based strategies. In the United States, experimental programs are attempting to reduce the emission of targeted pollutants and increase energy efficiency of the private sector. To date, there seems to be little experience with using voluntary agreements to deal with nonpoint pollutants or with the special problems of the service sector noted by Rejeski and Salzman (this volume, Chapter 2). The exception seems to be with ISO 14001 and other "green" labeling and certification efforts that are encouraged by consumer demand for environmentally friendly products, demands from large purchasers on their suppliers, or demands of import regulations in important markets. Although targeted at manufacturers, some of these have the potential of reducing nonpoint sources when products used by consumers are a source of environmental problems.

One of the central problems in dealing with nonpoint sources is that they are often composed of many tiny point sources. Households, small firms, or local governments may generate pollution almost inadvertently as a result of routine activities. Those carrying out these activities often have little awareness of the environmental consequences and little control over these impacts. The impacts may be embedded in the products or processes used, and the users may have little time, money, or technical capacity to develop for themselves or find in the market lower impact alternative products or practices. Thus policies targeting manufacturers and suppliers may be an appropriate strategy for dealing with nonpoint sources. Voluntary agreements may be an effective tool in changing the behavior of these corporate actors as an indirect way of addressing the nonpoint sources.

ATTENDING TO THE GLOBAL

Most work in this volume has focused on the U.S. experience. As noted at the beginning of Part Three, this is appropriate as a first step in developing an understanding of education, information, and voluntary measures that is useful for U.S. decision makers. But a continued neglect of experiences outside the United States would unnecessarily and inappropriately limit understanding of New Tools, for two reasons.

First, although there are many obstacles to generalizing across political and economic institutions and cultures, it is just such comparisons that clarify what is general and what is unique. Comparative analyses can lend substantial analytical power to research on voluntary agreements. As many industrialized nations experiment with new forms of policy, it will become even more important to understand what approaches may transfer to the U.S. context, what approaches will not work, and why. In addition, as consumption and production increase in nations that still are only beginning to evolve their environmental policy systems, comparative analyses are essential to provide useful guidance about the best ways to approach the challenges they will face. For example, approaches that are found effective in the U.S. context may not work well in a country with a different regulatory context or where a target industry has a different structure.[13]

Second, we must recognize that environmental policy is now inherently international. This holds for at least two reasons. First, a large portion of the firms regulated in any one nation will have trade with and often have production facilities in other nations and so must deal with a web of national regulations. As a result, voluntary agreements nearly always will operate in the context of multiple national laws and regulations. Second, a growing fraction of national environmental laws and regulations are developed as a result of participation in international regulatory regimes.[14] Randall's point (this volume, Chapter 19) that firms may act in anticipation of regulation can apply to expected international treaties as well as to uniquely national policies. So voluntary agreements are acting not just in the context of multiple national regulations and multiple national markets for products and supplies, but also in the context of global environmental treaties and the national laws and regulations that respond to them.

CONCLUSIONS

Voluntary agreements may have substantial potential to reduce the environmental impacts of production and consumption activities both directly and by changing organizational culture and capabilities. But as the chapters in Part 3 demonstrate, voluntary agreements come in many forms, and we do not understand many of them very well. In the worst case, it is not clear what changes in environmental impact have flowed from the voluntary agreements that have been implemented, in part because of limits in the design of the existing evaluations and in part because we have few evaluations of the less tangible impacts of voluntary agreements such as organizational changes. The participation of firms in voluntary agreements, including the effective implementation of environmental policies and procedures, depends on many factors that require further investigation. Some factors are intraorganizational: organizational structure and lines of control, culture and routines, available expertise, and available capital. Others depend on external factors: existing and anticipated regulations, the structure of the firm's industry, and the perceived interests and strategies of suppliers, customers, and competitors. To understand voluntary agreements, to evaluate their impacts, and especially to offer guidance on the design of future programs, substantial efforts at conceptualization, modeling, and empirical analysis are essential. The chapters here review the first efforts along these lines and point toward future research, but it is clear that the interest in voluntary agreements has run well ahead of our understanding of them.

Even less well researched is the potential of voluntary agreements to deal with the sources of environmental impacts that flow from nonpoint sources. In addition, we are only beginning to think through the implications of international environmental regulatory regimes for national and international voluntary agreements. It is almost certain that voluntary agreements will be an important part of environmental policy over the next decade. What is less certain is the benefits that will flow from them and the best strategies for developing and implementing them.

NOTES

1 Analyses of the international experience are beginning to appear (tenBrink, 2002).

2 One of the first efforts to apply program evaluation methods to an environmental policy is Poppitti and Dietz (1983). In examining the Resource Conservation and Recovery Act, we were surprised by how little evaluation research was available on environmental programs. This has been less true of energy and water conservation programs, where there is a long history of evaluation research using both experimental and quasi-experimental designs. See, for example, Dietz and Vine (1982), Kowalczyk et al. (1983), Harris and Blumstein (1984), Stern et al. (1986), and Vine and Crawley (1991).

3 The classic Campbell and Stanley (1963) volume documents the threats to valid causal inference in various research designs. Cook and Campbell (1979) update this discussion, while Achen (1986) discusses statistical issues in the analysis of quasi-experiments. Issues of evaluation research design also are discussed in the ecological literature (Manly, 2001: Chapter 6; Underwood, 1997).

4 This suggests that an important goal of research should be developing an understanding of the effects of organizational change on the environmental performance of firms.

5 Examples of methods for modeling the "spread" of a distribution include those described by Gould (1992, 1997). "Extreme values" or the tails of a distribution are of interest in many applications of statistics to environmental problems, where methods are steadily improving (Manly, 2001).

6 Axelrod (1997) considers the altruism problem the "Eshrecia coli" of the social sciences, while Dietz et al. (2002) refer to it as the "Drosophila melanogaster." Whichever analogy to biology is chosen, the argument remains the same: The problem of altruism, posed as Prisoner's Dilemma, the Tragedy of the Commons, or the Logic of Collective Action, presents a marvelous test bed for the theory and methods of the social sciences.

7 In addition to empirical studies of altruism, formal models of cultural evolution and of repeated Prisoner's Dilemma games give analytical insights as to why cooperation may not be as rare as earlier, simpler models suggested (Boyd and Richerson, 1985; Richerson and Boyd, in press; Richerson et al., 2002; Sober and Wilson, 1998).

8 ISO 14001 does not regulate production or products per se, but simply requires that environmental management systems are in place.

9 Such prohibitions differ radically across nations. For example, in Japan the keiretsu corporate groups are an important part of the national economic structure, and many firms have links through banks that hold large portions of their stock equity (Aoki, 1988). However, Nakamura et al. (2001) found that these links have little impact on firms' participation in the ISO 14001 certification process.

10 Laumann and Knoke (1987) and Dietz and Rycroft (1987) show that professional and trade associations are important participants in the energy and environmental risk policy networks.

11 In a series of papers, Wernick, Ausubel, and their collaborators document the reductions in environmental impact that have been achieved in a number of industries and suggest that better engineering could substantially alleviate environmental problems (Ausubel, 1996; Wernick, 1997; Wernick, 1994; Wernick et al., 2000). Mol and Sonnenfeld (2000) present arguments that this is a general trend, driven by societal concerns with the environment rather than just cost reduction. The literature on the environmental Kuznets curve makes a similar argument—increased affluence leads to reduced environmental impact per unit affluence (see Nordstrom and Vaughan, 1999, for a review). York et al., (2001) and Roberts and Grimes (1996) present skeptical views of the trend toward reduced environmental impact. At best, such reductions may occur with regard to local impacts, but there is little evidence of such a dynamic for global impacts such as greenhouse gas emissions. Voluntary agreements have been directed at both local and global pollutants.

12 For example, organizational ecologists in sociology do not necessarily assume strict profit maximization, but do treat organizations as unified wholes with regard to their strategies (Hannan and Freeman, 1989). This view is critiqued in McLaughlin (1996).

13 Because of possible differences among industries in their likelihood of developing effective voluntary agreements such as those noted by Nash (this volume, Chapter 14) and Furger (this volume, Chapter 17), it is important that comparative research be conducted both across countries and across industries.

14 These agreements are themselves voluntary agreements in that there is no mechanism to impose them on individual states. In addition to the importance of studying voluntary agreements in the context of international treaties on the environment, it seems likely that the literature on voluntary agreements and that on international regulatory regimes might provide theoretical and methodological guidance to each other.

REFERENCES

Achen, C.H.
    1986   *The Statistical Analysis of Quasi-Experiments.* Berkeley: University of California.
Andrews, C.
    1994   Policies to encourage clean technology. In *Industrial Ecology and Global Change,* R. Socolow, C. Andrews, F. Berkhout, and V. Thomas, eds. Cambridge, Eng.: Cambridge University Press.
Aoki, M.
    1988   *Information, Incentives and Bargaining in the Japanese Economy.* New York: Cambridge University Press.
Ausubel, J.H.
    1996   Can technology spare the earth? *American Scientist* 84:166-178.
Axelrod, R.
    1997   *The Complexity of Cooperation.* Princeton, NJ: Princeton University Press.
Boyd, R., and P.J. Richerson
    1985   *Culture and the Evolutionary Process.* Chicago: University of Chicago Press.

Campbell, D.T.
  1969 Reforms as experiments. *American Psychologist* 24:409-429.
Campbell, D.T., and J.C. Stanley
  1963 *Experimental and Quasi-Experimental Designs for Research.* Chicago: Rand McNally.
Cook, T.D., and D.T. Campbell
  1979 *Quasi-Experimentation.* Chicago: Rand McNally.
DeCanio, S.J., and W.E. Watkins
  1998 Investment in energy efficiency: Do the characteristics of firms matter? *Review of Economics and Statistics* 80:96-107.
Dietz, T., E.Ostrom, N. Dolšak, and P.C. Stern
  2002 The drama of the commons. Pp. 3-36 in National Research Council, *The Drama of the Commons,* Committee on the Human Dimensions of Global Change. E. Ostrom T. Dietz, N. Dolšak, P.C. Stern, S. Stonich, and E.U. Weber, eds. Washington, DC: National Academy Press.
Dietz, T., and R.W. Rycroft
  1987 *The Risk Professionals.* New York: Russell Sage Foundation.
Dietz, T., and E. Vine
  1982 Energy impacts of a municipal conservation policy. *Energy* 7:755-758.
Galbraith, J.K.
  1967 *The New Industrial State.* Boston: Houghton Mifflin.
Gould, W.W.
  1992 Quantile regression with bootstrapped standard errors. *Stata Technical Bulletin Reprints* 2:137-139.
  1997 Interquartile and simultaneous-quantile regression. *Stata Technical Bulletin Reprints* 7:167-176.
Gunderson, L.H., C.S. Holling, and S.S. Light, eds.
  1995 *Barriers and Bridges to the Renewal of Ecosystems and Institutions.* New York: Columbia University Press.
Hannan, M.T., and J. Freeman
  1989 *Organizational Ecology.* Cambridge, MA: Harvard University Press.
Harris, J., and C. Blumstein, eds.
  1984 *What Works: Documenting Energy Conservation in Buildings.* Washington, DC: American Council for an Energy Efficient Economy.
Heckman, J.
  1976 The common structure of statistical models of truncation, sample selection, and limited dependent variables. *The Annals of Economic and Social Measurement* 5:475-492.
  1979 Sample selection bias as a specification error. *Econometrica* 47:153-161.
Henriques, I., and P. Sadorsky
  1996 The determinants of an environmentally responsive firm: An empirical approach. *Journal of Environmental Economics and Management* 30:381-395.
Kowalczyk, D., J.C. Cramer, B. Hackett, P. Craig, T. Dietz, M. Levine, and E. Vine
  1983 Evaluation of a community-based electricity load management program. *Energy* 8:253-243.
Laumann, E.O., and D. Knoke
  1987 *The Organization State: Social Choice in National Policy Domains.* Madison: University of Wisconsin Press.
Manly, B.F.J.
  2001 *Statistics for Environmental Science and Management.* Boca Raton, FL: Chapman and Hall.
McLaughlin, P.
  1996 *Towards an Ecology of Social Action: Merging the Ecological and Constructivist Traditions.* Fairfax, VA: Human Ecology Research Group, George Mason University.

Mol, A.P., and D.A. Sonnenfeld, eds.
2000 *Ecological Modernisation Around the World.* London: Frank Cass Publishers.

Nakamura, M., T. Takhashi, and I. Vertinsky
2001 Why Japanese firms choose to certify: A study of managerial responses to environmental issues. *Journal of Environmental Economics and Management* 42: 23-52.

Nordstrom, H., and S. Vaughan
1999 *Trade and Environment.* Geneva, Switz.: World Trade Organization.

Olson, M.
1965 *The Logic of Collective Action: Public Goods and the Theory of Groups.* Cambridge, MA: Harvard University Press.

Poppitti, J., and T. Dietz
1983 Social impact evaluation of the U.S. Resource Conservation and Recovery Act. *Environmental Management* 7:501-504.

Rees, W.E., and M. Wackernagel
1994 Ecological footprints and appropriated carrying capacity: Measuring the natural capital requirements of the human capacity. Pp. 362-390 in *Investing in Natural Capital: The Ecological Economics Approach to Sustainability,* A.C. Jannson, M. Hammer, C. Folke, and R. Costanza, eds. Washington, DC: Island Press.

Rees, W.R.
1992 Ecological footprints and appropriated carrying capacity: What urban economics leaves out. *Environment and Urbanization* 4:121-130.

Richerson, P.J., and R. Boyd
in   The evolution of subjective commitment to groups: A tribal social instincts hypothesis.
press In *The Evolution of Subjective Commitment,* R.M. Neese, eds. New York: Russell Sage Foundation.

Richerson, P.J., R. Boyd, and B. Paciotti
2002 An evolutionary theory of commons management. Pp. 403-442 in National Research Council, *The Drama of the Commons,* Committee on Human Dimensions of Global Change. E. Ostrom, T. Dietz, N. Dolšak, P.C. Stern, S. Stonich, and E.U. Weber, eds. Washington, DC: National Academy Press.

Roberts, J.T., and P.E. Grimes
1996 *Social Roots of Global Environmental Change: A World-Systems Analysis of Carbon Dioxide Emissions.* New Orleans, LA: Department of Sociology, Tulane University.

Simon, H.A.
1947 *Administrative Behavior.* New York: Macmillan.

Sober, E., and D.S. Wilson
1998 *Unto Others: The Evolution and Psychology of Unselfish Behavior.* Cambridge, MA: Harvard University Press.

Socolow, R., C. Andrews, F. Berkhout, and V. Thomas
1994 *Industrial Ecology and Global Change.* New York: Cambridge University Press.

Stern, P.C., E. Aronson, J. Darley, D.H. Hill, E. Hirst, W. Kempton, and T.J. Wilbanks
1986 The effectiveness of incentives for residential energy conservation. *Evaluation Review* 10:147-176.

Underwood, A.J.
1997 *Experiments in Ecology: Their Logical Design and Interpretation Using Analysis of Variance.* Cambridge, Eng.: Cambridge University Press.

Vine, E., and D. Crawley
1991 *State of the Art Energy Efficiency: Future Directions.* Washington, DC: American Council for an Energy Efficient Economy.

Wackernagel, M., L. Onisto, P. Bello, A.C. Linares, I.S. Lopez Falfan, J.M. Garcia, A. Isabel, S. Guerrero, and M.G. Suarez Guerrero
    1999   National natural capital accounting with the ecological footprint concept. *Ecological Economics* 29:375-390.

Wackernagel, M., and W.O. Rees
    1996   *Our Ecological Footprint: Reducing the Human Impact on the Earth.* Gabriola Island, British Columbia, Can.: New Society Publishers.

Wernick, I.K.
    1994   Demarterialization and secondary materials recovery: A long run perspective. *Journal of the Minerals, Metals, and Materials Society* 46(4):39-42.
    1997   Consuming materials: The American way. Pp. 29-39 in National Research Council, *Environmentally Significant Consumption: Research Directions,* P.C. Stern, T. Dietz, V.W. Ruttan, R.H. Socolow, and J.L. Sweeney eds. Washington, DC: National Academy Press.

Wernick, I.K., R. Herman, S. Govind, and J.H. Ausubel
    1997   Materialization and dematerialization: Measures and trends. Pp. 135-156 in *Technological Trajectories and the Human Environment,* J.H. Ausubel, and H.D. Langford, eds. Washington, DC: National Academy Press.

Wernick, I.K., P.E. Waggoner, and J.H. Ausubel
    2000   The forester's lever: Industrial ecology and wood products. *Journal of Forestry* 98(10):8-14.

York, R., E. Rosa, and T. Dietz
    2001   Social Theories of Modernization and the Environment: An Empirical Analysis of the Human-Environment Relationship. Unpublished paper presented to the 2001 Symposium on the Political Economy of the World System, Anaheim, CA, August 17.

Young, O.
    2001   The effectiveness of international environmental regimes: A midterm report. In *Studies in the Human Dimensions of Global Change,* A. Diekemann, T. Dietz, C. Jaeger, and E. Rosa, eds. Cambridge, MA: MIT Press.

PART IV

CONCLUSION

21

New Tools for Environmental Protection: What We Know and Need to Know

Thomas J. Wilbanks and Paul C. Stern

Although the potentials of the tools discussed in the earlier chapters in this volume are intriguing, the main conclusions are a bit paradoxical. On the one hand, full information is one of the foundations of responsible citizenship, and voluntary action is increasingly important as a way to ensure environmental stewardship in the United States, in partnership with government-mandated rules and regulations (and often in preference to them). But in the case of information, it seems clear that many people possess far less than they need to have in order to determine what is responsible voluntary action. This suggests a powerful rationale for communication and diffusion instruments that emphasize education and information to support voluntary action. Yet in many cases, perhaps most, the effects of federal government information and education programs appear so far to have been rather modest (see Lutzenhiser, this volume, Chapter 3; Schultz, this volume, Chapter 4; Thøgersen, this volume, Chapter 5; Stern, this volume, Chapter 12).

Voluntary measures for firms and industries also have great potential in principle. They allow for a decentralization of decision making to actors who are in the best position to evaluate what works for them, thus potentially increasing efficiency as well as democratic control. But as with education and information, the effects of voluntary measures appear so far to have been rather modest. They are documented in only a few industries, and even there, much of the claimed effect cannot be attributed unequivocally to the programs (see Mazurek, this volume, Chapter 13; Nash, this volume, Chapter 14; Harrison, this volume, Chapter 16).

In exploring this paradox, this volume considers three central issues:

- *Why* (or why not) increase government support for education and information programs for individuals and households and for voluntary programs for firms and industries in support of environmental management?
- *How* can such programs be as effective as possible?
- What do we *need to know*—in order to do better—that we do not already know?

This chapter is one reading of "bottom line" answers to these questions, based on the chapters of this volume and the discussion at the workshop on which the volume is based.

EXPLORING THE RATIONALE FOR EDUCATION, INFORMATION, AND VOLUNTARY PROGRAMS

The growing attention to this topic, not only in scholarship but in policy-making, reflects the fact that the world of government is changing. While democratization has been spreading globally, for two decades in the United States we have been moving in the direction of less government, cheaper government, a devolution of government roles, and a tendency to question whether government regulation is the most appropriate and most effective way to reach social goals. This trend suggests that voluntary decision making will become ever more important for the foreseeable future and perhaps that the rationale for programs to support voluntary decision making will become more compelling for government than for the research community, which would be a reversal of the patterns of the past.

At the same time, while the context of voluntary decision making is changing, so are the problems to which decisions need to be applied and the tools that are available to assist (Rejeski and Salzman, this volume, Chapter 2). This suggests that education, information, and voluntary programs are best designed to be adaptive, so they can respond flexibly to shifting requirements. Given this framework for thought, there are a number of reasons for government to support education, information, and voluntary programs, but there are also several reasons to be cautious.

The Central Reasons in Favor

Government agencies often consider education, information, and voluntary programs for at least three reasons. First, education and information are intended to inform responsible citizenship: to help close a gap between what people know or are able to know on their own and what they need to know in order to make well-informed voluntary decisions. In this connection, programs may be intended to:

- Ensure information quality and reliability, especially if other sources are suspected by many citizens of being biased;
- Encourage broader citizen involvement;
- Catalyze and support voluntary actions, including correcting erroneous perceptions;
- Improve capacities to act effectively; or
- Encourage the establishment of voluntary partnerships and linkages across boundaries, for example, between national and local governments or between the public and private sectors.

Second, some of these programs respond to the citizen's right to know by ensuring that information to which the citizen has a legal and/or moral right is made available by:

- Requiring public notification: determining what information must be made available by whom, when, and how, as with the Toxics Release Inventory (see Herb et al., this volume, Chapter 15); and
- Removing constraints on access: for example, providing information labels when the citizen otherwise would have to exert a great deal of effort to find information that should be considered in making a decision (see, e.g., Thøgersen, this volume, Chapter 5).

Generally, the idea underlying education and information programs is that government should not shape the values of citizens, but that it has a duty to citizens to provide information that can reinforce values and relate them to actions, if that information is not likely to be made available otherwise in forms that would be considered credible and/or affordable. The intent is to have programs that empower, not coerce. Information and education programs such as the U.S. Environmental Protection Agency's (EPA's) Green Lights program respond to this public need by not only ensuring citizen awareness, but also by accelerating it and by helping to close gaps between awareness and appropriate action (see Valente and Schuster, this volume, Chapter 6).

Third, education, information, and voluntary programs are believed to increase the efficiency of consumers' and producers' responses to economic and other signals of the need to change behavior to reduce environmental costs. Both households and firms are in a much better position than the federal government to find the best ways to economize in their own situations, so informed, decentralized decision making is more efficient theoretically than central regulation. This improved efficiency, however, depends on the decentralized actors' access to accurate information about the nature and costs of their decision options. Education and information can, in principle, provide this information for consumers; the kinds of dialogue among firms and government involved in organizing and maintaining voluntary programs can, in principle, provide this information for producers.

Some Reasons to Be Cautious

As good as all this sounds, there are reasons to think carefully before investing in education, information, and voluntary programs. First, in our society, we tend to believe that government roles in shaping human behavior should be quite limited. Any indication of social engineering by government, for example, by "experimentally manipulating social norms" (Schultz, this volume, Chapter 4), is likely to be considered a threat to true democracy. This general philosophy, of course, is less of a limitation in some fields than others; for example, government invests in advertising campaigns to discourage smoking and to influence people in some other matters of public health or disaster preparation (see Valente and Schuster, this volume, Chapter 6; Mileti and Peek, this volume, Chapter 7). One possible reason is that most people consider government advocacy more appropriate where there is an obvious and pressing public benefit, or where policies already have been determined through democratic processes, than they do where policy objectives are still undecided.[1]

Second, there are serious questions about the cost-effectiveness of government education and information programs, for at least two reasons (see Rosenzweig, this volume, Chapter 8). The impact of such programs may be relatively modest compared with the costs, and education and information programs may not be more cost-effective than other policies for reaching the same goal. A relevant issue in both of these connections is that more information and knowledge may affect actions in some situations and contexts, but not others (Stern, this volume, Chapter 12).

There are also serious questions about the effectiveness of government support of voluntary programs for firms and industries. Here, effectiveness must be weighed not against cost in tax dollars, but in relation to the relaxation of regulatory oversight that is often part of the package of government support of these programs. Voluntary programs decentralize decision making, which has potential benefits, but they also put the decisions in the hands of actors whose objectives differ from the regulator's goal of providing public goods like environmental quality (see Prakash, this volume, Chapter 18). Randall (this volume, Chapter 19) concludes that voluntary programs "make a nice frosting on the regulatory cake. But the cake must be there."

Finally, education, information, and voluntary programs (and research on them) can be ways to avoid timely action, in essence passing the buck to citizens to deal with a policy problem that would be dealt with more appropriately by government itself. This problem sometimes has been noted in policy analyses (U.S. Environmental Protection Agency Science Advisory Board, 2001).

CONSIDERING HOW TO CARRY OUT EFFECTIVE EDUCATION
AND INFORMATION PROGRAMS

Where it makes sense to invest in education and information programs, the next question is how to assure that they yield environmental benefits and are as cost-effective as possible. "Information programs" can cover a wide spectrum of government actions, including regulatory initiatives and financial incentives that "send signals" for particular actions and thus have an information function, but the focus of this volume is on information and education as communication and diffusion instruments that are distinct from regulatory actions or financial inducements. In this more limited connection, the central predicament is that a particular government information program becomes one of a great many tiny tributaries feeding a virtual flood of information engulfing those citizens who are the intended audiences, like adding just one more ingredient to a complex "information soup" (Mileti and Peek, this volume, Chapter 7). The challenge is to navigate through this complexity in ways that get the desired messages across.

The contributors to this book identify five elements of effective education and information programs, beyond the imperative of pretesting any proposed approach before implementing it (see especially Valente and Schuster, this volume, Chapter 6; Mileti and Peek, this volume, Chapter 7, and Stern, this volume, Chapter 12):

- *Targeting* selected parts of a diverse audience and addressing their particular concerns. The objective should be either to reach a large part of the population or, if different audiences need different information, to focus on key groups by addressing the main questions on their minds: "Do I have to worry about this or not?" and if so, "What are the most important things to do about it?" Directing information to people who already have it or who do not need it is seldom a good use of resources. One aspect of determining the appropriate target is to assess whether the main voluntary actors are likely to be individuals or institutions and, again, focusing on the relevant concerns of the relevant target.
- *Personalizing* the process. The more individualized, less impersonal the communication, the more likely it is that the information will be transferred. This suggests several strategies, including: ensuring an ongoing communication process rather than single information provision acts or events; paying close attention to the levels of credibility and trust associated by the target audiences with different information sources; utilizing the most effective channels (which often are anchored in existing social networks); and inviting information *exchange,* not just information *provision*—in other words, incorporating interactive and experiential stakeholder involvement, perhaps after an initial awareness-raising stage.

- Assuring *multiple information sources/mechanisms.* Education and information programs are likely to be more effective if they incorporate an assortment of approaches: repeating and reinforcing the flow of information and telling people where they can get additional information; linking with other information and education efforts; paying attention to forces that might encourage partnerships; and in some cases, considering different mechanisms for different stages in the education and information process. For example, mass media approaches may be more effective in early stages, and interpersonal communication in later stages (Valente and Schuster, this volume, Chapter 6).

- Being prepared for *"windows of opportunity."* Given that the attention level of many citizens is related to crises of the moment, information programs can prepackage strategies and information to be brought out if and as events raise questions that are answered by the packages (Mileti and Peek, this volume, Chapter 7). For example, the response to the California energy crisis of 2001 was assisted by information packages that were on hand, advising people on ways to gain thermal comfort and other energy services with less electricity, to shift demand away from peak hours, and so forth.

- Making the right choices and picking the right combinations of policy tools. Designers of programs may have a larger menu of possible mechanisms at their disposal than they are aware. Identifying the full range, considering all the options, and making the right selections for the case at hand can make a difference in the effectiveness of a program. The chapters of this volume mention such possibilities as the following:

1. Partnerships. Partnerships involve government information (and/or public recognition) working directly in collaboration with nongovernmental voluntary action and education. One example is the Motor Challenge program of the Department of Energy (DOE), in which DOE invites private-sector firms to join in a partnership where the firm makes a commitment to use state-of-the-art, energy-efficient electric motors and drives where cost-effective; in turn, DOE provides full information about technology options, along with technical assistance and considerable public recognition for the partners.

2. Scorecards and benchmarks. Government provides ways to measure performance (scorecards) and to publicize the results of measurements, often associated with levels of performance that are among the best being achieved under current market and regulatory conditions (benchmarks) (Furger, this volume, Chapter 17).

3. Labeling. Government or another third party attaches a label or logo to consumer items to inform voluntary decisions about what to buy (e.g., appliance or automobile fuel efficiency labels, recycling symbols). The presumption is that voluntary actions will be different if consumers are

aware of their environmental implications (Thøgersen, this volume, Chapter 5).

4. Government purchasing. The federal government can use its enormous purchasing power to shape supplier characteristics. An example is the EPA's Energy Star computer program, which led to a government decision to buy only personal computers meeting energy-efficiency standards, making it unattractive for equipment suppliers to invest in producing equipment not meeting those standards. Large corporations also can use this strategy (see Rejeski and Salzman, this volume, Chapter 2).

5. State-of-the-art communication modes. Information providers can follow the example of the private sector in using the power of different communication modes for particular purposes, such as use of the growing arsenal of graphics tools emerging from the information technology revolution.

Sorting through all these choices is clearly a complicated business. It involves considering a variety of kinds of information that government program designers may not have at hand. It requires complicated operational decisions involving financial and human resources. It calls for cost-benefit estimation that only may be possible qualitatively. It also raises more fundamental issues. For example, in designing a public information program, who should decide what information is needed? Who decides what information is true? What if either or both of the decisions are wrong? Who is accountable? As an information program proceeds, how can it be determined when the information provided is enough? How does an information program handle uncertainties and possible surprises, especially if it is providing information about the future as well as the past and the present?

Nobody ever said it was going to be easy. But, at the same time, thoughtful applications of a rule of reason often can reduce the complications to a number where more careful program design is feasible. The problem is that in many cases, the detailed design stage is undermined by limitations on what even the nation's top experts know.

CONSIDERING HOW TO CARRY OUT EFFECTIVE VOLUNTARY PROGRAMS FOR THE PRIVATE SECTOR

The central policy question about voluntary measures is whether environmental objectives can be achieved more effectively or more cost-effectively if direct regulation is reduced in favor of policy instruments that enhance the power of market pressure, investor influence, public concern, reputation, and the like to press firms toward better environmental performance. The proper distinction is not between coerced and voluntary behavior. It is between direct regulation (regulatory demands to meet emission goals or adopt different technologies) and other instruments that may be perceived as less coercive (e.g., Andrews, 1998).

So-called self-regulation is perhaps the most obvious of the available possibilities. The preceding chapters suggest that this approach may be valuable under some conditions and in some industries where adequate incentives exist for firms to establish and maintain institutions that ratchet up environmental performance. Although more experimentation with this approach is warranted, the evidence strongly suggests that government should proceed very cautiously in the direction of relaxing regulations in the hope that self-regulation will take their place. It is not certain how much of the reported successes of industry self-regulation is real (Mazurek, this volume, Chapter 13; Nash, this volume, Chapter 14; Harrison, this volume, Chapter 16) or how much of the real improvement is attributable to the credible threat of regulation (Randall, this volume, Chapter 19)—the desire for a less painful way to meet current or potential regulatory demands.

The evidence indicates that successful self-regulation is most likely to occur in industries where three conditions exist: strong public concern about environmental damage from that industry; limited identification of this damage with specific firms; and industry leaders that are sufficiently large or well known that they have incentives to bear a disproportionate share of the costs of creating self-regulatory institutions (Nash, this volume, Chapter 14). An advantageous communication structure within the industry also may be necessary (Furger, this volume, Chapter 17). Even under these advantageous conditions, the industry's incentive is to produce a reputation for environmental stewardship, and this may be gained at lower cost by promoting a "green" image than by changing corporate environmental behavior. For this reason, the effectiveness of voluntary measures may be increased greatly by government-funded or-mandated programs that monitor actual environmental progress so that reputation can be tied to valid environmental indicators. The Toxics Release Inventory in the United States has this function, though it is vulnerable because its indicators are taken from firms' self-reports.

Little is known about how to make self-governance work in other kinds of industries. These include industries in which consumer products are tightly linked to brands so that the incentives fall on single firms rather than industries (e.g., pharmaceuticals), in which there are no industry leaders (e.g., dry cleaning), or in which environmental damage is not easily traceable to particular firms (e.g., trucking). It is reasonable to expect that industrywide self-governance is more difficult to achieve under these conditions, even though some individual firms may take voluntary action.

"Voluntary" strategies other than industry self-governance may have significant potential. Although there is little or no systematic knowledge about how to make them work, they are worthy of further attention. We mention three interesting mechanisms involving voluntary action as illustrative. One is the notion that information about the environmental performance of publicly traded firms may change their behavior through the influence of "green" investors (Herb et al., this volume, Chapter 15). A second mechanism is the use of consumer boycotts and other collective action to exert pressure on firms independent of regulation. An example was the consumer boycott and demonstrations directed at McDonalds restaurants that led to an agreement to end styrofoam packaging in 1991 (for an account, see Gardner and Stern, 1996). A third interesting mechanism of voluntary action involves arrangements between industries and nongovernmental organizations (NGOs) to support improved environmental performance. For example, some coffee marketers have been willing to pay extra for imported coffee they could certify as shade-grown if an NGO was willing to inspect the coffee plantations to provide legitimacy for the claim.

The knowledge base on voluntary measures is not as well developed as that for education and information, so it is early to draw conclusions about how to make these measures work best. Some insights about education and information may prove relevant to voluntary measures, though. In particular, targeting, the use of multiple mechanisms, preparation for windows of opportunity, and making the right choices of instruments all are likely to be important. Further insights probably can be gained from relevant theories in areas such as organizational behavior and collective action (e.g., Furger, this volume, Chapter 17; Prakash, this volume, Chapter 18). But much more knowledge is needed for voluntary measures to become a tool that can be used with precision, rather than just a promising idea.

WHAT DO WE NEED TO KNOW IN ORDER TO DO BETTER?

From the perspectives of the experts, including those represented in this volume, government program designers and decision makers are asking a number of critically important questions that cannot be answered with confidence from the existing knowledge base. The experts themselves need to know a lot more in order to be as helpful as the government needs, given the imperative of using taxpayers' money and government regulatory authority effectively and responsibly to aid voluntary decision making.

Most of what still needs to be learned, however, is not specific to governmental environmental education and information programs. It concerns broader issues for both government program effectiveness and the social and behavioral sciences at large. Based on the contributions to this volume, the highest priority questions to address for improving the knowledge base to support education and information program design are the following:

• When and how do people and organizations demand and use information? More needs to be known about information demand as well as supply, given the diversity of audiences and the need to target particular audiences. The challenges include understanding how information feeds into the voluntary decision making processes of organizations and individuals, how information demand varies across cultures, how organizations and individuals adapt to changes in information, and how organizations and individuals can become more adaptable in a changing, uncertain world (National Research Council, 1999).

• How can success be measured and documented? In an era of reengineered government, there is an urgent need to improve the capacity to evaluate all kinds of government-supported programs. The challenges include evaluating the success of efforts to *transmit information and learning* in stimulating voluntary actions and evaluating the *impacts of the voluntary actions,* both of which in many cases require establishing baselines against which to compare program-related outcomes. When effects are substantially lagged in time, the challenge is even greater. In these regards, there appear to be abundant opportunities to apply insights from evaluation research (Harrison, this volume, Chapter 16; Weiss, 1998) and learn from industry experience (Nash, this volume, Chapter 14).

• When and how does information lead to action? As indicated, we know far too little about how information relates to knowledge and how knowledge relates to action. We need to learn more about how different information processes may relate to underlying common issues, how a particular information program may reinforce or contradict other information processes, and how a variety of information programs may have cumulative impacts that add up to more than the sum of the parts (e.g., encouraging a stronger "environmental ethic"). In many cases, unraveling these questions calls for types of longitudinal studies, followup studies, and cumulative impact studies for which funding is exceedingly difficult to find.

• How can information infrastructures and programs be designed so they are more adaptive? Unless education and information programs are constructed so they can change as their contexts change, they are likely to become outdated quickly. The Toxics Release Inventory is an example (see Herb et al., this volume, Chapter 15). The challenge is to build adaptability into the structure from the beginning in the language and implementation of statutes and in relation to ongoing evaluation processes. This calls for communication between the executive and legislative branches and with the parties to which they listen.

• How can effects of the information technology revolution be harnessed in support of government education and information programs? Clearly, the world is being transformed rapidly through the tools available to facilitate communication. Electronic mail was uncommon a decade ago, use of the Internet is mushrooming, and graphics capabilities such as geographic information systems and hypermedia packages are growing rapidly. Such capabilities may make possible dramatic advances in instrumentation and measurement that can allow quick feedback about the effects of actions on environmental indicators. The ability to assess the potential and limitations of such new developments for information dissemination and for interaction between providers and receivers is still in its infancy.

THE BOTTOM LINE

Even though education and information programs are not the answer to every environmental policy need, they are fundamentally important in supporting responsible citizenship in a democracy. Voluntary programs in the private sector are also highly attractive from a governance perspective. The impact of these tools, however, often seems to have been modest at best. This record seems to call for increased attention to ways to make such programs more effective, which in turn calls for more attention to strengthening the knowledge base on which program planning and design is based.

The evidence strongly suggests that it is past time to move beyond debates about which tool is best for environmental protection—whether regulation or market-based approaches are better, whether it is good to increase voluntarism and decrease regulation, and so forth. Each tool has its place, not only because of the variety of policy targets, but also because each tool performs particular functions. The best policy normally uses a combination of tools, each serving its proper function (Stern, 2000; Stern, this volume, Chapter 12). For example, authors in this volume argue that voluntary measures in industries work best under the threat of regulation; that they depend on good information in the form of monitoring data on environmental performance; that they benefit from market forces that favor "green" performance; and that their success depends on whether an industry has agents (such as trade associations) to diffuse best practices. Thus, command and control, communication and diffusion, and market instruments all may help voluntary programs be more effective.

Much can be gained by developing better understanding of the functions performed by each type of policy tool so that policies can be designed to employ the tools in appropriate combinations. The need for new combinations—as well as for the new tools—is likely to increase as the nature of environmental problems and the identity of pollution sources changes (Rejeski and Salzman, this volume, Chapter 2).

NOTE

1 One should not presume that good information necessarily will reduce social and political conflict or differences of opinion about what kinds of voluntary actions make sense. Experience has taught that more information, far from generating agreement, can in fact strengthen the views of different constituencies in opposition to each other. The same body of information can be used by different parties to make opposing arguments and to justify opposing actions. "Information is power"; thus its contents and mechanisms are (or can be) politically sensitive. As one consequence, when education and information programs touch on controversial issues, some constituencies may oppose the programs because of the prospect that their content might be used effectively to support opposing views.

REFERENCES

Andrews, R.N.L.
   1998   Environmental regulation and business "self-regulation." *Policy Sciences* 31:177-197.
Gardner, G.T., and P.C. Stern.
   1996   *Environmental Problems and Human Behavior.* Needham Heights, MA: Allyn and Bacon.
National Research Council
   1999   *Our Common Journey: The Transition to Sustainability.* Board on Sustainable Development. Washington, DC: National Academy Press.
Stern, P.C.
   2000   Toward a coherent theory of environmentally significant behavior. *Journal of Social Issues* 56(3):407-424.
U.S. Environmental Protection Agency Science Advisory Board
   2001   *Improved Science-Based Environmental Stakeholder Processes.* Washington, DC: U.S. Environmental Protection Agency.
Weiss, C.H.
   1998   *Evaluation: Methods for Studying Programs and Policies.* Upper Saddle River, NJ: Prentice-Hall.

ABOUT THE CONTRIBUTORS

ELAINE ANDREWS is a member of the faculty at the University of Wisconsin Cooperative Extension–Environmental Resources Center, where she specializes in community environmental management and environmental education. Her leadership of numerous state and national grant projects has resulted in long-term education partnerships with more than 20 national organizations. The author of many publications, she authored the hazardous waste sections for the national pollution prevention education strategy, Farm*A*Syst/ Home*A*Syst, which won Vice President Gore's Hammer Award for reinventing government in 1998. She holds a B.A. in biology, an M.A. in secondary science education, and an M.S. in natural resource policy and management. She is also the executive director of the North American Association for Environmental Education.

ALAN BALCH holds a master's degree in environmental sciences and is a doctoral candidate in environmental studies at University of California, Santa Cruz. His dissertation work deals with sustainable resource use and the role that recycling and source reduction can play in that effort. He is coauthor of a chapter entitled "Biodiversity Politics and Policy in the U.S.," published in the forthcoming book, *The Loss of Biological Diversity,* (S. Spray and K. McGlothlin, editors, Rowman and Littlefield Publishing).

THOMAS DIETZ is College of Arts and Sciences Distinguished Professor and professor of environmental science and policy and sociology at George Mason University. He holds a bachelor of general studies from Kent State University and a Ph.D. in ecology from the University of California, Davis. He is a Fellow of the American Association for the Advancement of Science, a Danforth Fellow, and a past president of the Society for Human Ecology, and he has received the Distinguished Contribution Award from the Section on Environment, Technology and Society of the American Sociological Association. His research interests are in human ecology and cultural evolution. He has a longstanding program of scholarship on the relationship between science and democracy in environmental policy. Recent publications include *Environmentally Significant Consumption: Research Directions* (National Academy Press, 1997).

FRANCO FURGER is an associate research professor in the School of Public Policy at George Mason University. He received his M.S. in electrical engineering in 1982, and his Ph.D. in environmental sciences in 1992 from the Federal Institute of Technology, Zürich. Professor Furger's research interests include environmental policy, science and technology policy, economic sociology, and institutional economics. Before joining the School of Public Policy in 1994, he was visiting fellow at the Institute of Social Research in Frankfurt am Main (Germany), where he conducted extensive research on the social factors affecting the adoption of electronic systems of payment. He has conducted extensive work on voluntary environmental initiatives at the domestic and international levels. He has also done research in science and technology policy on issues such as the effectiveness of the National Science Foundation's Engineering Research Centers Program, the changing role of purchasing managers in R&D collaborations, and patterns of federal R&D spending. He is the author of several articles and two books.

KATHRYN HARRISON is an associate professor of political science at the University of British Columbia. She is the author of *Passing the Buck: Federalism and Canadian Environmental Policy* (University of British Columbia, 1996), coauthor with George Hoberg of *Risk, Science, and Politics: Regulating Toxic Substances in Canada and the United States* (McGill-Queens University press, 1994), and coeditor with Patrick Fafard of *Managing the Environmental Union* (Strathy Language Unit, 2000). She has recently published articles on comparative environmental policy and alternative policy instruments in *Governance, Policy Sciences, Environmental Politics,* and the *Journal of Industrial Ecology.*

SUSAN HELMS was an associate scientist in the risk analysis group at Tellus Institute. While at Tellus, she worked with government, businesses and nongovernmental organizations to develop a multimedia, integrated reporting system for environmental releases. She also codeveloped a materials accounting system for a major manufacturer, applied True Cost Assessment tools to inform decisions on wastewater treatment, and helped develop a strategy for thoroughly integrating pollution prevention into state regulations. Ms. Helms has presented on a range of topics, including integrated environmental reporting, measuring damages from methyl-mercury contamination, and assessing the extent of the market for existence values of natural resources. Prior to joining Tellus, Ms. Helms was an economist at Triangle Economic Research, where she conducted natural resource damage assessments used in settlements and litigation. Ms. Helms holds a master's degree in environmental studies from Yale University.

JEANNE HERB is manager of the Public Policy Program at Tellus Institute. In that role, she oversees Tellus' efforts to develop and study innovative policies related to environmental management and sustainable development. Prior to joining Tellus in 1998, she was the founding director of the Pollution Prevention (P2) Program at the New Jersey Department of Environmental Protection. She oversaw development of the state's P2 Act and regulations and managed its innovative P2-based multimedia permit program. She served as the agency's lead spokesperson on pollution prevention and as vice chairperson of the National Pollution Prevention Roundtable. She also oversaw research studies on topics including pollution prevention policies, industrial environmental reporting, risk communication, public perceptions of environmental policies and incentive-based environmental regulations. She holds a bachelor's degree in environmental studies from Rutgers University (Cook College) and a master's degree in environmental journalism from New York University.

HAROLD R. HUNGERFORD is a professor in the Department of Curriculum and Instruction at Southern Illinois University at Carbondale. He has played a defining role in environmental education for more than 30 years.

MICHAEL J. JENSEN is a research analyst at the Tellus Institute, working primarily within the business and sustainability group. His fields of interest include industrial ecology and cooperative environmental strategies and their resulting benefits and costs, particularly policies within both the private and public sectors that foster increased accountability for environmental decisions. While at Tellus he has worked on topics that include use of environmental information by equity investment services; regional solid waste planning; evaluation of the U.S. Environmental Protection Agency's policies; and the revision of the Global Reporting Initiative, a standard for corporate disclosure. As an undergraduate, he worked as a research assistant to the Industrial Environmental Management Program at the Yale University School of Forestry and Environmental Studies, where he coauthored two articles with Thomas Graedel. His joint senior thesis examined the integrated resource efficiency measures of the Ford Motor Company during the 1920s and 1930s. He holds a bachelor's degree in both history and studies in the environment from Yale University.

LOREN LUTZENHISER is associate professor of sociology and rural sociology at Washington State University. His research examines changing patterns of technology use (and understandings of technology) and their environmental consequences. A particular focus of his work has been consumer energy use and conservation practices. He is also interested in community sustainability and the emergence of market-based approaches to environmental policy. He has published widely in both the social science and policy-applied literatures and is the current chair of the American Sociological Association Section on Environment and Technology.

JANICE MAZUREK directs the Center for Innovation and Environment at the Progressive Policy Institute. Her work focuses on ways in which to update environmental management strategies to reflect industry restructuring, product time to market, and globalization. Previously, Mazurek advised the U.S. Environmental Protection Agency and the Organization for Economic Co-operation and Development on voluntary and market-based strategies to improve environmental performance. She has also worked as a researcher at Resources for the Future (RFF) and the National Academy of Public Administration. Mazurek is the author of *Making Microchips: Policy, Restructuring and Globalization in the Semiconductor Industry* (MIT Press, 1999) and the coauthor with J. Clarence Davies of *Pollution Control in the U.S.: Evaluating the System* (RFF/Johns Hopkins, 1998). She holds degrees in economics and regional planning from the University of California.

DENNIS S. MILETI is professor and chair of the department of sociology and director of the Natural Hazards Research and Applications Information Center at the University of Colorado at Boulder. He is the cofounder and coeditor-in-chief of the journal *Natural Hazards Review*. Professor Mileti is the author of more than 100 publications focusing on the societal aspects of mitigation and preparedness for natural hazards and disasters. He recently completed a 5-year national effort to assess knowledge, research, and policy needs for natural and related technological hazards in the United States. This work resulted in his most recent book, *Disasters by Design: A Reassessment of Natural Hazards in the United States* (Joseph Henry Press, 1999). He received his B.A. in sociology from the University of California at Los Angeles, his M.A. in sociology from California State University at Los Angeles, and his Ph.D. in sociology from the University of Colorado-Boulder.

JENNIFER NASH is director of the regulatory policy program at the John F. Kennedy School of Government at Harvard University. Her research investigates why firms adopt environmental practices that go beyond regulatory requirements and the effectiveness of new regulatory instruments in achieving policy goals. She is the coeditor, with Cary Coglianese, of the volume *Regulating from the Inside: Can Environmental Management Systems Achieve Policy Goals?* (Resources for the Future Press, 2001), the first book-length treatment of the public policy implications of the growing use of environmental management systems. She has published articles and book chapters on the roles of U.S. Environmental Protection Agency innovative programs, environmental standards, and trade association codes of practice in environmental protection. She received a B.A. from Bryn Mawr College in anthropology and a master of city planning degree from Michigan Institute of Technology.

LORI A. PEEK is a doctoral student in the department of sociology and a research assistant at the Natural Hazards Research and Applications Information Center at the University of Colorado at Boulder. She is also the assistant co-editor-in-chief of the multidisciplinary journal *Natural Hazards Review.* Over the past 3 years, Peek has presented and coauthored several articles in the areas of social vulnerability, environmental risk, and natural hazards and disasters. She is currently conducting dissertation research on the response of Arab and Muslim university students in the United States to the events of September 11, 2001. Peek received her B.A. in sociology from Ottawa University and her M.Ed. in human resource studies from Colorado State University.

ASEEM PRAKASH is assistant professor of political science at the University of Washington-Seattle. His research investigates two questions: first, how firms and governments can respond to the challenges posed by globalization; second, how policy instruments can be designed to induce businesses to adopt environmentally responsible policies. He is the author of *Greening the Firm: The Politics of Corporate Environmentalism* (Cambridge University Press, 2000) and the coeditor (with Jeffrey A. Hart) of *Globalization and Governance* (Routledge, 1999), *Coping with Globalization* (Routledge, 2000), and *Responding to Globalization* (Routledge, 2000). He received a B.A. from St. Stephen's College (India), an M.B.A. from the Indian Institute of Management at Ahmedabad (India), and a joint Ph.D. from the department of political science and the School of Public and Environmental Affairs, Indiana University, Bloomington.

DANIEL PRESS is associate professor of environmental studies at the University of California, Santa Cruz, where he teaches environmental politics and policy. He is the author of *Democratic Dilemmas in the Age of Ecology* (Duke University Press, 1994) and *Saving Open Space: The Politics of Local Open Space Preservation in California* (University of California Press, 2002).

JOHN RAMSEY is an associate professor of Science and Environmental Education, as well as Director of Teacher Education at the University of Houston. The North American Association of Environmental Education presented him with the 2001 Research Award in Environmental Education.

ALAN RANDALL is professor and chair of the department of agricultural, environmental, and development economics at the Ohio State University. He received his B.S. and M.S. in agricultural economics from the University of Sydney and his Ph.D. in agricultural economics from Oregon State University. His primary research areas are natural resource economics, project evaluation, and benefit cost analysis, including theory and methods of estimating environmental benefits and assessing environmental damages.

DAVID W. REJESKI is a resident scholar at the Woodrow Wilson Center, where he directs the center's project on foresight and governance. Most recently, he was a visiting fellow at Yale University's School of Forestry and Environmental Studies and an agency representative from the U.S. Environmental Protection Agency (EPA) to the White House Council on Environmental Quality (CEQ). Before moving to CEQ, he worked at the White House Office of Science and Technology Policy on a variety of technology and R&D issues. From 1990 to 1993, he headed the future studies unit at the EPA. He has written extensively on science, technology, and policy issues, in areas ranging from genetics to electronic commerce and pervasive computing. He has graduate degrees in public administration and environmental design from Harvard University and Yale University.

MARK R. ROSENZWEIG is the Walter H. Annenberg Professor in the Social Sciences and professor of economics at the University of Pennsylvania. He has published numerous articles on human capital and population in both the United States and in rural populations of low-income countries, including studies on schooling investments and returns, savings, agricultural investments, deforestation, migration, marriage, and fertility in environments characterized by uncertainty and technical change. Rosenzweig received his B.A. and Ph.D. degrees from Columbia University.

JAMES SALZMAN is a professor of law at the Washington College of Law, American University. With degrees from Yale College and Harvard Law School, he worked in the environment directorate of the Organization for Economic Co-operation and Development (OECD) in Paris from 1990 to 1992, directing the work programs on environmental labeling and the "green" consumer. From 1992 to 1995, he served as the European environmental manager for Johnson Wax, supervising environmental issues in 14 countries of operation. He has advised the OECD, the United Nations Environment Program, the U.S. Environmental Protection Agency, and the U.S. Trade Representative on a range of policy issues and served as a visiting professor at Stanford Law School and Harvard Law School. He has published on consumption, services, and other challenges to traditional approaches to environmental protection and coauthored the leading textbook on international environmental law.

P. WESLEY SCHULTZ is associate professor of psychology at California State University, San Marcos. His research interests are in applied social psychology, particularly in the area of sustainable behavior. Recent books include *Applied Social Psychology* (Prentice-Hall, 1998) and *Social Psychology: An Applied Perspective* (Prentice-Hall, 2000), and he is currently completing an edited book, *Psychology of Sustainable Development* (Kluwer, 2002). He earned a B.A. from the University of California, Irvine, an M.A. from the University of Maine, and a Ph.D. from Claremont Graduate University.

DARLEEN V. SCHUSTER is a doctoral student in preventive medicine (health behavior research) at the University of Southern California. She teaches graduate courses in health communication and health behavior theory at the University of Southern California, where she serves as the coordinator of the Master in Public Health program. She received a B.A. in political science from the University of California, Irvine, an M.A. in communications management from the University of Southern California Annenberg School for Communication, and an M.P.H. in community health education from California State University, Northridge. She is a certified health education specialist.

PAUL C. STERN is study director of two National Research Council committees: the Committee on the Human Dimensions of Global Change and the Committee to Review the Scientific Evidence on the Polygraph. His research interests include the determinants of environmentally significant behavior, particularly at the individual level, and participatory processes for informing environmental decision making. His recent books include *Environmental Problems and Human Behavior* (with G.T. Gardner, Allyn and Bacon, 1996), *Understanding Risk: Informing Decisions in a Democratic Society* (edited with H.V. Fineberg, National Academy Press, 1996), and *International Conflict Resolution after the Cold War* (edited with D. Druckman, National Academy Press, 2000). Stern received his B.A. from Amherst College and his M.A. and Ph.D. degrees from Clark University.

MARK STEVENS is a Ph.D. candidate in urban and regional planning and a Marie Christine Kohler fellow at the University of Wisconsin-Madison. His current research on perceptions of effectiveness in planning practice is part of his broader work on the social and environmental dynamics of problem solving. He received his B.S. in environmental conservation from the University of New Hampshire and his M.S. in land resource from the University of Wisconsin-Madison, Institute for Environmental Studies.

JOHN THØGERSEN is professor of economic psychology at the Aarhus School of Business. He has published extensively on environmental attitudes and behavior issues, including articles in refereed journals such as *Journal of Economic Psychology, Journal of Consumer Policy, Psychology and Marketing, Environment and Behavior, International Journal of Research in Marketing,* and

Business Strategy and the Environment. He is the coordinator of the business and environment research theme at the faculty of business administration, the Aarhus School of Business, and member of the board of the Centre for Transport Research on Environmental and Health Impacts and Policy.

THOMAS W. VALENTE is a professor and the director of the Master of Public Health program in the department of preventive medicine in the Keck School of Medicine at the University of Southern California. He specializes in health behavior, health promotion, health communication, social network analysis, and program evaluation. His writings include *Network Models of the Diffusion of Innovations* (Hampton Press, 1995), *Evaluating Health Promotion Programs* (Oxford University Press, 2002), and more than three dozen articles and chapters on family planning and reproductive health, tobacco use prevention, substance abuse prevention and treatment, and HIV/STD prevention. He received his B.S. in mathematics from Mary Washington College, his M.S. in mass communication from San Diego State University, and his Ph.D. from the Annenberg School for Communication at the University of Southern California.

THOMAS J. WILBANKS is a corporate research fellow at the Oak Ridge National Laboratory and leads the global change and developing country programs of the laboratory. He is also an associate of the Belfer Center for Science and International Affairs at Harvard University. His research has largely been concerned with energy and environmental policy, sustainable development, relationships between society and technology, institution building, and responses to concerns about global environmental change. A coauthored book, *Global Change and Local Places,* is forthcoming from Cambridge University Press. He received his B.A. from Trinity University and his M.A. and Ph.D. degrees from Syracuse University.

GREG WISE, an associate professor in the Department of Community Resource Development, is currently serving as Interim Assistant Vice Chancellor, University of Wisconsin-Extension. For more than 10 years, he served as a University of Wisconsin-Extension community development agent in Manitowoc and Sauk Counties. His community work has focused on assisting communities with strategic community development processes and leadership development. His most recent work has involved positioning UW-Extension as a leader in higher education in the definition and assessment of scholarly outreach efforts. He has presented that work at several national higher education conferences devoted to civic engagement. Prior to joining the UW-Extension, he worked for the U.S. Environmental Protection Agency. He holds an M.S. in Urban and Regional Planning, an M.A. in Public Policy and Administration, and a B.A. in Landscape Architecture, all from the University of Wisconsin-Madison.